Bioinformatics

High Performance Parallel Computer Architectures

Embedded Multi-Core Systems

Series Editors

Fayez Gebali and Haytham El Miligi
University of Victoria
Victoria, British Columbia

Multi-Core Embedded Systems, *Georgios Kornaros*

Bioinformatics: High Performance Parallel Computer Architectures, *edited by Bertil Schmidt*

Bioinformatics

High Performance Parallel Computer Architectures

Edited by

Bertil Schmidt

CRC Press
Taylor & Francis Group
Boca Raton London New York

CRC Press is an imprint of the
Taylor & Francis Group, an **informa** business

CRC Press
Taylor & Francis Group
6000 Broken Sound Parkway NW, Suite 300
Boca Raton, FL 33487-2742

Printed in the United States of America on acid-free paper
10 9 8 7 6 5 4 3 2 1

International Standard Book Number: 978-1-4398-1488-8 (Hardback)

Library of Congress Cataloging-in-Publication Data

Bioinformatics : high performance parallel computer architectures / edited by Bertil Schmidt.
p. cm.
Includes bibliographical references and index.
ISBN 978-1-4398-1488-8
1. Bioinformatics--Data processing. 2. Parallel processing (Electronic computers) I. Schmidt, Bertil. II. Title.

QH324.2.B5487 2010
572.80285--dc22 2010003953

Visit the Taylor & Francis Web site at
http://www.taylorandfrancis.com

and the CRC Press Web site at
http://www.crcpress.com

Contents

Preface

High-throughput techniques for DNA sequencing and gene expression analysis have led to a rapid growth in the amount of digital biological data. Prominent examples are the growth of DNA sequence information in NCBI's GenBank database and the growth of protein sequences in the UniProtKB/TrEMBL database. Furthermore, emerging next-generation sequencing technologies have broken many experimental barriers to genome scale sequencing, facilitating the extraction of huge quantities of sequences, which will further promote the future growth of biological databases.

Computer scientists and biomedical researchers face the challenge of transforming genomic data into biological understanding. Consequently, bioinformatics tools need to be scalable; that is, they need to deal with an ever growing amount of data. Unfortunately, the amount of publicly available sequence data grows faster than single processor core performance (which is stagnant at the moment). Thus, modern bioinformatics tools need to take advantage of parallel computing, which has always been a challenging task. In particular, modern multicore and many-core architectures are revolutionizing high-performance computing (HPC) in recent years. Incorporating multiple processor cores into a single silicon die has been the recent trend in improving performance by means of parallelism. As more and more processor cores are being incorporated into a single chip, the era of the many-core processor has begun. Thus, it is expected that future mainstream processors will be parallel systems, with their parallelism continuing to scale with Moore's law. The emergence of many-core architectures, such as general-purpose graphic processors (GPGPU), especially compute unified device architecture (CUDA)-enabled GPUs, and other accelerator technologies, such as field-programmable gate arrays (FPGAs) and the Cell/BE, provides the opportunity to significantly reduce the runtime of many biological algorithms on commonly available and inexpensive hardware with more powerful high-performance computing power, which are generally not provided by conventional general-purpose processors. These emerging parallel computer architectures therefore pose new challenges to the field of bioinformatics, since

- New bioinformatics algorithms and applications need to take advantage of these new architectures and
- Existing bioinformatics tools need to be ported efficiently to emerging parallel architectures.

This book consists of 14 chapters written by internationally recognized experts. To provide necessary background, it contains three introductory chapters on important bioinformatics algorithms, GPGPUs, and massively threaded programming and on reconfigurable computing with FPGAs. The major part of the book, consisting of 11 chapters, compiles recent approaches from prominent researchers in the field to parallelize bioinformatics applications on a variety of modern parallel architectures. The presented tools and algorithms include pairwise sequence alignment, multiple sequence alignment, BLAST, motif finding, pattern matching, sequence assembly, hidden Markov models, proteomics, and evolutionary tree reconstruction. Since both parallel computing and bioinformatics are two major technologies for a traditional and broad community, we envisage this book to be beneficial to researchers, graduate students, engineers, and teachers, who are actively involved in research and applications in the fields of HPC and bioinformatics.

The material of this book is organized as follows: Chapter 1 provides readers with background information on bioinformatics algorithms that is important to understand the remaining chapters of this book. This chapter is at an introductory level and suitable for readers who are new to the field of bioinformatics. In Chapter 2, Rob Farber gives an introduction to GPGPU technology and the associated massively threaded CUDA programming model. An overview of FPGA architecture and programming is provided by Douglas Maskell in Chapter 3. In Chapter 4, Sarje and Aluru present several parallel algorithms for computing alignments on the Cell/BE architecture. This includes linear-space pairwise alignment, syntenic alignment, and spliced alignment. In Chapter 5, Stamatakis focuses on computational aspects of phylogenetic inference. He reviews underlying concepts, current developments, and advances in orchestrating the phylogenetic likelihood function on parallel computer architectures ranging from FPGAs up to the IBM BlueGene/L supercomputer. Chapter 6, by Liu, Schmidt, and Maskell, covers several effective techniques to fully exploit the computing capability of many-core CUDA-enabled GPUs to accelerate protein sequence database searching, multiple sequence alignment, and motif finding. Second-generation sequencing machines are able to produce a huge amount of high-throughput short-read (HTSR) data. In Chapter 7, Shi, Liu, and Schmidt present a parallel CUDA-based method for correcting sequencing base-pair errors in HTSR data. Chapter 8, by Jacob, Lancaster, Buhler, and Chamberlain, deals with an FPGA accelerator for BLASTN and BLASTP that exploits the characteristics of the streaming model. In Chapter 9, Lavenier and Nguyen present a parallel seed-based algorithm called PLAST for comparing protein banks, and its instantiation into two technologies (GPU boards and reconfigurable accelerators). The algorithm has been thought to express the maximum of parallelism and to be easily speeded up by specific hardware platforms. In Chapter 10, Walters, Chaudhary, and Schmidt show how the Viterbi algorithm for database searching with profile-hidden Markov models can be efficiently parallelized on reconfigurable hardware (FPGAs) as well as on many-core architectures (GPUs) with the CUDA programming model. The FPGA-based massively parallel COPACOBANA architecture is the subject of Chapter 11. Schimmler, Wienbrandt, Güneysu, and Bissel demonstrate how its usage, originally designed for cryptanalysis, can be extended to bioinformatics applications. In Chapter 12, Dandass describes techniques for accelerating the performance of string set matching solutions using an implementation of the Aho-Corasick algorithm on FPGA devices, with particular emphasis on applications in computational proteomics. Chapter 13, by Lin and Stepanova, is devoted to a two-phase neural system for recognition of dimeric DNA motifs. The authors demonstrate its power by applying a hybrid system into genome-wide identification of hormone response elements on DNA. Finally, in Chapter 14, Coca, Bogdan, and Beynon advocate the use of FPGAs as an alternative approach to conventional HPC for protein identification based on mass spectrometry using database searching.

Last but not least, I want to gratefully thank all the authors for their valuable contributions.

Editor

Bertil Schmidt is associate professor at the School of Computer Engineering at Nanyang Technological University in Singapore. Earlier, he held a faculty position at the University of New South Wales and was a senior researcher at the University of Melbourne. He received his undergraduate degree in computer science (Diplom-Informatiker) from the University of Kiel (Germany) in 1995 and his PhD degree from Loughborough University (UK) in 1999.

Dr. Schmidt has been involved in the design and implementation of scalable algorithms for over a decade. He has worked extensively with fine-grained, coarse-grained, and hybrid parallel architectures. He has successfully applied these technologies to various domains, including bioinformatics, cryptography, computational science, and data compression. He is currently the principal investigator (PI) of an AcRF Tier-1 project and the GPU-enabled genomics project funded by NVIDA as well as Co-PI of an AcRF Tier-2 project. Previously, he has successfully completed an A-Star BMRC-funded project as well as the industry-funded hybrid computing project. He has published extensively in premium and leading journals such as *Parallel Computing, Journal of VLSI Signal Processing, Microelectronic Engineering, IEEE Transactions on Circuits and Systems II, IEEE Transactions on Parallel and Distributed Systems, IEEE Transactions on IT in Biomedicine, Journal of Parallel and Distributed Computing, Concurrency and Computation: Practice and Experience, Future Generation Computer Systems, Bioinformatics, BMC Bioinformatics,* and *Autoimmunity.*

Contributors

Srinivas Aluru
Department of Electrical and Computer Engineering
State University
Ames, Iowa

Robert J. Beynon
Department of Veterinary Preclinical Sciences
University of Liverpool
Liverpool, United Kingdom

Jost Bissel
Department of Computer Science
Christian-Albrechts-University of Kiel
Kiel, Germany

Istvan Bogdan
Department of Automatic Control & Systems Engineering
University of Sheffield
Sheffield, United Kingdom

Jeremy D. Buhler
Computer Science & Engineering
Washington University in St. Louis
St. Louis, Missouri

Roger D. Chamberlain
Computer Science & Engineering
Washington University in St. Louis
St. Louis, Missouri

Vipin Chaudhary
Department of Computer Science and Engineering
University of Buffalo, SUNY
Buffalo, New York

Daniel Coca
Department of Automatic Control & Systems Engineering
University of Sheffield
Sheffield, United Kingdom

Yoginder S. Dandass
Department of Computer Science and Engineering
Mississippi State University
Starkville, Mississippi

Robert M. Farber
Pacific Northwest National Laboratory (PNNL)
Richland, Washington

Tim Güneysu
Horst Görtz Institute for IT-Security
Ruhr University Bochum
Bochum, Germany

Arpith C. Jacob
Computer Science & Engineering
Washington University in St. Louis
St. Louis, Missouri

Joseph M. Lancaster
Computer Science & Engineering
Washington University in St. Louis
St. Louis, Missouri

Dominique Lavenier
Informatique et Télécommunication
ENS Cachan
Rennes, France

Feng Lin
School of Computer Engineering
Nanyang Technological University
Singapore

Weiguo Liu
School of Computer Engineering
Nanyang Technological University
Singapore

Yongchao Liu
School of Computer Engineering
Nanyang Technological University
Singapore

Douglas Maskell
School of Computer Engineering
Nanyang Technological University
Singapore

Van-Hoa Nguyen
EPI Symbiose
National Institute for Research in Computer Science and Control
Rennes, France

Abhinav Sarje
Department of Electrical and Computer Engineering
Iowa State University
Ames, Iowa

Manfred Schimmler
Department of Computer Science
Christian-Albrechts-University of Kiel
Kiel, Germany

Bertil Schmidt
School of Computer Engineering
Nanyang Technological University
Singapore

Haixiang Shi
School of Computer Engineering
Nanyang Technological University
Singapore

Alexandros Stamatakis
Department of Computer Science
Technical University of Munich
Munich, Germany

Maria Stepanova
School of Computer Engineering
Nanyang Technological University
Singapore

John Paul Walters
University of Southern California Information Sciences Institute East
Arlington, Virginia

Lars Wienbrandt
Department of Computer Science
Christian-Albrechts-University of Kiel
Kiel, Germany

1

Algorithms for Bioinformatics

Bertil Schmidt

1.1 Introduction

In this chapter we provide some important background on bioinformatics algorithms that is important to understand the remaining chapters of this book. This chapter is at an introductory level and is suitable for readers who are new to the field of bioinformatics. Attention has been paid to provide a sufficient number of examples and illustrations to explain concepts and ideas.

We start with the most basic bioinformatics algorithm in Section 1.1: pairwise sequence alignment. This includes global and local pairwise alignment as well as linear and affine gap penalties, which can all be computed in quadratic time and space using dynamic programming (DP). Furthermore,

Hirschberg's divide-and-conquer approach, which reduces the space complexity from quadratic to linear by just doubling the amount of computation, is presented.

Section 1.2 explains how the pairwise alignment problem can be extended to multiple sequence alignment (MSA). Unfortunately, the straightforward extension of the DP approach leads to an exponential time complexity. Therefore, heuristics are commonly used to compute multiple alignments in practice. We first present the simple star alignment heuristic and then show how this is extended to the progressive alignment approach used in ClustalW (which is one of the most popular multiple alignment tools with more than 26,000 citations in the ISI web of science).

The most popular bioinformatics tool is undoubtedly the basic local alignment search tool (BLAST). BLAST is a sequence database search method, where BLASTN is for DNA database search and BLASTP is for protein database search. We explain the filtration approach that is the basis of BLAST in Section 1.3. We further describe a number of efficient data structure for exact string matching, namely, the suffix tree and the suffix array.

Of course there are many more bioinformatics algorithms that are not discussed in this chapter. The interested reader is referred to corresponding books [1–4].

1.2 Pairwise Sequence Alignment

1.2.1 Definitions and Notations

Consider a sequence S of length l over the alphabet Σ. We use the following notations.

- $S[i\ldots j]$ denotes the substring of S starting at position i and ending at position j; that is, $S = S[0\ldots l\text{–}1]$.
- $S[i]$ denotes the letter at position i in S.
- $|S|$ denotes the length of string S; that is, $|S| = l$.
- The string of length zero is called the empty string and is denoted as ε.
- The gap symbol is denoted as $-$, where $-\notin\Sigma$.

Alphabets used in bioinformatics are often the DNA alphabet with four nucleotides (i.e., $\Sigma = \{A, C, G, T\}$) and the protein alphabet with the 20 standard amino acids (i.e., $\Sigma = \{A, C, D, E, F, G, H, I, K, L, M, N, P, Q, R, S, T, V, W, Y\}$).

- Given are two sequences S_0 and S_1 over the alphabet Σ of length l_0 and l_1, respectively. We define a *global pairwise sequence alignment* of

M_1

	0	1	2	3	4	5	6	7	8	9
0	T	A	–	G	A	C	T	A	–	G
1	–	A	C	G	–	–	T	A	T	G

M_2

	0	1	2	3	4	5	6	7	8
0	T	A	G	A	C	T	A	G	–
1	A	C	G	–	–	T	A	T	G

FIGURE 1.1
Two possible global alignments M1 and M2 of S0 = TAGACTAG and S1 = ACGGTATG.

S_0 and S_1 as a matrix M of size $2 \times n$ with $n \geq max\{l_0, l_1\}$ with the following properties for all $0 \leq k \leq 1$ and $0 \leq i \leq n-1$.

- $M[k][i] = -$ or $M[k][i] = S_k[p]$ for some $p \in \{0,\ldots,l_k-1\}$.
- If $M[k][i] = -$ then $M[1-k][i] \neq -$.
- If $M[k][i] = S_k[p]$ and $M[k][j] = S_k[q]$, then $p < q$, for all $i < j$.
- It exists $j \in \{0,\ldots,n-1\}$ with $M[k][j] = S_k[p]$ for all $0 \leq p \leq l_k-1$.

In other words, in a global pairwise alignment all letters of both sequences occur in the corresponding row of the alignment matrix in the same order as in the original sequence, possibly interspersed by gaps. Furthermore, it is not allowed to have two gaps in a column of the alignment matrix. An example of two possible global alignments of a pair of DNA sequences is shown in Figure 1.1.

For a pair of given sequences there is a very large (exponential) number of possible global alignments. Thus, we are only interested in certain alignments with high alignment scores. Therefore, a pairwise alignment scoring method needs to be defined. A frequently used method is to score every column of an alignment independently and then add up the individual column scores. This is known as a *linear scoring scheme.* Given a global pairwise alignment M of length n, the linear score of M is defined by Equation 1.1, where $\delta_{pair}(M[0][i],M[1][i])$ is the score of the ith alignment column in M.

$$\text{score}_{\text{linear}}(M) = \sum_{i=0}^{n-1} \delta_{\text{pair}}(M[0][i], M[1][i]) \quad (1.1)$$

The classification of each alignment columns as

- An *insertion* or a *deletion* (or *indel* for short) if it includes a gap
- A *match,* if it consists of two equal letters
- A *mismatch,* if it consists of two unequal letters

can be used for a possible definition of δ_{pair} as shown in Equation 1.2, where g (gap penalty), α (match score), and β (mismatch penalty) are parameters of the scoring scheme.

	A	R	N	D	C	Q	E	G	H	I	L	K	M	F	P	S	T	W	Y	V
A	4																			
R	−1	5																		
N	−2	0	6																	
D	−2	−2	1	6																
C	0	−3	−3	−3	9															
Q	−1	1	0	0	−3	5														
E	−1	0	0	2	−4	2	5													
G	0	−2	0	−1	−3	−2	−2	6												
H	−2	0	1	−1	−3	0	0	−2	8											
I	−1	−3	−3	−3	−1	−3	−3	−4	−3	4										
L	−1	−2	−3	−4	−1	−2	−3	−4	−3	2	4									
K	−1	2	0	−1	−3	1	1	−2	−1	−3	−2	5								
M	−1	−1	−2	−3	−1	0	−2	−3	−2	1	2	−1	5							
F	−2	−3	−3	−3	−2	−3	−3	−3	−1	0	0	−3	0	6						
P	−1	−2	−2	−1	−3	−1	−1	−2	−2	−3	−3	−1	−2	−4	7					
S	1	−1	1	0	−1	0	0	0	−1	−2	−2	0	−1	−2	−1	4				
T	0	−1	0	−1	−1	−1	−1	−2	−2	−1	−1	−1	−1	−2	−1	1	5			
W	−3	−3	−4	−4	−2	−2	−3	−2	−2	−3	−2	−3	−1	1	−4	−3	−2	11		
Y	−2	−2	−2	−3	−2	−1	−2	−3	2	−1	−1	−2	−1	3	−3	−2	−2	2	7	
V	0	−3	−3	−3	−1	−2	−2	−3	−3	3	1	−2	1	−1	−2	−2	0	−3	−1	4

FIGURE 1.2
The BLOSUM62 substitution matrix for amino acids. Because the matrix is symmetric, we only show the lower triangular part (positive values are shaded).

$$\delta_{\text{pair}}(M[0][i], M[1][i]) = \begin{cases} g & \text{if } M[0][i] = - \text{ or } M[1][i] = - \text{ (indel)} \\ \alpha & \text{if } M[0][i] = M[1][i] \text{ (match)} \\ \beta & \text{if } M[0][i] \neq M[1][i] \text{ (mismatch)} \end{cases} \quad (1.2)$$

Typically, g and β are negative while α is positive. Using this scheme with the parameters $g = -1$, $\alpha = +2$, and $\beta = -2$ results in the following scores for the alignment shown in Figure 1.1: $\text{score}_{\text{linear}}(M_1) = +5$ and $\text{score}_{\text{linear}}(M_2) = -3$. In practice, the usage of a single score for match (α) and mismatch (β) is often replaced by a more general substitution matrix *sbt* of size $|\Sigma| \times |\Sigma|$. The reason for using substitution matrices instead of match/mismatch is that they can model evolutionary events more accurately (i.e., the mutation of one amino acid to another amino acid), since they include individual scores for each pair of letters. An example of a frequently used substitution matrix for amino acids is BLOSUM62 [5], which is shown in Figure 1.2.

M_3

	0	1	2	3	4	5	6	7
0	A	T	C	T	T	C	T	G
1	A	–	–	T	T	–	–	G

M_4

	0	1	2	3	4	5	6	7
0	A	T	C	T	T	C	T	G
1	A	T	–	–	–	–	T	G

FIGURE 1.3
Examples of two global alignments that have the same score under a linear scoring scheme. However, under an affine scoring scheme alignment M_4 would be preferred.

When using a substitution matrix *sbt*, the definition of δ_{pair} is modified to Equation 1.3.

$$\delta_{\text{pair}}(M[0][i], M[1][i]) = \begin{cases} g & \text{if } M[0][i] = - \text{ or } M[1][i] = - \\ sbt[M[0][i]][M[1][i]] & \text{otherwise} \end{cases} \quad (1.3)$$

Consider the two global alignments M_3 and M_4 shown in Figure 1.3. Both alignments have an identical score under a linear scoring scheme; for example, for $g = -1$, $\alpha = +2$, and $\beta = -2$, $\text{score}_{\text{linear}}(M_3) = +4 = \text{score}_{\text{linear}}(M_4)$. However, from a biological perspective, M_3 would correspond to two evolutionary events (two indels of length two each) while M_4 corresponds to only a single event (one indel of length four). Hence, M_4 should have a higher score than M_3. Biologists therefore often prefer an *affine scoring scheme* rather than the simple linear scoring scheme used so far.

In an affine scoring scheme there are two values for scoring indels:

- *go*: the gap opening penalty, and
- *ge*: the gap extension penalty.

Under affine scoring, a continuous indel of length k is charged $go + k \cdot ge$ rather than $k \cdot g$ (used in linear scoring). For example, using $go = -3$, $ge = -1$, $\alpha = +2$, and $\beta = -2$, $\text{score}_{\text{affine}}(M_3) = -2$ and $\text{score}_{\text{affine}}(M_4) = +1$. More formally, given a global pairwise alignment M of length n, go, ge, and sbt, the affine score of M is defined by Equation 1.4, where $\delta_{\text{pair}}(M[0][i], M[1][i])$ is the same as in Equation 1.2 or 1.3 with ge used for g, and the Δ is the number of gap openings in M; that is, $\Delta = \left|\{i \mid \exists k : M[k][i] = - \text{ and } (M[k][i-1] \neq - \text{ or } i = 0)\}\right|$.

$$\text{score}_{\text{affine}}(M) = \left(\sum_{i=0}^{n-1} \delta_{\text{pair}}(M[0][i], M[1][i])\right) - \Delta \cdot go \quad (1.4)$$

The linear scoring is also referred to as global pairwise alignment with *linear gap penalty function* and the affine scoring to global pairwise alignment with *affine gap penalty function*.

Another important pairwise alignment is local pairwise alignment. Given are two sequences S_0 and S_1 over the alphabet Σ of length l_0 and l_1, respectively. A *local pairwise sequence alignment* of S_0 and S_1 is a global alignment of two substrings $S_0[i_0 \ldots j_0]$ and $S_1[i_1 \ldots j_1]$ of S_0 and S_1 for any $0 \le i_0 \le j_0 \le l_0$ and $0 \le i_1 \le j_1 \le l_1$. The score of a local alignment is the score of the associated global substring alignment.

1.2.2 DP for Optimal Pairwise Alignment with Linear Gap Penalty Function

In this section we describe how an optimal global pairwise alignment (i.e., a pairwise global alignment with maximum score) under a linear scoring scheme can be computed with a *DP* approach with time and space complexity of order $O(l_0 \cdot l_1)$. Afterward, we show how this DP approach can be easily modified to compute an optimal local pairwise alignment. The former is also known as the Needleman–Wunsch algorithm [6] and the latter, as the Smith–Waterman algorithm [7]. The generalization to optimal alignment with an affine scoring scheme is described in Section 1.2.3.

Given are two sequences S_0 and S_1 of length l_0 and l_1 and a linear scoring scheme (consisting either of g and *sbt* or of g, α, and β). Let $H[i][j]$ denote the score of an optimal global pairwise alignment of the prefixes $S_0[0 \ldots i-1]$ and $S_1[0 \ldots j-1]$. For $i \ge 1$ and $j \ge 1$, only the following three cases are possible for the last column of an associated alignment.

1. $S_0[i-1]$ and $S_1[j-1]$ are aligned.
2. $S_0[i-1]$ is aligned with a gap.
3. $S_1[j-1]$ is aligned with a gap.

For each case the optimal global alignment score can be computed as follows:

1. $\delta(S_0[i-1], S_1[j-1])$ plus the optimal alignment global score of the prefixes $S_0[0 \ldots i-2]$ and $S_1[0 \ldots j-2]$ (which is stored in $H[i-1][j-1]$).
2. $\delta(S_0[i-1], -)$ plus the optimal alignment global score of the prefixes $S_0[0 \ldots i-2]$ and $S_1[0 \ldots j-1]$ (which is stored in $H[i-1][j]$).
3. $\delta(-, S_1[j-1])$ plus the optimal alignment global score of the prefixes $S_0[0 \ldots i-1]$ and $S_1[0 \ldots j-2]$ (which is stored in $H[i][j-1]$).

The optimal global alignment score is then the maximum of these three values. Thus, $H[i][j]$ can be computed using the recurrence relation in Equation 1.5 for the linear scoring scheme given by g, α, and β.

$$H[i][j] = \max \begin{cases} H[i-1][j-1] + \begin{cases} \alpha & \text{if } S_0[i-1] = S_1[j-1] \\ \beta & \text{if } S_0[i-1] \neq S_1[j-1] \end{cases} \\ H[i-1][j] + g \\ H[i][j-1] + g \end{cases} \quad \text{, for all } 1 \le i \le l_0 \text{ and } 1 \le j \le l_1 \tag{1.5}$$

For the linear scoring scheme given by g and $sbt[][]$, the recurrence relation for $H[i][j]$ is given by Equation 1.6.

$$H[i][j] = \max \begin{cases} H[i-1][j-1] + sbt[S_0[i-1]][S_1[j-1]] \\ H[i-1][j] + g \\ H[i][j-1] + g \end{cases} \quad \text{, for all } 1 \le i \le l_0 \text{ and } 1 \le j \le l_1 \tag{1.6}$$

Any value in the first column of H, that is, $H[0][j]$ for all $1 \le j \ge l_1$, simply refers to the optimal global alignment score for aligning the empty string ($\varepsilon = S_0[0\ldots-1]$) to the prefix $S_1[0\ldots j-1]$. There is only one possible alignment for this case (aligning $S_1[0\ldots j-1]$ to gaps) with the score $g \cdot j$. Therefore, the initial conditions for the first row and the first column of H are given by Equation 1.7.

$$H[i][j] = \begin{cases} 0 & \text{if } i = 0 \text{ and } j = 0 \\ g \cdot j & \text{if } i = 0 \text{ and } j > 0 \\ g \cdot i & \text{if } i > 0 \text{ and } j = 0 \end{cases} \tag{1.7}$$

The complete matrix $H[i][j]$ of size $(l_0 + 1) \times (l_1 + 1)$ can now be computed (in a row major order) using the aforementioned recurrences for $i = 0\ldots l_0$ and $j = 0\ldots l_1$. The optimal global pairwise alignment score is then $H[l_0][l_1]$. Figure 1.4 shows all values of $H[][]$ for the alignment of S_0 = AGT and S_1 = AAGT using $g = -2$, $\alpha = +1$, and $\beta = -1$.

So far we have computed only the optimal global alignment score, but not the actual alignment. The actual alignment can be computed from the DP matrix H by a *trace-back procedure*. The trace-back starts at the lower right matrix cell $H[l_0][l_1]$ and traverses H until it reaches the upper-left matrix cell $H[0][0]$. For each matrix cell $H[i][j]$ that the trace-back traverses, it is checked whether the value $H[i][j]$ has been formed from the upper-left neighbor (i.e., $H[i][j] = H[i-1][j-1] + sbt[S_0[i-1]][S_1[j-1]]$), the left neighbor (i.e., $H[i][j] = H[i][j-1] + g$), or the upper neighbor (i.e., $H[i][j] = H[i-1][j] + g$). The trace-back

		A	A	G	T
	0	−2	−4	−6	−8
A	−2	1	−1	−3	−5
G	−4	−1	0	0	−2
T	−6	−3	−2	−1	1

FIGURE 1.4
The matrix $H[][]$ of size 4 × 5 for the global alignment of the input DNA sequences S_0 = AGT and S_1 = AAGT using $g = -2$, $\alpha = +1$, and $\beta = -1$. The optimal global pairwise alignment score is $H[3][4] = +1$. $S_0[i]$ and $S_1[j]$ are also displayed for each row and each column.

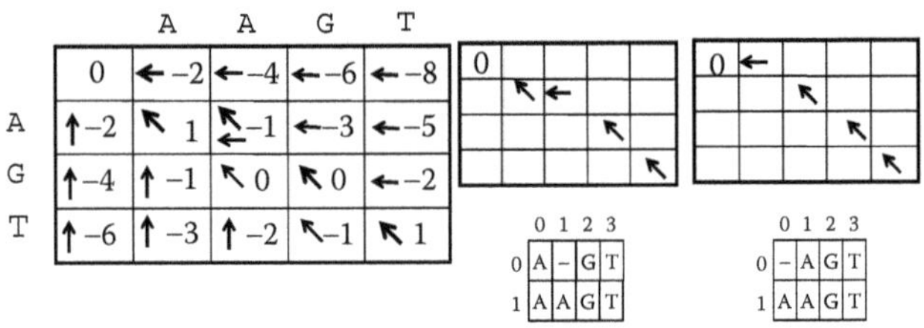

FIGURE 1.5
The matrix H with trace-back pointers for the example shown in Figure 1.4 (left). The two possible trace-back paths from $H[l_0][l_1]$ to $H[0][0]$ (center and right). The optimal alignments corresponding to each trace-back path are also shown.

then moves to the corresponding cell. For cells in the first row (column), the trace-back always moves to the left (up). One way to implement the trace-back procedure is to use trace-back pointers in each matrix cell $H[i][j]$. A trace-back pointer can have one of three values (up-left (↖), left (←), up (↑)) depending on which neighbor the value $H[i][j]$ has been formed from. Note that, in case of a tie between two or three neighbors, several trace-back pointers can be stored in a matrix cell. As a consequence, there can be several possible trace-back paths from $H[l_0][l_1]$ to $H[0][0]$, where each distinct path corresponds to an optimal global alignment. Figure 1.5 shows all trace-back pointers for the example shown in Figure 1.4 as well as the two possible trace-back paths.

The actual alignment can be constructed from a trace-back path using the following three rules for each traversed matrix cell $H[i][j]$ and the column k of the alignment matrix M.

- If an up-left pointer is used, then $M[0][k] = S_0[i]$ and $M[0][k] = S_1[j]$, move to $H[i-1][j-1]$, and move to column k–1 in M.
- If a left pointer is used, then $M[0][k]$ = – and $M[0][k] = S_1[j]$, move to $H[i][j-1]$, and move to column k–1 in M.
- If an up pointer is used, then $M[0][k] = S_0[i]$ and $M[0][k]$ = –, move to $H[i-1][j]$, and column k–1 in M.

The two optimal alignments corresponding to the two trace-back paths in Figure 1.4 are also shown in Figure 1.5.

We now modify the presented DP algorithm to the computation of optimal local alignments. This can be done by a slight modification of the definition of $H[i][j]$. For local alignment, $H[i][j]$ denotes the score of the best optimal global pairwise alignment of any two suffixes of $S_0[0 \ldots i-1]$ and $S_1[0 \ldots j-1]$; that is, the maximum score from all optimal global alignments of $S_0[p \ldots i-1]$ and $S_1[q \ldots j-1]$ for all $0 \le p \le i$ and $0 \le q \le j$. The following four cases are possible for the last column of an associated alignment:

1. $S_0[i-1]$ and $S_1[j-1]$ are aligned.
2. $S_0[i-1]$ is aligned with a gap.
3. $S_1[j-1]$ is aligned with a gap.
4. Both suffixes are empty strings (in which the alignment has no column).

The scores for the first three cases are computed in the same way as for the global alignment recurrence relation. The score for Case 4 is always zero (i.e., an optimal local alignment can never have a negative score). The recurrence relation for local alignment with the linear scoring scheme given by g, α, and β is defined by Equation 1.8.

$$H[i][j] = \max \begin{cases} H[i-1][j-1] + \begin{cases} \alpha & \text{if } S_0[i-1] = S_1[j-1] \\ \beta & \text{if } S_0[i-1] \neq S_1[j-1] \end{cases} \\ H[i-1][j] + g \\ H[i][j-1] + g \\ 0 \end{cases}, \text{ for all } 1 \le i \le l_0 \text{ and } 1 \le j \le l_1 \tag{1.8}$$

For the linear scoring scheme given by g and $sbt[][]$ the recurrence is defined by Equation 1.9.

$$H[i][j] = \max \begin{cases} H[i-1][j-1] + sbt[S_0[i-1]][S_1[j-1]] \\ H[i-1][j] + g \\ H[i][j-1] + g \\ 0 \end{cases}, \text{for all } 1 \le i \le l_0 \text{ and } 1 \le j \le l_1 \tag{1.9}$$

The initial conditions are of $H[0][0] = H[i][0] = H[0][j] = 0$ for all $1 \le i \le l_0$ and $1 \le j \le l_1$. Since a local alignment considers any two substrings of S_0 and S_1, the

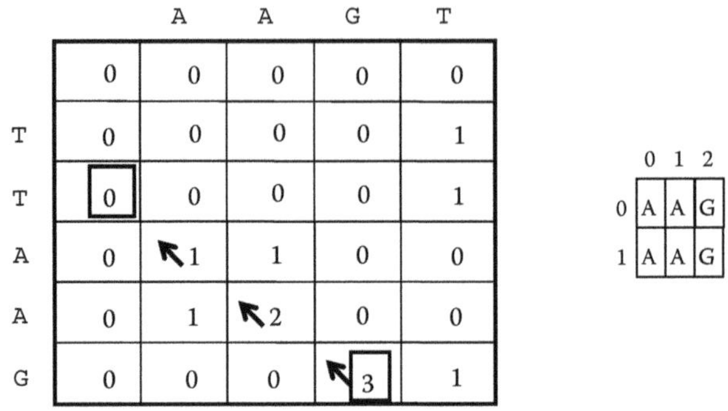

FIGURE 1.6
The matrix H of size 6×5 for the local alignment computation of sequences S_0 = TTAAG and S_1 = AAGT using $g = -2$, $\alpha = +1$, and $\beta = -1$. The optimal alignment trace-back path from $H[5][3]$ to $H[2][0]$ and the actual alignment are also shown.

score of the optimal local alignment is the maximum score in matrix H. Let $H[i_{max}][j_{max}]$ be a matrix cell with a maximum score. An actual alignment can now be found by executing a trace-back procedure, which ends at the first matrix cell with the score zero. Figure 1.6 shows an example of an optimal local pairwise alignment computation with trace-back.

1.2.3 DP for Optimal Pairwise Alignment with Affine Gap Penalty Function

In this section we extend the previously described DP algorithms for optimal global and local alignment with a linear scoring scheme to an affine scoring scheme. The recurrence relations for affine scoring need to consider whether a gap is opened (for which the penalty is $go + ge$) or a gap is extended (for which the penalty is ge). To model these two situations, three DP matrices are used instead of only one. Given two sequences S_0 and S_1 of length l_0 and l_1 and an affine scoring scheme (consisting of go, ge, sbt or of go, ge, α, β), they are defined as follows for global alignment.

- $H[i][j]$: Score of an optimal global pairwise alignment of $S_0[0\ldots i-1]$ and $S_1[0\ldots j-1]$.
- $E[i][j]$: Score of an optimal global pairwise alignment of $S_0[0\ldots i-1]$ and $S_1[0\ldots j-1]$, which ends with $S_0[i-1]$ aligned to a gap.
- $F[i][j]$: Score of an optimal global pairwise alignment of $S_0[0\ldots i-1]$ and $S_1[0\ldots j-1]$, which ends with a gap aligned to $S_1[j-1]$.

Thus, the two different gap penalties are used for the calculation of $E[i][j]$ (and $F[i][j]$). The value of $E[i][j]$ ends either with a gap extension (in which case $E[i][j] = E[i-1][j] + ge$) or with a gap opening (in which case

$E[i][j] = H[i-1] j] + go + ge$). The calculation of $F[i][j]$ is similar. The value of $H[i][j]$ is then $E[i][j]$ ($S_0[i-1]$ is aligned with a gap), $F[i][j]$ ($S_1[j-1]$ is aligned with a gap), or $H[i][j] + sbt[S_0[i-1]][S_1[j-1]]$ ($S_0[i-1]$ and $S_1[j-1]$ are aligned). In summary, the recurrence relations for all $1 \le i \le l_1$ and $1 \le j \le l_2$ for global alignment with the affine scoring scheme *go*, *ge*, *sbt*[][] are shown in Equations 1.10.

$$H[i][j] = \max\begin{cases} H[i-1][j-1] + sbt[S_0[i-1]][S_1[j-1]] \\ E[i][j] \\ F[i][j] \end{cases}$$
$$E[i][j] = \max\begin{cases} H[i-1][j] + go + ge \\ E[i-1][j] + ge \end{cases}$$
$$F[i][j] = \max\begin{cases} H[i][j-1] + go + ge \\ F[i][j-1] + ge \end{cases} \tag{1.10}$$

The recurrence relations for the scoring scheme *go*, *ge*, α, β are the same except that $H[i][j] + sbt[][]$ is changed to $H[i][j] + \alpha$ (if $S_0[i-1] = S_1[j-1]$) and $H[i][j] + \beta$ (otherwise). The initial conditions for all $1 \le i \le l_1$ and $1 \le j \le l_2$ are given by

- $H[0][0] = 0$; $H[i][0] = go + i \cdot ge$; $H[0][j] = go + j \cdot ge$;
- $E[0][0] = -\infty$; $E[i][0] = go + i \cdot ge$; $E[0][j] = -\infty$;
- $F[0][0] = -\infty$; $F[i][0] = -\infty$; $F[0][j] = go + j \cdot ge$.

An example is shown in Figure 1.7.

An alignment corresponding to the optimal global alignment score can again be found by a trace-back procedure from $H[l_0][l_1]$ to $H[0][0]$. The rules for the trace-back are as follows:

- From $H[i][j]$, move to $H[i-1][j-1]$ (if $H[i][j] = H[i-1][j-1] + sbt[S_0[i-1]][S_1[j-1]]$), to $E[i][j]$ (if $H[i][j] = E[i][j]$), or to $F[i][j]$ (if $H[i][j] = F[i][j]$).
- From $E[i][j]$, move to $E[i-1][j]$ (if $E[i][j] = E[i-1][j] + ge$), or to $H[i-1][j]$ (if $E[i][j] = H[i-1][j] + ge + go$).
- From $F[i][j]$, move to $F[i][j-1]$ (if $F[i][j] = F[i][j-1] + ge$), or to $H[i][j-1]$ (if $F[i][j] = H[i][j-1] + ge + go$).

The recurrence relations for optimal local pairwise alignment can be constructed by adding the term zero to the maximum computation of $H[][]$ and changing the initial conditions. In summary, the recurrence relations for all $1 \le i \le l_1$ and $1 \le j \le l_2$ for local alignment with the affine scoring scheme *go*, *ge*, *sbt*[][] are given by Equations 1.11.

H[][]		V	L	S	P	A
	(0)	−8	−9	−10	−11	−12
V	−8	↖2	−6	−7	−8	−9
S	−9	−6	↖1	−4	↖−8	−9
A	−10	−7	−7	0	−5	↖(−6)

E[][]		V	L	S	P	A
	−∞	−∞	−∞	−∞	−∞	−∞
V	−8	−16	−17	−18	−19	−20
S	−9	−6	−14	−15	−16	−17
A	−10	−7	−7	−12	−16	−17

F[][]		V	L	S	P	A
	−∞	−8	−9	−10	−11	−12
V	−∞	−16	←−6	←−7	−8	−9
S	−∞	−17	−14	←−7	←−8	−9
A	−∞	−18	−15	−15	−8	−9

	0	1	2	3	4
0	V	S	–	–	A
1	V	L	S	P	A

	0	1	2	3	4
0	V	–	–	S	A
1	V	L	S	P	A

FIGURE 1.7
Matrices $H[][]$, $E[][]$, and $F[][]$ for global pairwise alignment of S_0 = VSA and S_1 = VLSPA using the affine scoring scheme $\alpha = +2$, $\beta = -1$, $go = -7$, and $ge = -1$. The two trace-back paths from $H[3][5]$ to $H[0][0]$ for the two optimal alignments shown on the right are also indicated by trace-back pointers.

$$H[i][j] = \max\begin{cases} H[i-1][j-1] + sbt[S_0[i-1]][S_1[j-1]] \\ E[i][j] \\ F[i][j] \\ 0 \end{cases}$$
$$E[i][j] = \max\begin{cases} H[i-1][j] + go + ge \\ E[i-1][j] + ge \end{cases} \tag{1.11}$$
$$F[i][j] = \max\begin{cases} H[i][j-1] + go + ge \\ F[i][j-1] + ge \end{cases}$$

The initial conditions are $H[0][0] = H[i][0] = H[0][j] = 0$; $E[0][0] = -\infty$; $E[i][0] = go + i \cdot ge$; $E[0][j] = -\infty$; $F[0][0] = -\infty$; $F[i][0] = -\infty$; $F[0][j] = go + j \cdot ge$; for $1 \le i \le l_1$ and $1 \le j \le l_2$.

1.2.4 Computing Alignments in Linear Space Using Divide and Conquer

A critical resource in the described DP alignment algorithm is memory. Assume we want to align two sequences of length one million each. Then the DP matrix would have one trillion entries, leading to a memory requirement of four terabytes (assuming four bytes per matrix cell). An important improvement is therefore the space-saving divide-and-conquer method that was first

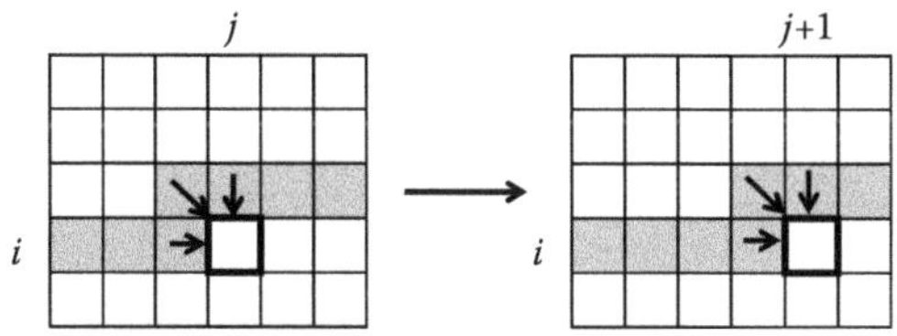

FIGURE 1.8
Linear–space score-only alignment computation. The values stored in the one-dimensional vector *HH* at iteration step (*i,j*) are shaded (left); that is, $HH[k] = H[i][k]$ for $0 \le k \le j-1$ and $HH[k + 1] = H[i-1][k]$ for $j-1 \le k \le l_1$. Afterward, step $(i, j + 1)$ is performed.

introduced by Hirschberg [8] and later applied to bioinformatics by Myers and Miller [9]. The method reduced the required memory from quadratic to linear; that is, for the aforementioned example we would need only a few megabytes instead of a few terabytes. The linear-space method is described for optimal global pairwise alignment with linear scoring in the following text.

We first note that the optimal alignment *score* can easily be computed in linear space. Consider two input sequences S_0 and S_1 of length l_0 and l_1. For the computation of the matrix cell $H[i][j]$, only the values $H[i-1][j-1]$, $H[i][j-1]$, and $H[i-1][j]$ are required. When iteratively computing $H[][]$ in row-major order, only the values $H[i-1][j-1]$ to $H[i-1][l_1]$ and $H[i][0]$ to $H[i][j-1]$ need to be kept at any point (i, j) in the iteration. After the calculation of $H[i][j]$, the $H[i-1][j-1]$ is not required anymore. Thus, instead of using a two-dimensional DP matrix H of size $(l_0 + 1)\times(l_1 + 1)$, it suffices to use a one-dimensional vector of size $(l_1 + 2)$ for score-only computation. Figure 1.8 illustrates the linear–space score-only computation.

However, to get an actual optimal alignment and not just the score, a trace-back path needs to be established. If we want to use only linear space, the trace-back procedure described in the previous section has to be modified. Let us assume that together with the linear–space score-only alignment matrix computation we can identify an *optimal midpoint* in the middle row $l_0/2$, denoted as $om(l_0/2)$. In general, $om(i)$, an optimal midpoint in row i, is defined as an intersection of the trace-back path with row i; that is, the trace-back path of an optimal global alignment goes through the cell $H[i][om(i)]$. The identified optimal midpoint $om(l_0/2)$ divides the DP matrix H into four quadrants:

A. $[0 \ldots l_0/2][0 \ldots om(i)]$.
B. $[0 \ldots l_0/2][om(i) \ldots l_1]$.
C. $[l_0/2 \ldots l_0][0 \ldots om(i)]$.
D. $[l_0/2 \ldots l_0][om(i) \ldots l_1]$.

Owing to the trace-back path properties (only left, up, or up-left), we know that the path can pass through only quadrants A and D. Therefore, quadrants II and III can be eliminated from further consideration. The procedure is then applied to identify $om(l_0/4)$ in quadrant A and $om(3 \cdot l_0/4)$ in quadrant D, resulting in

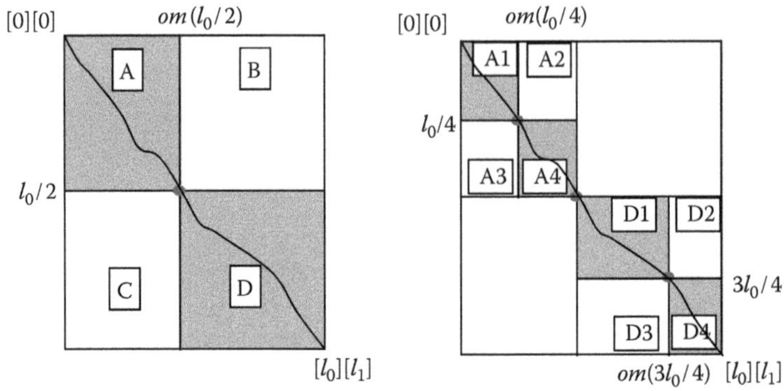

FIGURE 1.9
Two steps of the divide-and-conquer algorithm to find an optimal global alignment in linear space.

the four quadrants A1, A4, D1, and D4, which the optimal alignment traverses (see Figure 1.9). It is recursively applied until the resulting quadrants are small enough to compute an actual alignment in quadratic in quadratic space.

It remains to be shown how to find an optimal midpoint in linear space. For the two given sequences S_0 and S_1 of length l_0 and l_1, an optimal midpoint $om(^{l_0}/_2)$ can be found by calculating *score(i)* for all $0 \le i \le l_1$, where *score(i)* is defined as the score of the best global alignment of S_0 and S_1 passing through the matrix cell $[^{l_0}/_2][i]$. An optimal midpoint can then be found by $om(^{l_0}/_2) = \text{argmax}_{0 \le i \le l_1}\{score(i)\}$. We can calculate the *score(i)* values by *score(i)* = *upper(i)* + *lower(i)*; where *upper(i)* is the optimal global alignment score of $S_0[0\ldots{}^{l_0}/_2{-}1]$ and $S_1[0\ldots i{-}1]$ and *lower(i)* is the optimal global alignment score of $S_0^R[^{l_0}/_2\ldots l_0{-}1]$ and $S_1^R[i\ldots l_1{-}1]$ (where S_0^R and S_1^R denote *reverse* sequences). The values *upper(i)* (*lower[i]*) can simply be computed by executing the linear–space score-only alignment computation for $S_0[0\ldots{}^{l_0}/_2{-}1]$ and $S_1[0\ldots l_1{-}1]$ (for $S_0^R[^{l_0}/_2\ldots l_0{-}1]$ and $S_1^R[0\ldots l_1{-}1]$). The concept is illustrated in Figure 1.10.

Overall, the linear-space divide-and-conquer method needs to compute double the amount of DP-matrix cells compared with to the square-space method. However, the massive saving of space clearly outweighs this drawback. The presented linear-space method can also be extended to local alignment as well as to affine scoring schemes. The details are omitted here and the interested reader is referred to literature [3, 9].

1.3 Multiple Sequence Alignment

1.3.1 Background

In this subsection, the definitions and scoring schemes for pairwise alignment are generalized for *MSA*. Afterward, we describe how DP can be used

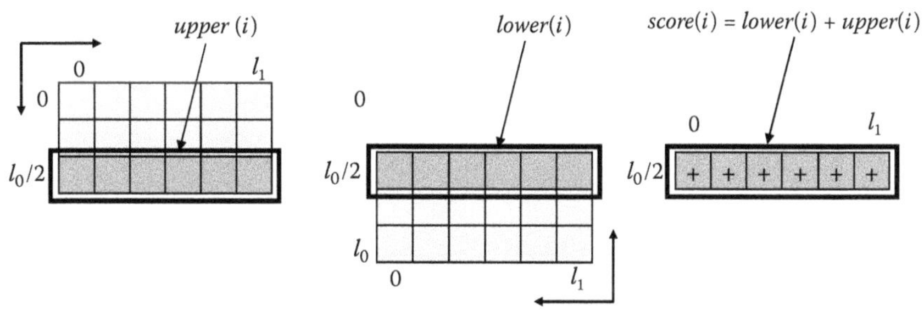

FIGURE 1.10
Optimal midpoint computation in linear space. The rows *upper(i)* and *lower(i)* are computed by executing the score-only linear-space algorithm for the upper half and the lower half of the DP matrix. Then, the row *score(i)* = *lower(i)* + *upper(i)* is calculated, which is used to identify an optimal midpoint.

to compute optimal global MSAs. Unfortunately, this approach leads to an exponential runtime in terms of the number of input sequences and is therefore unpractical. Consequently, heuristic approaches that run in polynomial time are used in practice. A very popular heuristic is the progressive alignment approach, which is described in the following subsection.

Given is a set $\mathcal{S} = \{S_0,\ldots,S_{t-1}\}$ of t sequences over the alphabet Σ with $|S_i| = l_i$ for $i \in \{0,\ldots,t-1\}$. We define a *global MSA* of these sequences as a matrix $M[][]$ of size $t \times n$ with $n \geq max\{l_0,\ldots, l_{t-1}\}$ that has the following properties for all $0 \leq k \leq t-1$.

- $M[k][i] = -$ or $M[k][i] = S_k[p]$ for some $p \in \{0,\ldots,l_k-1\}$ for all $0 \leq i \leq n-1$.
- For all $0 \leq i \leq n-1$, exists $r \in \{0,\ldots,t-1\}$ with $M[r][i] \neq -$.
- If $M[k][i] = S_k[p]$ and $M[k][j] = S_k[q]$, then $p < q$, for all $0 \leq i < j \leq n-1$.
- It exists $i \in \{0,\ldots,n-1\}$ with $M[k][i] = S_k[p]$ for all $0 \leq p \leq l_k-1$.

In other words, in a global MSA all letters of each sequence occur in the corresponding row of the alignment matrix in the same order as in the original sequence, possibly interspersed by gaps. Furthermore, it is not allowed to have only gaps in any alignment column. An example of a global MSA of four protein sequences is shown in Figure 1.11. A *local MSA* of $\mathcal{S} = \{S_0,\ldots,S_{t-1}\}$ is a global MSA of k substrings $S_k[i_k\ldots j_k]$ for any $0 \leq i_k \leq j_k \leq l_k$ and for each $k \in \{0,\ldots,t-1\}$.

We now need to define how a MSA can be scored. Given a global MSA M of a set of t sequences with n columns, the score of M is defined by Equation 1.12.

$$score(M) = \sum_{i=0}^{n-1} \delta_{\text{obj}}(M[0][i],\ldots, M[t-1][i]) \tag{1.12}$$

In Equation 1.12, $\delta_{\text{obj}}(M[0][i],\ldots,M[t-1][i])$ is the score of the ith alignment column in the matrix M. The function δ_{obj} is called the *objective function*.

M_3

	0	1	2	3	4	5	6	7	8	9	10	11
0	P	Y	R	F	T	–	–	–	I	K	S	M
1	P	Y	K	F	–	–	–	S	I	K	S	M
2	P	Y	M	Y	–	–	–	S	S	E	S	M
3	P	M	D	D	N	P	F	S	F	Q	S	M

FIGURE 1.11
A global MSA of S = {PYRFTIKSM, PYKFSIKSM, PYMYSSESM, PMDDNPFSFQSM}.

Several types of objective functions for MSA have been used. Two popular objective functions are

- Sum-of-pairs function, and
- Consensus function.

Given is an associated linear pairwise alignment scoring scheme δ_{pair}. The *sum-of-pairs score* of the ith alignment column is defined by Equation 1.13.

$$\delta_{obj}(M[0][i],...,M[t-1][i]) = \sum_{r=0}^{k-2}\sum_{s=r+1}^{k-1}\delta_{pair}(M[r][i],M[s][i]) \tag{1.13}$$

For example, for the pairwise scoring scheme $\alpha = +1$, $\beta = -1$, and $g = -2$, the sum-of-pairs-score of the MSA M_3 shown in Figure 1.11 is score(M_3) = 6 + 0 – 6 – 4 – 9 – 6 – 6 – 3 – 4 – 4 + 6 + 6 = –24. An important feature of the sum-of-pairs score of an MSA is that it is equal to the sum of all pairwise global alignments induced by the given MSA. A global MSA of t sequences induces a global pairwise alignment for each pair of sequences i, j; that is, $t \cdot (t-1)/2$ in total. The pairwise alignment of the sequences $S_i, S_j \in \mathcal{S}$, with $i < j$, is defined as the pairwise alignment matrix given by taking the rows i and j in the given MSA matrix and removing all columns containing two gaps. Figure 1.12 shows all induced pairwise alignments for the example shown in Figure 1.11. Using the pairwise scoring scheme $\alpha = +1$, $\beta = -1$, and $g = -2$, it can be seen that the sum of all induced pairwise alignments is: 2 – 4 – 9 + 1 – 7 – 7 = –24. Please note that an induced pairwise global alignment is not necessarily also an optimal pairwise global alignment.

Given is an associated linear pairwise alignment scoring scheme δ_{pair}. The *consensus score* of the ith alignment column of the MSA matrix M is defined by Equation 1.14, where *cons*(i) is the consensus character for column i.

$$\delta_{obj}(M[0][i],...,M[t-1][i]) = \sum_{r=0}^{k-1}\delta_{pair}(M[r][i], cons(i)) \tag{1.14}$$

$M_{0,1}$
```
P Y R F T - I K S M
P Y K F - S I K S M
```

$M_{0,2}$
```
P Y R F T - I K S M
P Y M Y - S S E S M
```

$M_{0,3}$
```
P Y R F T - - - I K S M
P M D D N P F S F Q S M
```

M
```
P Y K F S I K S M
P Y M Y S S E S M
```

$M_{1,3}$
```
P Y K F - - - S I K S M
P M D D N P F S F Q S M
```

$M_{2,3}$
```
P Y M Y - - - S S E S M
P M D D N P F S F Q S M
```

FIGURE 1.12
The six global pairwise alignments induced by M_3 (see Figure 1.11). $M_{i,j}$ denotes the induced pairwise alignment for S_i and S_j.

The consensus character is defined as the character from $\Sigma \cup \{-\}$ that maximizes the overall similarity for column i (see Equation 1.15).

$$cons(i) = \underset{c \in \Sigma \cup \{-\}}{\arg\max} \left(\sum_{r=0}^{k-1} \delta_{\text{pair}}(M[r][i], c) \right) \tag{1.15}$$

For the pairwise scoring scheme $\alpha = +1$, $\beta = -1$, and $g = -2$, the consensus sequence of the MSA M_3 (see Figure 1.11) consisting of the consensus characters of each column is PYKF–SIKSM. The corresponding consensus score is then $score(M_3) = 4 + 2 - 2 + 0 - 4 - 2 - 2 + 1 + 0 + 0 + 4 + 4 = +5$.

For a given set of three sequences $\mathcal{S} = \{S_0, S_1, S_2\}$ of length l_0, l_1, l_2 and an objective δ_{obj}, the optimal global MSA score can be computed by a straightforward extension of the pairwise DP approach. Let $H[i][j][k]$ denote the score of an optimal global MSA of the prefixes $S_0[0\ldots i-1]$, $S_1[0\ldots j-1]$, and $S_2[0\ldots k-1]$. For $i \geq 1$, $j \geq 1$, and $k \geq 1$ only the following seven cases are possible for the last column of an associated alignment.

1. $S_0[i-1]$, $S_1[j-1]$, and $S_2[k-1]$ are aligned.
2. $S_0[i-1]$, $S_1[j-1]$, and *gap* are aligned.
3. $S_0[i-1]$, *gap*, and $S_2[k-1]$ are aligned.
4. *gap*, $S_1[j-1]$, and $S_2[k-1]$ are aligned.
5. $S_0[i-1]$, *gap*, and *gap* are aligned.
6. *gap*, $S_1[j-1]$, and *gap* are aligned.
7. *gap*, *gap*, and $S_2[k-1]$ are aligned.

For each case the optimal global alignment score can be computed by the recurrence relation shown in Equation 1.16 for all $1 \leq i \leq l_0$, $1 \leq j \leq l_1$, and $1 \leq k \leq l_2$.

$$H[i][j][k] = \max \begin{cases} H[i-1][j-1][k-1] + \delta_{\text{obj}}(S_0[i], S_1[j], S_2[k]) \\ H[i-1][j-1][k] + \delta_{\text{obj}}(S_0[i], S_1[j], -) \\ H[i-1][j][k-1] + \delta_{\text{obj}}(S_0[i], -, S_2[k]) \\ H[i][j-1][k-1] + \delta_{\text{obj}}(-, S_1[j], S_2[k]) \\ H[i-1][j][k] + \delta_{\text{obj}}(S_0[i], -, -) \\ H[i][j-1][k] + \delta_{\text{obj}}(-, S_1[j], -) \\ H[i[j][k-1] + \delta_{\text{obj}}(-, -, S_2[k]) \end{cases} \tag{1.16}$$

Initial conditions are given by Equation 1.17.

$$H[i][j][k] = \begin{cases} g \cdot k & \text{if } i = 0, j = 0, k \geq 0 \\ g \cdot j & \text{if } i = 0, j > 0, k = 0 \\ g \cdot i & \text{if } i > 0, j = 0, k = 0 \end{cases} \tag{1.17}$$

Obviously, the DP matrix for three sequences has $l_0 \cdot l_1 \cdot l_2$ cells. In general, the DP approach to MSA for k input sequences requires $O(l_{\text{ave}}^k)$ DP matrix cells, where l_{ave} is the average sequence length. Furthermore, each inner cell depends on $O(2^k)$ other cells. Assuming that δ_{obj} for a single cell can be calculated in $O(k)$ time, this leads to an overall complexity of $O(k \cdot 2^k \cdot l_{\text{ave}}^k)$, where l_{ave} is the average sequence length. Obviously, this complexity leads to prohibitive runtimes even for small values of k. The Carillo–Lipman bound [10] allows to reduce the number of computed in the DP matrix by determining areas where the optimal alignment path cannot pass through. Even though this technique reduces the time and space complexity somewhat, overall runtimes still remain prohibitive in practice. As a consequence, heuristic methods that are suboptimal but run in polynomial time (usually between $O(k^2 \cdot l_{\text{ave}})$ and $O(k^2 \cdot l_{\text{ave}}^2)$) are used for MSA. We describe the popular progressive alignment approach in the next subsection.

1.3.2 Progressive Alignment

The main idea of the *star alignment* approach to global MSA is to compute the optimal global alignments for each pair of sequences. They are then used to determine a *center* sequence, which is the sequence with the largest overall similarity. The MSA is then built by combining all optimal pairwise alignments to the center sequence. Obviously, this method is suboptimal since the pairwise alignments induced by the constructed MSA from two sequences different from the center sequence can be incompatible. Figure 1.13 gives an example of the MSA computation by the star method.

The star alignment method uses a star topology to progressively align sequences to a growing multiple alignment. This can be generalized to progressively aligning to a growing multiple alignment along a so-called *guided*

	S_0	S_1	S_2	S_3	S_4	**Sum**
S_0		+1	+2	+1	–1	**+3**
S_1	+1		–1	+2	–2	**0**
S_2	+2	–1		+1	0	**+2**
S_3	+1	+2	+1		0	**+4**
S_4	–1	–2	0	0		**–3**

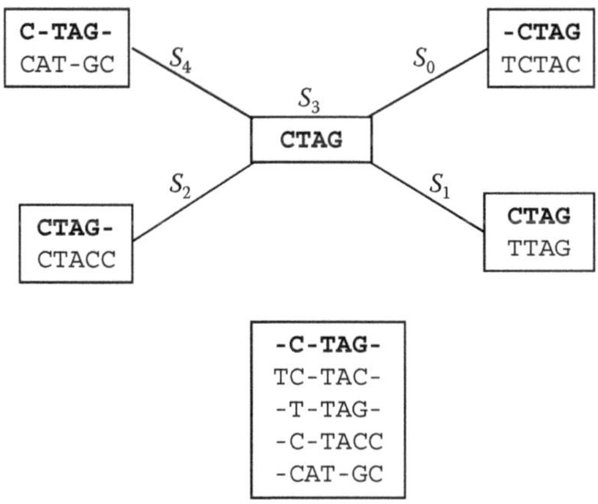

FIGURE 1.13
Star alignment using the input sequence S_0 = TCTAC, S_1 = TTAG, S_2 = CTACC, S_3 = CTAG, S_4 = CATGC and the scoring scheme $\alpha = +1$, $\beta = -1$, and $g = -1$. Top: The matrix with the optimal global alignment scores for each pair of sequences. The sequence with the highest sum of all pairwise scores for each sequence is taken as the center (i.e., S_3). Middle: All optimal global pairwise alignments to the center (S_3) are taken to get the final MSA (bottom).

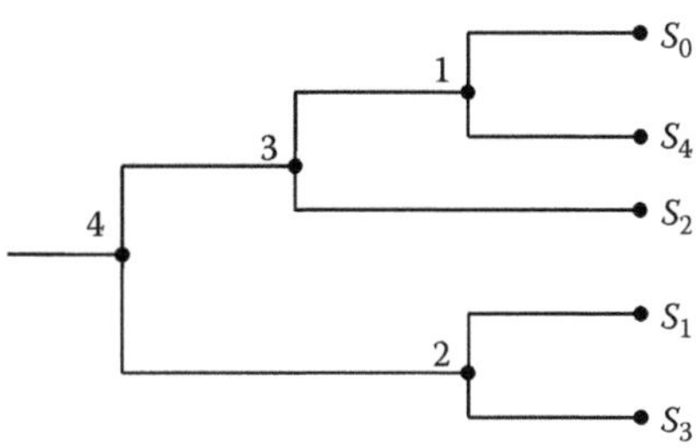

FIGURE 1.14
A guided tree for five input sequences $S_0, \ldots, S_4$.

tree. The guided tree is a rooted tree with each leaf node labeled by a unique sequence from the input set.

Figure 1.14 shows an example of a guided tree for five input sequences. The internal nodes are labeled 1–4. Each internal corresponds to one of the following three cases:

- A pairwise sequence alignment (e.g., internal node 1 and 2 corresponding to the alignment of S_0–S_4 and S_1–S_3, respectively). This results in an alignment profile, which we denote as *prof(A,B)* (e.g., nodes 1 and 2 represent $prof(S_0,S_4)$ and $prof(S_1,S_3)$).
- A pairwise sequence–profile alignment (e.g., internal node 3 corresponds to the alignment of S_2 to $prof(S_0,S_4)$).
- A pairwise profile–profile alignment (e.g., internal node 4 corresponds to the alignment of $prof(prof(S_0,S_4),S_2)$ to $prof(S_1,S_3)$).

The guided tree in Figure 1.14 therefore produces the final MSA $prof(prof(prof(S_0,S_4),S_2), prof(S_1,S_3))$.

The guided tree is usually computed using a distance-based clustering method like UPGMA [11] or neighbor-joining (NJ) [12, 13]. In the following section we describe the *NJ* method, which is used in the popular ClustalW tool [14]. Input to NJ is a distance matrix *D* of size $k \times k$ for a set of *k* input sequences $\mathcal{S} = \{S_0, \ldots, S_{k-1}\}$, where the value $D[i][j]$ denotes the distance between S_i and S_j. There are many ways to compute a distance value between two sequences. For example, ClustalW computes the distance between two protein sequences by the number of exact matches in optimal local alignment trace-back path of S_i and S_j divided by $\min\{l_i,l_j\}$. NJ then iteratively selects an entry in *D*. The corresponding two sequences are then connected with a node in the guided tree. The corresponding two rows in *D* are then merged to produce a new smaller matrix for the next iteration step. Iterations consist of the following steps:

1. Compute the rate-corrected distance matrix *DR* from *D* for all $i \neq j$ using Equation 1.18.

$$DR[i][j] = D[i][j] - \left(r[i] + r[j]\right) \text{ with } r[q] = \frac{1}{k-2}\sum_{p=0}^{K-1} D[q][p] \tag{1.18}$$

2. Find the minimum entry $DR[i_{\min}][j_{\min}]$ in *DR*.
3. Create a new node *N* in the guided tree that joins the two entries corresponding to $i_{\min}$ and $j_{\min}$ and calculate all distances to *N* as shown in Equations 1.19.

$$\begin{aligned} D[i_{\min}][N] &= D[i_{\min}][j_{\min}] + r[i_{\min}] - r[j_{\min}] \\ D[j_{\min}][N] &= D[i_{\min}][j_{\min}] - D[i_{\min}][N] \\ D[x][N] &= D[i_{\min}][x] + D[j_{\min}][x] - D[i_{\min}][j_{\min}], \text{ for all } x \in \{0,\ldots,k-1\} \setminus \{i_{\min}, j_{\min}\} \end{aligned} \tag{1.19}$$

4. Replace the rows and columns $i_{\min}$ and $j_{\min}$ in *D* by a new row and column representing the distances to *N*. The resulting distance matrix is used for the next iteration step.

Figure 1.15 shows an example of a single iteration step of NJ. An iteration of NJ has the complexity $O(k^2)$. Therefore, the overall NJ complexity is $O(k^3)$.

$D[\,][\,]$

A	0	3	6	9	12
B	3	0	5	9	11
C	6	5	0	10	13
D	9	9	10	0	19
E	12	11	13	19	0

$r[\,]$

10
9.3
11.3
15.7
18.3

$DR[\,][\,]$

0				
−16.3	0			
−15.3	−15.7	0		
−16.7	−16.0	**−17.0**	0	
−16.3	−16.7	−16.7	−15.0	0

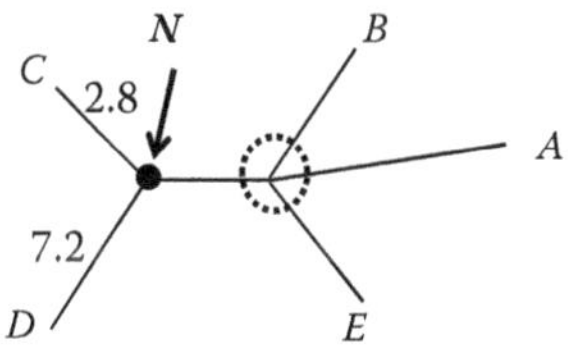

$D[\,][\,]$

A	0	3	2.5	12
B	3	0	2	11
N	2.5	2	0	11
C	12	11	11	0

FIGURE 1.15
Single NJ iteration step. Starting from the matrix *D* on the upper left, the rate-corrected matrix *DR* is calculated using the vector *r*. The minimum in *DR* is detected and the corresponding nodes *C* and *D* are joined in the guided tree by the newly created node *N*. All distances to *N* are calculated, resulting in an updated smaller matrix *D*.

As mentioned earlier, sequence–profile alignments and profile–profile alignments are frequently used in progressive alignment to build up an MSA. In the following section we briefly describe how a sequence can be aligned to a profile using DP. Profile–profile alignment is then a straightforward extension of the sequence–profile alignment. Given are a sequence S of length l, an MSA matrix M of size $k \times n$, and a linear scoring scheme (e.g., α, β, and g). For each column $0 \le j \le n-1$ in M, and each letter $c \in \Sigma \cup \{-\}$, we now build a letter frequency matrix P as shown in Equation 1.20.

$$P[c][j] = \frac{\left|\left\{i \mid M[i][j] = c\right\}\right|}{k} \tag{1.20}$$

On the basis of the letter frequency matrix, we can now define the score of a letter $c \in \Sigma \cup \{-\}$ with profile column j in Equation 1.21.

$$t(c,j) = \begin{cases} \alpha \cdot P[c][j] + \beta \cdot \sum\limits_{b \in \Sigma \setminus \{c\}} P[b][j] + g \cdot P[-][j] & \text{if } c \neq - \\ g \cdot \sum\limits_{b \in \Sigma} P[b][j] & \text{if } c = - \end{cases} \tag{1.21}$$

M_1

A	G	C	-	A
A	G	A	G	A
A	T	C	G	-
C	G	-	G	C

P

A	0.75	0.00	0.25	0.00	0.50
C	0.25	0.00	0.50	0.00	0.25
G	0.00	0.75	0.00	0.75	0.00
T	0.00	0.25	0.00	0.00	0.00
–	0.00	0.00	0.25	0.25	0.25

M_2

A	A	C	-	G	C
1	-	2	3	4	5

$$\begin{aligned}\text{score}(M_2) &= t(\text{A},1) + g + t(\text{C},2) + t(-,3) + t(\text{G},4) + t(\text{C},5)\\ &= (2\cdot 0.75 - 3\cdot 0.25) + (-1) + (-3\cdot(0.75 + 0.35)) + (-1\cdot(0.25 + 0.75)) + (2\cdot 0.75 - 1\cdot 0.25) +\\ &\quad (2\cdot 0.25 - 3\cdot 0.5 - 1\cdot 0.25) = \mathbf{-3.4875}\end{aligned}$$

FIGURE 1.16
A sequence–profile alignment M_2 of the DNA sequence AACGC to the letter frequency matrix P of the MSA M_1. The score of M_2 is –3.4875 using the scoring scheme $\alpha = +3$, $\beta = -3$, and $g = -1$.

The score of a sequence–profile alignment is then simply the sum of all column scores. An example of a letter frequency matrix for a given MSA and the score of a sequence–profile alignment is shown in Figure 1.16.

The optimal global sequence–profile alignment can be computed by DP using the recurrence relation in Equation 1.22.

$$H[i][j] = \max\begin{cases} H[i-1][j-1] + t(S_0[i-1], j-1) \\ H[i-1][j] + g \\ H[i][j-1] + t(-, j-1) \end{cases}, \text{ for all } 1 \le i \le l \text{ and } 1 \le j \le n \tag{1.22}$$

The initial conditions are $H[i][0] = i \cdot g$ for $0 \le i \le l$ and $H[0][j] = H[0][j-1] + t(-,j-1)$ for $0 \le j \le n$. The time complexity of the sequence–profile DP algorithm is $O(l \cdot n \cdot |\Sigma|)$. The four steps of the ClustalW progressive alignment method can be summarized as follows:

1. Distance matrix calculation using pairwise alignments.
2. Guided tree computation using neighbor-joining.
3. Rooting the guided tree and calculating sequence weights.
4. Progressively building the MSA following the branching order of the rooted guided tree.

1.4 Database Search and Exact Matching

1.4.1 Filtration

Given a query sequence S of length l_0 and a sequence D of length l_1, where D is constructed by concatenating all sequences of a sequence database,

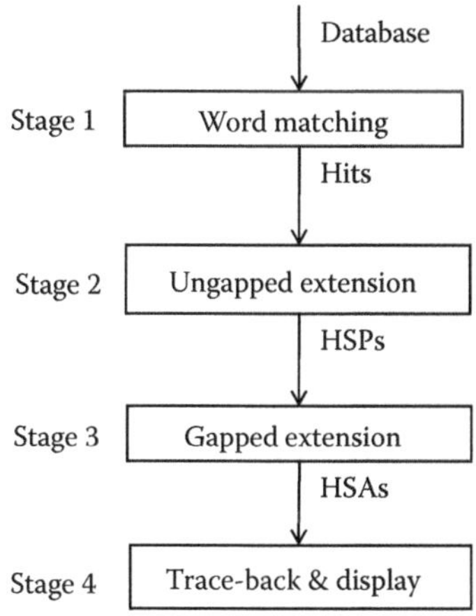

FIGURE 1.17
Stages of the BLAST pipeline.

the task of a database search tool is to find all significant local alignments between S and D. Using a DP-only approach for this problem would lead to a complexity of $O(l_0 \cdot l_1)$. Unfortunately, this would lead to prohibitive runtimes for large databases such as GenBank.

The basic idea for fast sequence database search is therefore *filtration*. Filtration assumes that good alignments usually contain short exact matches. Such matches can be quickly computed by using data structures such as lookup tables. Identified matches are then used as seeds for further detailed analysis. The analysis pipeline of the popular BLAST algorithm [15, 16] is shown in Figure 1.17.

We briefly describe each step of the pipeline for BLASTP, which is the version of BLAST for searching protein sequence databases.

- **Stage 1:** This stage identifies *hits* (or seeds). Each hit is defined as an offset pair (i,j) for which $\sum_{k=0}^{w-1} sbt(Q[i+k],D[j+k]) \geq T$, where sbt is a amino acid substitution matrix, w is the user-defined word length, and T is a user-defined threshold. BLASTP implements this stage by preprocessing Q as follows. For each position i of Q the neighborhood $N(Q[i \ldots i + w - 1],T)$ is computed consisting of all w-mers p for which $\sum_{k=0}^{w-1} sbt(Q[i+k],p[k]) \geq T$. The complete neighborhood of a query is typically stored in an efficient data structure (e.g., lookup table, finite-state automaton, or keyword tree). Default parameter values are $w = 3$ and $T = 11$.
- **Stage 2**: Stage 2 outputs high-scoring segment pairs (*HSPs*). HSPs are identified by performing an ungapped extension on a diagonal d that contains a nonoverlapping hit pair (i_1,j_1), (i_2,j_2) within a window A; that is, $d = i_1 - j_1 = i_2 - j_2$ and $w \leq i_2 - i_1 \leq A$. The identification

of pairs is also known as the *two-hit algorithm*. If the resulting ungapped alignment scores above a certain threshold it is passed to Stage 3.

- **Stage 3:** This stage outputs *HSAs*. HSAs are identified by performing a seeded banded gapped pairwise DP alignment algorithm using the previously identified HSPs as seeds. Alignments that score above a certain threshold are then passed to the final stage.
- **Stage 4:** The final alignments of the highest-scoring sequences are calculated and displayed to the user. This requires the computation of the trace-back path using the local pairwise DP approach in linear space.

Note that each stage of the pipeline progressively reduces the search space in the database for significant alignment. In addition, the computational cost associated with each stage and the sensitivity also progressively increases. However, since the amount of input data is becoming significantly smaller for each stage, the actual runtimes generally do not increase.

A notable difference of BLASTN (for searching DNA databases) compared to BLASTP is that the length of hits identified in Stage 1 is significantly longer (e.g., $w = 7$, 11, or 15). The PatternHunter [17] database search tool introduced the concept of *spaced seeds*. A spaced seed is similar to a hit in Stage 1 of BLASTN, but allows that it is interspersed by mismatches at certain positions.

It should also be mentioned that the size of the hits in Stage 1 (i.e., w in BLAST) affects sensitivity and runtime of a filtration method. In the case of BLASTP, a small value of w (e.g., $w = 2$) leads to more hits in Stages 1 and 2. This in turn leads to higher sensitivity, since less significant alignment might be missed. However, it also increases the runtime, since more data needs to be processed in Stages 2 and 3. On the other hand, a large value of w (e.g., $w = 4$) leads to less hits in Stages 1 and 2, which in turn leads to lower sensitivity due to the fact that significant alignment might be missed. On the positive, it reduces the runtime since less data is processed in Stages 2 and 3.

1.4.2 Suffix Trees and Suffix Arrays

Many sequence analysis tools in bioinformatics (such as database searching or read mapping) require some form of exact pattern matching. In the exact pattern matching problem, we are given a sequence T (called the *text*) of length n and a sequence P (called the *pattern*) of length m. We are required to find all exact occurrences of P in T; that is, all $i \in \{0,\ldots,n-m\}$ with $T[i\ldots i+m-1] = P$. Performing this task very fast is crucial to many bioinformatics tools, with examples including BLAST [15, 16], MUMmer [18], and Bowtie [19]. Approaches for exact matching can be classified into

1. Hashing
2. Preprocessing of the pattern(s) (e.g., with a keyword tree [19])
3. Preprocessing of the text (e.g., with a suffix tree or a suffix array).

The third approach is particularly interesting, since the text is usually a database or a genome. Thus, in many situations the text can be preprocessed offline in a suffix tree or a suffix array data structure. This preprocessing only takes $O(n)$ time; for example, by using Ukkonen's linear-time suffix tree construction algorithm [20]. Afterward, the actual pattern matching can be done in time $O(m)$. This approach is very efficient, since m is usually much smaller than n. In the following section we briefly describe how exact matching with a suffix tree and a suffix array can be done. More details can be found in the books by Gusfield [3] and Aluru [1].

The suffix tree of sequence T of length m has the following features:

- It has a rooted directed tree with exactly m leaves numbered from 0 to m–1.
- Each internal node, other than the root, has at least two children.
- Each edge is labeled with a nonempty substring of T.
- No two edges out of the same node have edge labels beginning with the same letter.
- For any leaf i, the concatenation of edge labels on the path from the root to leaf i exactly spells out the suffix $T[i \ldots m–1]$.

The suffix tree of the sequence T = TACTA$ is shown in Figure 1.18. Note that a unique termination symbol is commonly appended to the input sequence to guarantee that every suffix ends in a leaf.

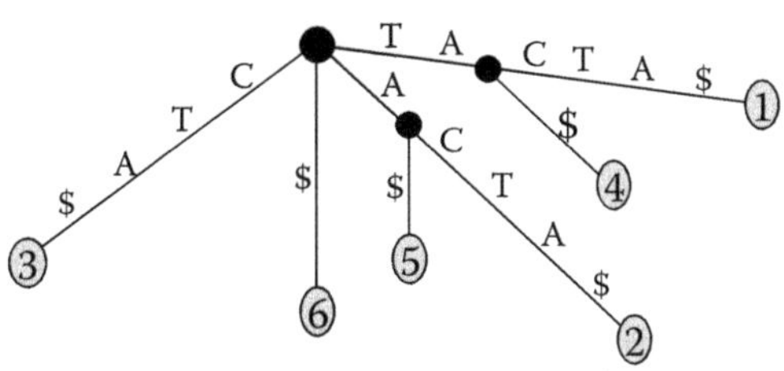

6	5	2	3	4	1

FIGURE 1.18
Suffix tree and suffix array of the T = TACTA$.

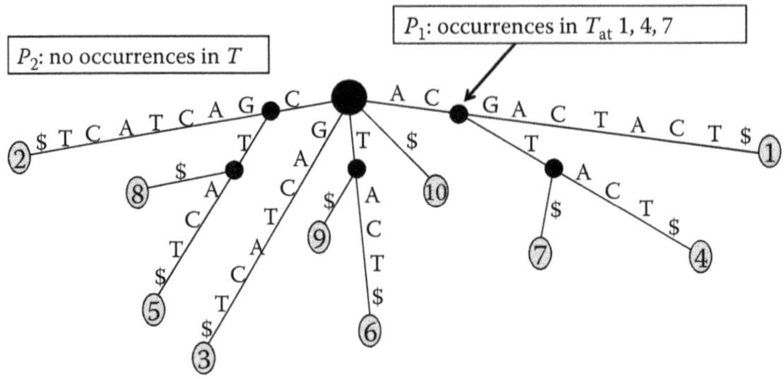

FIGURE 1.19
Matching the two patterns P_1 = AC and P_2 = CA against the suffix tree of T = ACGACTACT$.

All exact occurrences of the pattern *P* in the text *T* can be found using the suffix tree of *T* by matching *P* against the suffix tree starting at the root. If *P* can be completely matched, then every number of all leaf nodes below the final matching position in the tree is a starting position of occurrences of *P* in *T*. If *P* cannot be completely matched, then P does not occur in *T*. Figure 1.19 illustrates an example.

Besides exact pattern matching, suffix trees can also be used for the efficient solution of other string-based problems. Examples of problems include the maximal computation of repeat problems, longest common substrings, all-pairs suffix-prefix matching, Ziv–Lempel decomposition, common substrings of multiple sequences, exact set matching, and matching statistics [3]. To solve some of these problems, the suffix tree might be traversed top-down, bottom-up, or with suffix links.

Because of space and cache efficiency reasons, suffix arrays are sometimes preferred to suffix trees. A suffix array of a string *T* of length *n* is defined as an array of integer *n* specifying the lexicographical order of the *m* suffixes of *T*. The suffix array of *T* = TACTA$ is also shown in Figure 1.18, where we assume that $ is the lexicographically smallest letter. Suffix arrays can also be constructed efficiently [21]. Most string problems that can be efficiently solved with suffix trees can also be efficiently implemented using a suffix array [22].

1.5 References

1. Aluru, S. 2006. *Handbook of Computational Molecular Biology*. Chapman & Hall/CRC Press.
2 Durbin, R., Eddy, S., Krogh, A. and Mitchison, G. 1998. *Biological Sequence Analysis: Probabilistic Models of Protein and Nucleic Acids*. Cambridge University Press.
3. Gusfield, D. 1999. *Algorithms on Trees, Strings, and Sequences*. Cambridge University Press.

4. Jones, N.C. and Pevzner, P.A. *An Introduction to Bioinformatics Algorithms.* MIT Press, 2004.
5. Henikoff, S. and Henikoff, J.G. 1992. Amino acid substitution matrices from protein blocks. *Proceedings of the National Academy of Science of the USA* 89(22), 10915–10919.
6. Needleman, S.B. and Wunsch, C.D. 1970. A general method applicable to the search for similarities in the amino acid sequence of two proteins. *Journal of Molecular Biology* 48(3), 443–453.
7. Smith, T.F. and Waterman, M.S. 1981. Identification of common molecular subsequences. *Journal of Molecular Biology* 147(1), 195–197.
8. Hirschberg, D.S. 1975. A linear-space algorithm for computing maximal common subsequences. *Communication of the ACM* 18(6), 341–342.
9. Myers, E.W. and Miller, W. 1988. Optimal alignment in linear space. *Computer Applications in the Biosciences* 4(1), 11–17.
10. Carillo, H. and Lipman, D. 1988. The multiple sequence alignment problem in biology. *SIAM Journal on Applied Mathematics* 48(5), 1073–1082.
11. Michener, C.D. and Sokal, R.R. 1957. A quantitative approach to a problem of classification. *Evolution* 11, 130–162.
12. Saitou, N. and Nei, M. 1987. The neighbor-joining method: a new method for reconstructing phylogenetic trees. *Molecular Biology and Evolution* 4(4), 406–425.7.
13. Studier, J.A. and Keppler, K.J. 1988. A note on the neighbor-joining algorithm of Saitou and Nei. *Molecular Biology and Evolution* 5(6), 729–731.
14. Thompson, J.D., Higgins, D.G. and Gibson T.J. 1994. CLUSTAL W: improving the sensitivity of progressive multiple sequence alignment through sequence weighting, position-specific gap penalties and weight matrix choice. *Nucleic Acids Research* 11(22), 4673–4680.
15. Altschul, S.F., Gish, W., Miller, W., Myers, E.W. and Lipman, D.J. 1990. Basic local alignment search tool. *Journal of Molecular Biology* 215(3), 403–410.
16. Altschul, S.F., Madden, T.L., Schäffer, A.A., Zhang, J., Zhang, Z., Miller, W. and Lipman, D.J. 1997. Gapped BLAST and PSI-BLAST: a new generation of protein database search programs. *Nucleic Acids Research* 25(17), 3389–3402.
17. Ma, B., Tromp, J. and Li, M. 2002. PatternHunter: faster and more sensitive homology search. *Bioinformatics* 18(3), 440–445.
18. Kurtz, S., Phillippy, A., Delcher, A.L., Smoot, M., Shumway, M., Antonescu, C. and Salzberg, S.L. 2004. Versatile and open software for comparing large genomes. *Genome Biology* 5, R12.
19. Langmead, B., Trapnell, C., Pop, M. and Salzberg, S.L. 2009. Ultrafast and memory-efficient alignment of short DNA sequences to the human genome. *Genome Biology* 10, R25.
20. Aho, A.V. and Corasick, M.J. 1975. Efficient string matching: an aid to bibliographic search. *Communications of the ACM* 18(6), 333–340.
21. Ukkonen, E. 1995. On-line construction of suffix trees. *Algorithmica* 14(3), 249–260.
22. Puglisi, S.J., Smyth, W.F. and Turpin, A. 2007. A taxonomy of suffix array construction algorithms. *ACM Computing Surveys* 39(2), 1–31.
23. Abouelhoda, M.I., Kurtz, S. and Ohlebusch, E. 2004. Replacing suffix trees with enhanced suffix arrays. *Journal of Discrete Algorithms* 2(1), 53–86.

2

Introduction to GPGPUs and Massively Threaded Programming

Robert M. Farber

2.1 Introduction

Today, science and technology are inextricably linked. Human insight in bioinformatics, in particular, is driven by the vast amounts of data that can be collected with automated instruments coupled with sufficient computational capability to extract, analyze, model, and visualize results.

As they say, "the proof of the pudding is in the tasting," and computationally driven advances are very enticing. For example, the Harvard Connectome project is in the process of creating a complete "wiring-diagram" of a rat brain at a 3 nm/pixel resolution using automated slicing and data collection instruments [1]. Just as with the invention of the microscope, projects such as this will let biologists see into the structure of the brain and potentially revolutionize the entire field of study.

Of course any recipe for success can fail because of the lack of any single ingredient. Unfortunately, many important problems have remained intractable because there were no computers powerful enough or because scientists simply could not afford to access machines with the necessary capabilities.

Remarkably, the current revolution in scientific computation is happening because intense competition in computer graphics, mainly driven by the computer gaming industry, has evolved graphics processors into extremely capable yet low-cost general-purpose computational platforms. These general-purpose graphics processor units (GPGPUs) are C-language programmable computers that are capable of delivering well over a teraflop (a teraflop represents one trillion floating-point operations per second) of floating-point performance.

NVIDIA coined the phrase "supercomputing for the masses" to convey the catalytic effect generally available massively threaded GPGPU technology has had on high-performance computing.

To put this in perspective, Sandia National Laboratory in Albuquerque, NM, announced in December 1996 that their ASCI Red supercomputer was the first to exceed a trillion floating-point operations per second. It is truly amazing that roughly 10 years later any student or scientist could go to his or her favorite electronics store and purchase a teraflop capable graphics processing unit (GPU) for a nominal amount.

This computational bonanza is starting to bear fruit, as the scientific and technical literature over the past couple of years contains an explosion of GPGPU-enabled applications and algorithms in an astounding number of algorithmic and scientific application areas. In other words, it is clear that many scientists and programmers, using existing tools, are able to achieve one to two orders of magnitude, 10×–100×, of performance increase over conventional hardware when running their applications on GPGPUs. (Some researchers have even reported three orders of magnitude, or 1,000×, of faster performance when running algorithms that heavily utilize the NVIDIA GPU special processing units for transcendental functions.)

In comparing commodity processors versus graphics processors, I noted in my *Scientific Computing* column "GPGPUs: Neat Idea or Disruptive Technology?" [2] that newer dual- and quad-processor commodity workstations provided incremental 2×–4× performance gains. While this level of performance increase is nice, it does not fundamentally change how people work.

A 10× performance increase is a significant advance but does not necessarily represent a fundamental change. Machines with this level of performance make the computational workflow more interactive because computational tasks that previously took hours now take minutes and extended computational work that previously took days can now occur overnight.

Computer hardware that delivers 100× of faster performance is disruptive and has the potential to fundamentally affect scientific research by removing time-to-discovery barriers. Algorithms and computational tasks that previously would have required a year to complete can finish in days on the new hardware. Better scientific insight becomes possible because scientists can work with more data; utilize more accurate yet computationally expensive approximations; and work with larger, more realistic simulations and systems of equations. For the experimentalist, the results of newer high-throughput instruments (or collections of many instruments) can be utilized to create higher resolution and more informative pictures of what is occurring in nature. It's like transitioning from light-based microscopy to a powerful new electron microscope that allows one to see more and in much greater detail.

2.2 Massive Multithreading Is the Key

Massive multithreading (using hundreds to thousands of simultaneous *threads*) is the key to harnessing computational power of GPGPUs because it provides a common paradigm that both programmers and hardware designers can exploit to attain the highest possible performance.

Essentially *threads* are individual pieces of the same program that can execute simultaneously. For example, the vector multiply shown in Example 2.1 can be broken into N separate threads, where each thread simultaneously calculates vector c for each element index i.

Example 2.1: A Simple Vector Multiply

```
for(i = 0; i < N; i++) c[i] = a[i] * b[i]
```

This example also illustrates the importance of floating-point performance relative to memory bandwidth, as three memory operations are required for every floating-point multiply. Assuming that single-precision (32 bit or 4 byte) floating-point values are being used, the memory subsystem of a teraflop capable computer would need to provide 12 terabytes per second of memory bandwidth for this vector multiply example to run at full speed! This is roughly 50×–100× the capability of current GPU technology and roughly 375× more than the latest generation of high-end commodity processors. When the extra precision of 64-bit (8-byte) floating-point arithmetic is required, the reader can double these numbers (or halve the effective computational rate).

Massive multithreading (coupled with other architectural features of GPGPU hardware) permits graphics processors to achieve extremely high floating-point performance because the latency of memory accesses can be hidden and the full bandwidth of the memory subsystem can be utilized. An extremely low-latency hardware thread scheduler is an essential ingredient in this recipe for success.

Roughly speaking, graphics processors can be considered "streaming processors" because best performance is achieved when coalesced memory operations are used to simultaneously stream data from all of the on-board graphics memory banks. (A *coalesced* memory operation combines simultaneous memory accesses by multiple threads into a single memory transaction. This is in contrast to a *bank conflict*, which occurs when multiple memory requests fall in the same memory bank and causes the competing accesses to be serialized.)

It is easy to see that a linear increase in memory bandwidth can be achieved by simultaneously fetching data from multiple memory banks (or chips). From our simple vector multiply example, we can see that tying together two memory banks will double both memory bandwidth and floating-point performance. Similarly, tying together four banks of memory would result in a 4× speedup, and so on. Progressive generations of both graphics and conventional processors have used this technique to increase memory bandwidth.

Programming with a large number of threads allows the hardware thread scheduler to fully utilize the capabilities of the GPU because it can pick and choose amongst the active threads to

- Fully utilize all the internal resources of the hardware (floating-point, integer, or special function units) by scheduling those threads that do not have to wait on the memory subsystem to use whatever internal resources that happen to be available at that moment.
- Maximize the memory bandwidth of the memory subsystem by working together with internal coalescing units to stream data to/from all the memory banks at the highest possible data rate.
- Minimize the time taken. From a programmer's point of view, this thread scheduling occurs so quickly that it effectively takes no time and happens for free.

Now comes the important part: teaching scientists and programmers how to write (or rewrite) portions of their software to exploit the remarkable capabilities possible because of massive multithreading. C-language programmers utilizing NVIDIA's compute unified device architecture (CUDA) in particular should find the transition straightforward as they only need to understand a few key concepts and the mechanics of some simple additions (e.g., keywords and pragmas) to the C-language. In this way they can exploit the superb performance and scaling behavior that GPGPUs can deliver to advance scientific discovery.

While the concepts discussed in this chapter are general, specific capabilities provided by the runtime application programming interface (API) of the massively threaded CUDA C-language compiler, libraries, and software development kit (SDK) will be used as examples. This software can be freely downloaded off the NVIDIA web site and can be used on all NVIDIA CUDA-enabled graphics processors or run in emulation mode on conventional processors [3]. Please note that the CUDA emulator is not optimized for speed.

Other development platforms exist aside from CUDA that can be used to create programs for heterogeneous platforms consisting of CPUs, GPUs from multiple vendors, and other processors. OpenCL is a new technology that holds promise.

Several companies are developing C and FORTRAN compilers to support GPU computing. The Portland Group is one such company. Another company, CAPS Enterprise, is taking the innovative approach of generating hardware-specific *codelets* for C and FORTRAN code through the use of *compiler directives*. These codelets can then be used as is to run on GPUs and other architectures or they can be hand optimized to deliver the best possible performance. Their HMPP compiler also supports the ubiquitous message passing interface (MPI) that is heavily utilized in distributed scientific computing.

Many will also discover that structuring code to efficiently run on a massively threaded GPGPU has the added bonus of increasing performance and scalability on existing multicore processors. This can be an important step in "future-proofing" applications because multicore workstations (using four to eight processing cores) will have to become many-core systems (containing tens to hundreds of cores) to compete in the future.

This trend appears to be inevitable because manufacturers must now add cores to their processor chips rather than increase clock speed to remain competitive. In the past, manufacturers could introduce new generations of single-core processors with higher clock speeds, which in turn would entice customers to upgrade. If they could, manufacturers would be delighted to continue with this same business model. Switching to multicore processors affects the entire computer industry: it is disruptive to customers, requires that customers change how they design their programs, and forces existing software to be rewritten to use the extra processing cores.

Dennard's scaling laws are at the heart of the change. Effectively they say that power density will remain constant even as the number of transistors and their switching speed increases. For that relationship to hold, voltages need to be reduced in proportion to the linear dimensions of the transistor. Fabrication techniques have reduced the size of transistors to the point that manufacturers are no longer able to lower operating voltages sufficiently to match the performance gains that can be achieved by simply adding more computational cores to the processor chip. In a competitive market, minor changes in processor performance do not translate into increased sales for CPU manufacturers—so we now have multicore processors. Many in the computing industry believe that this trend will continue and the number of cores per processor will increase.

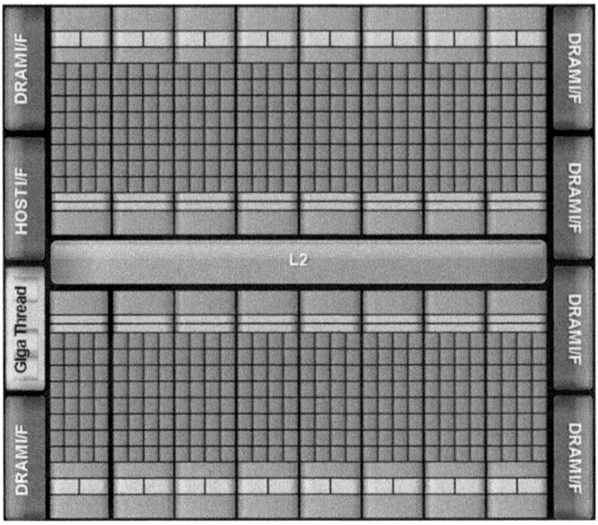

FIGURE 2.1
Block diagram showing 16 multiprocessors each with 32 threads. (From NVIDIA, Tesla Fermi launch—FINAL (press kit), p. 4, 2009. With permission.)

GPGPUs on the other hand evolved in a large and competitive market where massive parallelism is the evolutionary pathway to success because computer graphics operations that "push pixels" are inherently parallel. Simply stated, manufacturers were able to increase performance by adding pipelines and shaders to meet the intense market need for ever faster photo-realistic games and imaging software.

The advent of *programmable shaders* allowed each pixel or vertex to be processed by a short program. With the addition of floating-point math and looping capability, GPU hardware effectively evolved into single program multiple data (SPMD) massively parallel computers with hundreds of processing cores. CUDA was created to take advantage of this GPGPU technology by enabling the development of higher-level language (e.g., C-language) programs for this massively multithreaded hardware environment.

From a hardware standpoint, massively parallel multithreading is achieved through the use of a common architectural building block called a *multiprocessor* that can be replicated as required to provide the largest number of processing cores for a given price point. This massive replication can be seen in Figure 2.1 in the block diagram of the latest 20-series Fermi GPUs. NVIDIA generally bundles 32 threads into a *warp,* which runs single instruction multiple data fashion (SIMD) on each multiprocessor. SIMD execution allows instructions to be broadcast inexpensively and quickly within the multiprocessor but requires that all parallel processing happen in lock-step. SIMD execution is very efficient, but be aware that conditionals (if statements) can degrade performance because the multiprocessor must evaluate each branch of every conditional operation. Because GPGPUs are built with many multiprocessing units, they are considered SPMD rather than SIMD computers. Look to Flynn's taxonomy for a better understanding of the classification of computer architectures.

2.3 CUDA Simplifies the Creation of Massively Threaded Software

Writing massively threaded applications is greatly simplified because CUDA and the GPGPU hardware work together to manage threads for the programmer. Instead of explicitly creating threads as one would do on a conventional processor using a thread library such as *pthreads*, a CUDA developer writes a *kernel* that will run on the GPU. In reality, a kernel is nothing more that a C-language subroutine that runs and utilizes variables that reside on the GPU. In addition to the parameters, a call to a kernel includes the specification of an *execution configuration* that defines how the threads will be mapped to the GPGPU hardware.

Syntactically, the call to a CUDA kernel looks like a C-language subroutine call except the execution configuration is added between triple angle brackets "<<<" and ">>>" as is seen in Example 2.2.

Example 2.2: Example CUDA Kernel Call from the Host Code

```
cudaKernel <<< nBlocks,nThreadsPerBlock >>> (a_d, b_d, c_d)
```

The first two parameters between the angle brackets define the number of thread blocks, **nBlocks**, and the number of threads within a thread block, **nThreadsPerBlock**. The total number of simultaneously running threads for a given kernel is the product of these two parameters.

A key feature of a *thread block*, defined in the previous example by **nThreadsPerBlock**, is that only threads within the block can communicate with each other via high-speed shared memory. Otherwise threads cannot effectively communicate with each other outside their thread block unless the programmer is willing to pay a significant performance penalty to use global memory. Without going into further detail, other parameters in the execution configuration can be used to define 2D and 3D topologies and shared-memory allocations.

Unlike a C-language subroutine call, kernels are launched asynchronously, which means that the host processor merely queues the kernel in a pipeline to be launched when the hardware is ready. By changing the execution configuration, the programmer can easily specify a few or many thousands of threads. When not resource constrained, the NVIDIA documentation recommends using a large numbers of threads—on order of thousands—to future proof your code to maintain high performance on future generations of GPU products.

From the programmer's point of view, the CUDA kernel running on the GPU acts as if it were contained within the scope of a loop over the total number of threads—except each loop iteration runs in a separate thread of execution. Within each thread, the programmer has all the information needed to distinguish each thread from all other threads such as thread id within a thread block, the number of thread blocks, and coordinates within the execution configuration grid. With this information, the developer can then program each thread to perform the appropriate work and with the relevant data.

Arguably, we are watching an example of convergent evolution in computer technology as both GPUs and conventional processors adapt according to market pressures and technology constraints to become many-core processors. At this time, GPGPUs are currently the price and performance leader in this evolutionary race that delivers hardware platforms with hundreds of processing cores (currently 512 at the high end) and teraflops of performance at prices most scientists and students can afford.

Regardless of the process, the end result is clear and massively multithreaded programming models will remain an essential part of developing software for these evolving hardware platforms.

Physically, the current high-end graphics processors are peripheral cards that plug into a host computer via an industry standard PCI express (PCIe) slot. Figure 2.2 shows an example of the latest Fermi C2050 GPU. Many host computers can support multiple cards. Using four of the current high-end NVIDIA GTX 295 GPU cards that cost less that $500 each, for example, it is possible to give a student a *dedicated* workstation that can deliver over 7 teraflops of single-precision floating-point capability. In contrast the MPP2 supercomputer recently decommissioned in December 2008 was *shared* by the entire user community at Pacific Northwest National Laboratory. Built with Itanium processors, it occupied a large computer room and was only rated at 11.8 teraflops.

As can be seen in Table 2.1, there is a significant performance difference between the memory bandwidth of global memory (the main memory of the GPU) and the PCIe bus. For this reason, many people refer to the PCIe bus as a "PCIe straw" because accessing data across the bus is much like sipping through a very small straw—it is not possible to move much data quickly through the straw. Because of the relatively poor performance of the PCIe bus, it is essential from a performance standpoint to get the data into the memory space (and onto the memory subsystem) of the graphics processor.

Figure 2.2
An NVIDIA C2050 GPU. (From NVIDIA, 2009. With permission.)

TABLE 2.1
Transfer Speeds

Data Subsystem	Gigabytes Per Second (GiB/s)
NVIDIA GTX 295 Global Memory Bandwidth	223.8
PCIe ×16 version 2.0	8 unidirectional/16 bidirectional

Experience has shown that writing or porting software to graphics processors consists of three steps:

1. Getting (and keeping) the data on the GPU to eliminate the PCIe memory bandwidth bottleneck
2. Maximizing the amount of work performed per call to the GPU to eliminate the latency incurred when passing even short commands and small amounts of data to the GPU over the PCIe bus
3. Exploiting internal resources on the GPU (such as registers, shared memory, etc.) to bypass internal memory bottlenecks and maximize performance.

While the concepts and methods discussed in the remainder of this chapter are generally applicable, references to the CUDA runtime API will be used to illustrate many of the points.

TABLE 2.2

Examples Showing Data Movement with **CUDAMemcpy()**

PCIe Transfer	CUDA Runtime Call
Host to Device	*cudaMemcpy*(**a_d, a_h, nBytes, cudaMemcpyHostToDevice**)
Device to Device	*cudaMemcpy*(**b_d, a_d, nBytes, cudaMemcpyDeviceToDevice**)
Device to Host	*cudaMemcpy*(**a_h, b_d, nBytes, cudaMemcpyDeviceToHost**)

2.3.1 Step 1: Getting (and Keeping) the Data on the GPU

The CUDA runtime provides two main methods for transferring data between the host memory space and the graphics processor:

1. Explicit programmer-initiated data transfers
2. Mapped memory data transfers

Explicit transfers can be initiated through the use of *cudaMemcpy*(). As can be seen in Table 2.2, the CUDA runtime methods to transfer data between the host, graphics processor, and internally within the GPU are straightforward and closely resemble the C-language *memcpy*() routine. By convention, the variables **a_d** and **b_d** are assumed to reside on the GPU device and **a_h** is assumed to reside in the memory of the host computer.

C-programmers will also find the method used to allocate memory on the GPU, *cudaMalloc*(), is as familiar to use as the standard *malloc*() routine. When allocating regions of memory that will primarily be used to transfer data to/from the GPU, it is best to use pinned memory (meaning the memory cannot be swapped out on a virtual memory machine) allocated with *cudaHostAlloc*() so PCIe transfers can occur at the highest possible speed.

Transfers initiated with *cudaMemcpy*() are synchronous, meaning the call blocks until the data transfer is complete. Asynchronous data movement is important for speed because computation and communication can be overlapped so more work per unit time can be performed. Applications in CUDA manage concurrent data movement with the *streams* runtime API. Please consult the CUDA documentation for more information.

Program initiated transfers via mapped, pinned memory is a relatively new feature introduced in the CUDA 2.2 release. It provides an important (and convenient) capability to keep data synchronized between the host and GPU memory spaces via bidirectional asynchronous PCIe data transfers.

Having transparently synchronized memory eases the work involved in rewriting the computationally intensive portions of an application to run on the GPU. Essentially, the developer:

1. Profiles the existing code to find the computationally intensive routines.

2. Maps all the variables used by the computationally intensive sections of code and writes CUDA methods to perform the calculation on the GPU. Because mapped, pinned memory keeps the host and GPU data synchronized, the developer can focus on writing the GPU code and not on data movement.
3. Eventually moves enough of the calculation to the GPU so it no longer becomes necessary to keep data synchronized with the host. Unnecessary mappings can then be disabled or removed—thus creating a GPU-based version of the code that will not be affected by the PCI bottleneck.

Mapped, pinned memory provides a wonderful capability to facilitate *regression testing* when porting legacy application code to CUDA. (*Regressions* are unintended consequences and errors that result from program modifications.) Without question, regression testing is an essential software practice that is vital to the creation and verification of correctly working software!

When porting legacy software, the developer is provided with an existing known working code base that can be used for comparison against GPU generated results to identify errors. By transparently maintaining a synchronized version of data between both the host and device memory spaces, mapped memory allows comparison between results calculated on the GPU and host. By using *memcmp*() or other simple functions, developers can quickly search for the first appearance of a discrepancy between the original host and new GPU code. With care, the programmer can exploit the transparent synchronization provided by mapped, pinned memory throughout the entire porting project. As a result, there will always be a known working version that can be used to compare all calculations performed by new code running on the GPU.

2.3.2 Step 2: Maximizing the Amount of Work Performed per Call to the GPU

Graphics processors generally run relatively small pieces of computationally intensive code—otherwise known as *kernels*. Essentially, a kernel is a C-language subroutine that runs on each thread of the graphics processor. In CUDA, kernel calls are asynchronous so they can be pipelined to maximize performance. This also means that a kernel cannot act as a function because asynchronous execution precludes returning function values to the host code.

The current generation of NVIDIA GT200 graphics processors limits kernel sizes to roughly two million instructions per kernel. Of course, an application can have many kernels so the size restriction is not really a limiting factor. However, it does emphasis the computationally intensive off-load nature of graphics processors. Most applications will be hybrid codes that run the massively parallel sections of code on the GPU and the remaining serial or small-scale parallel sections on the host processor(s).

The adage, "be careful what you ask for because you might get it," applies to GPU computing. Graphics processors are very fast, which is extremely desirable but conversely raises the challenge of giving them enough work to do.

Even with pipelined asynchronous calls, there is still some overhead incurred in transferring parameters across the PCIe bus, setting up the grid in the GPU, and performing other housekeeping chores. To get a sense of the numbers, let's assume this overhead is 4 μsec for 1 teraflop GPU that takes four cycles to perform a floating-point operation. To achieve peak performance, each kernel must perform roughly one million floating-point operations or the GPU will stall waiting for the next kernel to start. If the kernel only takes 2 μsec to complete, then 50% of the GPU cycles will be wasted.

Most computationally oriented scientists and programmers are familiar with the basic linear algebra subprograms (BLAS) package, which is the de facto programming interface for basic linear algebra operations. NVIDIA supports this interface with their own library for the GPU called *CUBLAS*.

BLAS is structured according to three different levels with increasing data and runtime requirements. (The following discussion uses Big-O notation, which is a convenient way to describe how the size of an input affects the consumption by an algorithm of some resource such as time or memory.)

- *Level-1:* Vector-vector operations that require $O(n)$ data and $O(n)$ work. Examples include taking the inner product of two vectors or scaling a vector by a constant multiplier.
- *Level-2:* Matrix-vector operations that require $O(n^2)$ data and $O(n^2)$ work. Examples include matrix-vector multiplication or a single right-hand-side triangular solve.
- *Level-3:* Matrix-vector operations that require $\mathbf{O}(n^2)$ data and $O(n^3)$ work. Examples include dense matrix-matrix multiplication.

Table 2.3 illustrates the amount of work that the GPU will be required to perform for each floating-point value transferred to the GPU.

Effectively, Table 2.3 tells us that Level-3 BLAS operations should run efficiently on graphics processors because they perform many floating-point operations of work for every floating-point value transferred to the GPU. In addition, it is necessary to put as many level-1 and level-2 types of BLAS

TABLE 2.3

Work per Datum According to BLAS Level

BLAS Level	Data	Work	Work Per Datum
1	$O(n)$	$O(n)$	$O(1)$
2	$O(n^2)$	$O(n^2)$	$O(1)$
3	$O(n^2)$	$O(n^3)$	$O(n)$

operations into a single kernel call or risk having the GPU stall, as described earlier, because of kernel launch overhead.

The same work-per-datum analysis—and potential to stall the GPU—applies to non–BLAS-related computational problems as well. NIVDIA provides performance profiling tools (such as the CUDA profiler, cuda-prof, and Nexus) to assist developers in identifying inefficiencies in their code. Experience has shown that incorporating multiple smaller work-per-datum tasks together into a single GPU kernel can greatly boost efficiency.

Unfortunately, some problems may just be too small to justify the costs associated with transferring data to/from the GPU. The current generation of conventional processors from Intel and AMD has both large caches and decent memory bandwidth per processing core, which makes them ideal for small-scale parallel work. For example, the CUFFT library is a highly optimized fast fourier transform (FFT) library for NVIDIA CUDA-enabled GPUs. While this library can provide excellent performance, there are a number of studies in the literature and on the internet that show that it is not worth paying the data transfer overhead for smaller problems. However, GPU hardware and the CUFFT library—as well as conventional processors and libraries such as FFTW—are all evolving quickly, so please check the latest literature to determine where the break-even point might be for your problems. Please note that many FFT-intensive calculations such as Car–Parrinello quantum chemistry applications demonstrate excellent performance when running on graphics processors.

Previously, we assumed that all the kernel calls could be asynchronously queued up in the pipeline. In many cases, it is necessary to pass some value or data from the GPU to the host before a calculation can proceed. If this data transfer must occur synchronously—which is common for reduction operations like calculating a sum—then the programmer must plan on the GPU stalling for even longer periods of time. Unfortunately, synchronous data transfers will break the asynchronous pipelining of kernel launches.

In many cases, synchronization is necessary for correct program execution. For instance, CUDA can be used to generate or modify data that will be rendered by the GPU. (Mixing CUDA computational kernels and 3D rendering in the same application can deliver very high performance for scientific visualization.) This requires that the host processor wait until the GPU kernel has finished before issuing appropriate rendering commands. Otherwise, the results of a partially completed calculation might be accidentally rendered to the screen. When synchronization is required, the CUDA runtime method *cudaSynchronizeThreads*() can be called to ensure that all the kernels in the pipeline have completed.

2.3.3 Step 3: Exploiting Internal Resources on the GPU

So far, we have discussed only *global memory* on graphics processors. This is by far the largest type of memory on GPUs with capacities measured in gigabytes (a gigabyte equals one billion bytes). Most data will reside in global memory. It is also the source and destination for most data transfers.

Data reuse is a key characteristic of application kernels that achieve high performance on graphics processors. Our simple vector multiply example will not perform well as it demonstrates no data reuse—each element of vectors **a**, **b**, and **c** are used only once. As a result, global memory bandwidth becomes the rate-limiting factor for computational throughput.

Conventional processors also rely on data reuse to achieve high-computational efficiency. For smaller problems, internal processor caches can transparently buffer application data. In this way, any data reuse can be exploited allowing the processor to deliver a high computational throughput on smaller problems or those that heavily utilize a smaller amount of data that fits well in cache.

These caches are an important reason why it is currently advantageous to run some problems, such as smaller FFTs, on the local host processor(s) rather than on the GPGPU. Once cache utilization starts to drop, conventional processors also become memory bandwidth limited and application performance can precipitously drop.

Graphics processor hardware is evolving quickly. NVIDIA is attempting to double the performance of their hardware roughly every 18 months. They recently announced the newest generation of GPGPU architecture, Fermi, which incorporates a local cache on the multiprocessors. The advent of local multiprocessor cache has the potential to greatly expand the domain of problems that run efficiently on GPGPU hardware as well as ease the development effort for programmers. As mentioned in the FFT discussion, please look to the latest performance studies for information that can help decide how to most effectively allocate work between the host and GPGPU devices.

GPGPU programmers also have the ability to declare variables in several CUDA memory types to both reuse data and exploit various performance characteristics.

2.3.3.1 Register and Shared Memory

Obviously, the on-chip register and shared-memory types are highly desirable from a performance standpoint because they have single-cycle access times. Because they are located on the multiprocessor chip, they are also a very limited resource. NVIDA provides the CUDA occupancy calculator in the form of an Excel spreadsheet to help developers calculate how to best utilize these scarce resources. This spreadsheet is freely downloadable from the NVIDIA web site, and predefines the various multiprocessor register and shared-memory capacities for each generation of NVIDIA GPGPU architecture.

The CUDA occupancy calculator is an indispensible tool that allows GPGPU developers to balance their use of these scarce on-chip memories and greatly increase the performance of their application kernels. Using too many registers, for example, will force the compiler to utilize local memory for register storage. As can be seen in Table 2.4, local memory is significantly

TABLE 2.4
CUDA Memory Types

Memory	Location	Cached	Access	Scope	Access Latency
Register	On-chip	No	Read/write	One thread	Single cycle
Shared	On-chip	N/A	Read/write	All threads in a block	Single cycle
Constant	Off-chip	Yes	Read	All threads + host	One to hundreds of cycles depending on cache locality
Texture	Off-chip	Yes	Read/write (CUDA 2.2 and later)	All threads + host	One to hundreds of cycles depending on cache locality
Global	Off-chip	No	Read/write	All threads + host	*slow* (400–600 cycles)
Local	Off-chip	No	Read/write	One thread	*slow* (400–600 cycles)

slower than register memory, which implies dire performance consequences for those application kernels that spill registers to local memory.

Shared memory is effectively *the* high-speed pathway to share information between the threads of a thread block. CUDA developers generally spend much effort in partitioning problems to eliminate communications between threads that do not belong to the same thread block. Otherwise, slower global memory must be used with potentially drastic performance implications. Please note that shared memory—like global memory—is also subject to bank conflicts and the resulting serialization of accesses will quickly reduce performance.

2.3.3.2 Constant Memory

Constant memory is hardware optimized for the case when all threads read from the same location. Essentially, data can be broadcast from constant memory to all threads with one cycle of latency when there is a cache hit. This is remarkable given that constant memory resides in global memory. The constant memory cache includes an intelligent prefetch mechanism so the first access incurs only one cycle of latency. Constant memory can only be written via a data transfer from the host with the CUDA runtime method *cudaMemcpyToSymbol*(). It is also persistent across kernel calls within the same application, which allows constants to be loaded at startup and utilized throughout the application lifetime.

2.3.3.3 Texture Memory

From a C-programmer's perspective, texture memory provides an unusual combination of cache memory (separate from register, global, and shared memory), local processing capability (separate from the thread processors), and a necessary path for CUDA to interact with the display capabilities of the GPU. It is possible, through the judicious use of the caching behavior of texture memory to avoid the bandwidth limitations of global memory and greatly accelerate GPU application performance.

The easiest way to think of texture memory is as a simple hardware interface with limited processing capability that the CUDA programmer can bind to arbitrary regions of the global memory. GPGPUs have multiple texture units, each of which

- Has roughly 8 KB (a KB, or kilobyte, is a thousand bytes) of local memory per multiprocessor to prefetch data from global memory.
- Is optimized for 2D spatial locality and can provide a performance boost when all the threads in a warp access nearby locations in the texture according to this expectation of locality.

Also, please note that texture memory can be used as a limited form of high-performance random-access memory! One ingenious mapping of a random-access bloom filer to texture memory has been implemented within the CUDA-EC software [4].

Potentially the most important capability texture memory provides is the ability to mix CUDA and visualization code (either OpenGL or DirectX based) within the same application. Very high performance can be achieved because data never needs to be moved off the GPU—thus avoiding both host computer and PCIe bottlenecks. As a result, mixed CUDA and graphics programs can perform complex data creation, modification, and rendering at hundreds of frames per second on the current generation of mid- and high-end CUDA-enabled graphics processors such as the GeForce GTX 285.

Other performance benefits of texture memory include the following:

- Packed data may be broadcast to separate variables in a single operation.
- 8-bit and 16-bit integer input data may be optionally converted to 32-bit floating-point values in the range [0.0, 1.0] or [–1.0, 1.0] by the texture unit hardware.
- Linear, bilinear, and trilinear interpolation is performed using dedicated hardware separate from the thread processors.

2.3.3.4 Global Memory

Global memory is by far the largest memory space, with capacities measured in gigabytes. Briefly, global memory

- Is potentially 150× slower than register or shared memory. Data reuse enabled by other GPGPU memory types is required to prevent thread data starvation and achieve high performance.

- Requires coalesced operations to deliver the highest performance.
- Has the lifetime of the application.
- Is accessible from either the host or the device.

Much of the discussion in this chapter has centered on the memory bandwidth and latency issues associated with global memory, which also reflects much of where the effort is spent when programming GPGPUs. Without question, data reuse is a mandatory requirement to achieve high-GPGPU kernel performance. Massive multithreading helps because it provides the large thread count that utilizes the capability of each multiprocessor to support many outstanding load requests. As a result, load-to-use latency can be hidden when accessing data stored in global memory.

The recently announced Fermi architecture, or 20-series of NVIDIA hardware, provides two important advances related to global memory:

1. Local multiprocessor caches facilitate data reuse and enable more applications to achieve high performance.
2. It breaks the 4 GB global memory barrier imposed by the 32-bit addressing space utilized in previous generations of GPGPUs.

2.3.3.5 Local Memory

Local memory is actually a memory abstraction utilized exclusively by the compiler that implies "local in the scope of each thread." It is not an actual hardware component but rather is allocated and resides in global memory. Normally, automatic variables declared in a kernel reside in registers, which provide the fastest form of memory access. Performance can drastically drop when registers are spilled to local memory, so it is important to understand the circumstances that might cause the compiler to place automatic variables in local memory. They include the following:

- There are too many register variables.
- A structure would consume too much register space.
- The compiler cannot determine if an array is indexed with constant quantities.

2.4 Visualization

Amusingly, the focus of this chapter has been on utilizing the massively threaded computational capabilities provided by graphics processors—while the remarkable ability of this same hardware to render graphics has been largely ignored. Suffice it to say that GPGPUs exhibit extraordinary capabilities to display information as well.

Through the use of CUDA, high-performance visualization is possible without requiring that data be moved on and off the GPU. As with computational kernels, eliminating data transfers over the PCI bus removes a significant bottleneck and can dramatically speed the rendering process. Much like combining calculations in a single kernel, newer OpenGL constructs such as *primitive restart* provide significant performance benefits for the CUDA programmer by effectively combining multiple rendering operations into a single call. Unlike other OpenGL calls, primitive restart only utilizes information defined and kept on the GPU and does not require transferring any index or length arrays from the host to the GPU. A working example of primitive restart is provided in article sixteen of my *Doctor Dobb's Journal* series of online tutorials [5].

CUDA enables the efficient interoperability with graphics by mapping a pointer from the graphics buffer(s) to CUDA. Once provided with the mapped pointer(s), CUDA programmers are then free to exploit their knowledge of CUDA to write fast and efficient kernels to operate on the graphics buffers. The separation is distinct because graphical manipulation of buffers currently mapped by CUDA is not allowed.

There are two very clear benefits of the separation (yet efficient interoperability) between CUDA and graphics:

- *From a programming view:* When not mapped into the CUDA memory space, graphics gurus are free to exploit existing legacy code bases, their expertise and the full power of all the tools available to them.
- *From an investment view:* Probably the most important benefit of this separated approach is the efficient exploitation of existing legacy visualization software investments. Essentially, CUDA code can be incrementally added to existing visualization software without assuming significant risk or requiring that substantial portions of the code be rewritten.

2.5 Conclusion

It is possible today to purchase a commodity graphics processor at your favorite electronics store that nearly doubles the world record–breaking teraflop supercomputer performance announced in December 1996 by Sandia National Laboratory. Combining four of these same graphics processors that retail for less than $500 in a *dedicated* workstation can provide a student with over 7 teraflops of dedicated computational capability, which compares favorably with the recently decommissioned 11.8 teraflop MPP2 supercomputer provided by the Pacific Northwest National Laboratory for *shared* use by their entire world-wide user community. The latest generation of GPGPUs, announced but as yet untested, tout significantly greater computational capabilities and should be available in the first half of 2010.

Massively threaded programming models are the key to exploiting the capabilities of this massively parallel hardware technology. The myriad of scientific applications that now utilize GPGPU technology provide ample proof that scientists and engineers find the existing CUDA programming model, compiler, and development tools to be adequate. Newer models and programming pragmas are being advanced to make massively threaded programming even more general and familiar to C- and FORTRAN programmers.

With so many GPGPU massively threaded applications delivering orders of magnitude greater performance than multithreaded software running on conventional hardware, it is clear that extraordinary scientific advances are possible. Technology is only a tool, and scientists who combine knowledge of massively threaded programming with perceptiveness and insight are the ones who will likely make the most significant advances in the coming years. In a very real sense, computational biology and bioinformatics are transitioning from the "light-based microscopy" of conventional computing to the powerful new GPGPU "electron microscopes" of commodity massively parallel computing.

Several excellent sources of information exist on the internet that provide more detailed information about GPGPU computing and CUDA. I recommend visiting the NVIDIA CUDA Zone web site for the latest information on both CUDA software and hardware [6]. Those who wish to learn CUDA should consult my "Supercomputing for the Masses" series of tutorials that are freely available online at the Doctor Dobb's journal web site [5].

2.6 References

1. The Connectome Project. http://iic.harvard.edu/research/connectome. Accessed February 15, 2010.
2. R. Farber. GPGPUs: Neat idea or disruptive technology. *Scientific Computing*, 24(11), (January 2008).
3. NVIDIA Cuda Zone. http://nvidia.com/cuda http://iic.harvard.edu/research/connectome. Accessed February 15, 2010.

4. H. Shi, B. Schmidt, W. Liu, W. Mueller-Wittig. Accelerating error correction in high-throughput short-read DNA sequencing data with CUDA, Proceedings of the 2009 IEEE International Symposium on Parallel and Distributed Processing (IPDPS).
5. *Doctor Dobb's Journal.* http://www.ddj.com http://iic.harvard.edu/research/connectome Accessed February 15, 2010.
6. R. Farber. Using vertex buffer objects with CUDA and OpenGL, 2009, to be published online.

3

FPGA: Architecture and Programming

Douglas Maskell

3.1 Introduction

The history of the field-programmable gate array (FPGA) dates back to the 1970s with the commercial development of programmable logic array (PLA) and programmable array logic (PAL) devices. While PLA and PAL devices have evolved into today's complex programmable logic devices (CPLD), FPGA development took a slightly different route more akin to gate array technology. The first commercially successful FPGA, the Xilinx XC2064 [1], was developed by Ross Freeman and Bernard Vonderschmitt in 1985. This device, called a *logic cell array*, consisted of three different types of user configurable elements: configurable logic blocks (CLB), configurable I/O blocks (IOB), and programmable interconnect. The schematic layout of the XC2064 logic cell array and the CLB structure is shown in Figure 3.1. The configuration of these elements was achieved by writing data to the configuration memory to establish the various logic functions and connections. However, it was not until 1989 when Stan Baker of EETimes came up with the term FPGA that these devices became known as field-programmable gate arrays or FPGAs.

Since their introduction, FPGAs have evolved from relatively small homogeneous arrays of configurable elements to large heterogeneous computing systems consisting of a programmable fabric, an embedded memory, and a range of hardwired functional units, such as multipliers, communications

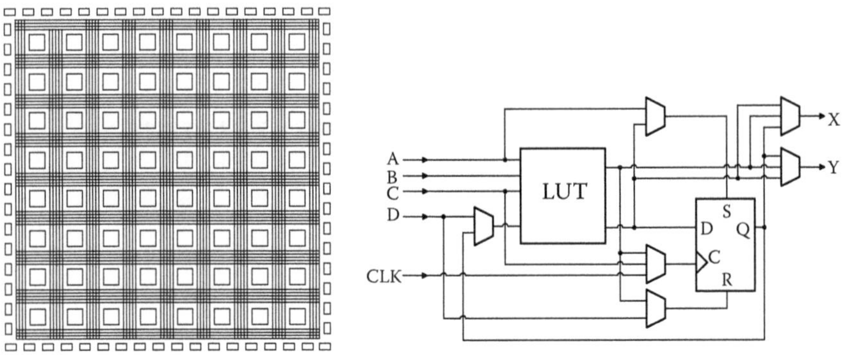

(a) (b)

FIGURE 3.1
The XC2064 logic cell array [1], (a) schematic layout, showing the array of CLBs with the configurable IOBs around the outside and the programmable interconnect, and (b) schematic of a CLB.

transceivers, digital signal processing (DSP) blocks, and even complete microprocessors. The major FPGA vendors also provide a range of intellectual property (IP) soft cores, including soft-core processors and peripherals, and a comprehensive development environment to compose complete FPGA systems. Modern FPGAs provide multimillion equivalent gates with embedded computing capability and are large enough for entire systems to be implemented onto a single chip. These devices are sometimes called *system on programmable chip* (SoPC) or platform FPGA to parallel the terminology used in application-specific integrated circuit (ASIC) design.

3.2 The Need for FPGA Computing

As the demand for computing resources increases, central processing unit (CPU) development has relied on a combination of clock speed increases and changes to the architecture to improve instruction level parallelism. However, there is a growing performance gap between the capabilities of the microprocessor and the data transfer rate of memory. To fill this gap, techniques such as caching, pipelining, out-of-order execution, and branch prediction are being pushed to their limits. These microarchitectural changes come at a cost, as running a system with extra circuitry and at a higher frequency consumes more power and dissipates more heat. As a result, CPU designers have moved toward multicore and many-core technologies to improve performance while keeping thermal dissipation within manageable limits. Unfortunately, multicore architectures have not really addressed the memory bottleneck issue and have introduced a completely new set of problems. To tap their full potential, multicore systems require more sophisticated operating systems and better parallelizing compilers, while applications developers require time to adapt and to become proficient with both the tools and the platform itself. The bottom line is that when it comes to accelerating algorithms, systems based on processors alone do not scale well with increasing frequency. Instead, other technologies, which can either work in conjunction with existing computing platforms or even replace them, need to be developed. One such technology is the FPGA-based reconfigurable computing accelerator.

Virtually since the FPGA's inception, it has been used for application acceleration, particularly in the DSP and digital communications areas. More recently, as FPGA devices have become larger, they are being increasingly seen as enabling technology, used alongside conventional high-performance computing systems, to accelerate computational tasks. Unlike conventional microprocessors, which must transfer data from/to registers, or memory, as part of the computation cycle, FPGAs can stream data between different processing elements eliminating the bulk of the data transfers. FPGA computing can achieve higher performance by exploiting the massive inherent parallelism and by using flexibility to specialize the architecture. A significant advantage over CPU devices is achieved because FPGA resources are configurable. Precomputed lookup tables are used in place of complex mathematical functions. Issues of memory bandwidth are alleviated by having access to multiple memory banks in parallel. The use of custom circuits means that the application can be better matched to the architecture. For example, in bioinformatics research, the algorithms for gene and protein sequence matching do not require the full integer precision available in modern CPUs, let alone needing floating point arithmetic. Therefore, adapting the application so that it exploits its inherent precision and parallelism, results in better utilization of the available resources even though the hardware may be clocked at a significantly lower frequency than that of the latest CPUs. This approach not only achieves algorithm acceleration, but also usually results in a reduced power consumption and heat dissipation.

The fact that the FPGA can be reprogrammed at will and with little effort makes it very attractive for developing efficient implementations of algorithms. Because many of these algorithms often originate in the form of a software application that requires acceleration, converting the implementation from software to hardware requires a fair amount of tweaking and experimentation. The FPGA offers precisely the right combination of resources and flexibility to help develop, test, and evaluate the performance of an algorithm's implementation. The developer can, and must, choose which portions of the algorithm to implement in FPGA hardware and which should be left to execute on a CPU. This process is known as *hardware-software partitioning*. Co-synthesis techniques that apply a hardware-software partitioning algorithm to automatically generate a software model that can be cross-compiled and a hardware model that can be synthesized are fairly mature [2–4]. The end result is that a sequence of instructions is converted into a hardware circuit that is functionally equivalent.

3.3 FPGA Computing Architectures

CPUs are designed to handle a rich mix of operations, while FPGAs are able to accelerate a well-defined set of repetitive operations. In many software applications, most of the computation time is attributable to only a small portion of the application, with the execution of the larger part of the code, which is necessary for completeness, having little effect on the performance. Consequently, an interesting hybrid computing approach couples a CPU with an FPGA fabric [5]. This hybrid approach has seen several systems proposed that couple a general-purpose CPU with a reconfigurable array. An early example of this is the dynamic instruction set computer (DISC) [6]. Later examples include Garp [7] and CHIMAERA [8]. This was followed by commercial hybrid platforms, including the Triscend A7 and E5 CSoC Families [9, 10], the QuickSilver Technology ADAPT2400 ACM Architecture [11], and the Stretch S5 and S6 SCP Engines [12].

An increasing number of new chips include a portion of FPGA-like fabric to add flexibility to adapt to a number of applications. This trend is due to companies such as M2000 (now Abound Logic) who market FPGA fabric for integration into new ASIC devices [13]. While many hybrid architectures have been proposed, most have had little commercial success. As of writing the only commercial hybrid device still existent in the market is the Stretch S6. The mix of paradigms has the effect of mixing the complexities of application design for software and hardware. To adopt such a device, a company has to use specialized compiler tools that are as yet unproven. Thus, the benefit of using such a device does not outweigh the risk inherent in complex systems that are only supported by a single vendor.

The FPGA vendors have had slightly more success introducing processors embedded inside their FPGA fabrics. Examples of commercially available platforms are the Xilinx Virtex-2 Pro, Virtex-4 FX, and Virtex-5 FXT FPGAs with an embedded PowerPC [14], and the Altera Excalibur with an embedded ARM [15]. In contrast to the hybrid architecture vendors mentioned previously, the FPGA development environments of Altera and Xilinx are well established. They have built on this advantage and integrated processor cores that have well-established programming environments. What is interesting is that Xilinx has no plans to incorporate a hard-core processor into its latest generation Virtex-6 devices, while Altera after exiting the hard-core processor arena in 2002 has recently licensed the MIPS32 RISC architecture from MIPS Technologies [16]. Alternatively, soft-core processors offer a cheaper and lower risk alternative to hybrid architectures, albeit at a lower speed. Soft-core processors use standard FPGA resources to build microprocessor architectures. In addition to the vendor-specific cores, such as the Xilinx Microblaze [14] and Altera NIOS-2 [15], there are a range of both commercial and open-source cores on offer. Commercial cores include ARM's FPGA-optimized Cortex M1 processor [17] and Freescale's V1 Coldfire processor

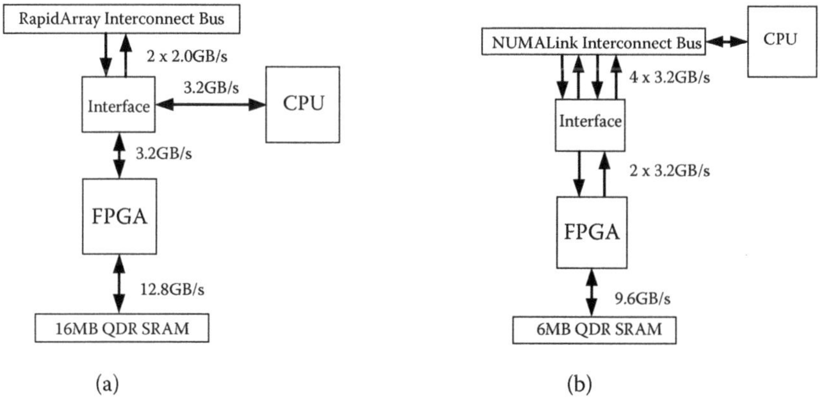

FIGURE 3.2
(a) XD1 architecture (Modified from Cray, Inc., http://www.cray.com/), and (b) the SCI RASC architecture (Modified from SGI, http://www.sgi.com/).

[18] while open-source cores include the LEON SPARC [19] and Opencores OpenRISC [20].

As FPGA technology has developed, its use as a computing platform has moved from academic research into the commercial domain. Many companies offer PCI bus FPGA accelerators with the necessary IP to develop hardware accelerated solutions. High-performance computing platforms based on FPGA initially developed as research platforms [21, 22] are now a commercial reality, with companies such as Starbridge Systems [23], Nallatech [24], and SRC Computers [25] offering high-performance FPGA-based computing solutions. FPGA technology was also briefly adopted by high-profile supercomputing companies such as Cray with their XD1 platform [26], and SGI with their RASC platform [27], although both appear to have discontinued these product lines. The architecture of the Cray XD1 and the SGI RASC platforms showing the FPGA accelerator is given in Figure 3.2.

The mainstream microprocessor manufacturers are also showing interest in FPGA accelerators. What started with AMD's strategy to improve the potential of their Opteron processor-based platforms by opening up the HyperTransport specification [28] has resulted in a number of tightly coupled FPGA accelerators targeted at both AMD and Intel processors. In opening up HyperTransport, AMD had hoped that the resulting application-specific accelerators, closely coupled with their processors, would provide platforms that outperform their competitors, thus giving them a competitive edge. Initially, two companies developed FPGA platforms that slot into a available Opteron socket on a multiple processor motherboard. DRC Computer offers a reconfigurable Xilinx Virtex-4 platform [29] and XtremeData offers an Altera Stratix-2 platform [30]. Placing the FPGA at this point in the system provides a 5.4 GB/s link with up to 4 GB of memory and a 1.6 GB/s link to the other AMD

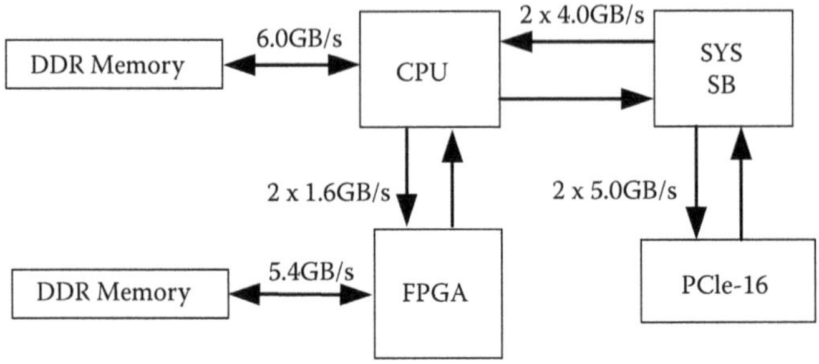

FIGURE 3.3
The XtremeData XD1000 architecture (Modified from XtremeData, Inc., http://www.xtremedatainc.com/).

Opteron processors on the motherboard, as shown in Figure 3.3. In response, Intel has also provided the ability to connect to its Xeon processor front side bus (FSB) and the QuickAssist accelerator abstraction layer. XtremeData offers an Altera Stratix III FPGA-based module, which targets the Intel FSB at 1066 MHz, while Nallatech [24] offers a stackable Vertix-5-based solution. Both Nallatech and Xtreme Data have recently announced that they are developing FPGA accelerators for Intel's new high-speed QuickPath interconnect.

3.4 FPGA Development Tools

Similar to software designs, FPGA design descriptions must be optimized and translated to a form usable at the physical hardware level. The resulting configuration data, called a *bitstream,* is then loaded into the FPGA to form the connections to make the required hardware circuit. Traditionally, an FPGA designer would develop a behavioral register transfer level (RTL) description of the required circuit using a hardware description language (HDL), such as VHDL or Verilog. This behavioral RTL representation would then be input into an FPGA compiler tool, such as Xilinx ISE [14] or Altera Quartus [15], for synthesis, logic optimization, and finally mapping to a specific FPGA.

Both VHDL and Verilog are well-established HDLs, and allow the definition of both high-level algorithms and low-level optimizations in the same language. The resultant code is reasonably straightforward for a software programmer to interpret, provided that the languages built in concurrency are understood. However, as designs become larger and more complicated, it becomes more difficult to manage the complexity at the HDL level. To meet these design challenges, automated EDA tools have been developed to make system-level design easier. System-level design tools that can effectively integrate the different design strategies within domains, so as to better leverage FPGA resources, are available from both FPGA vendor and third party

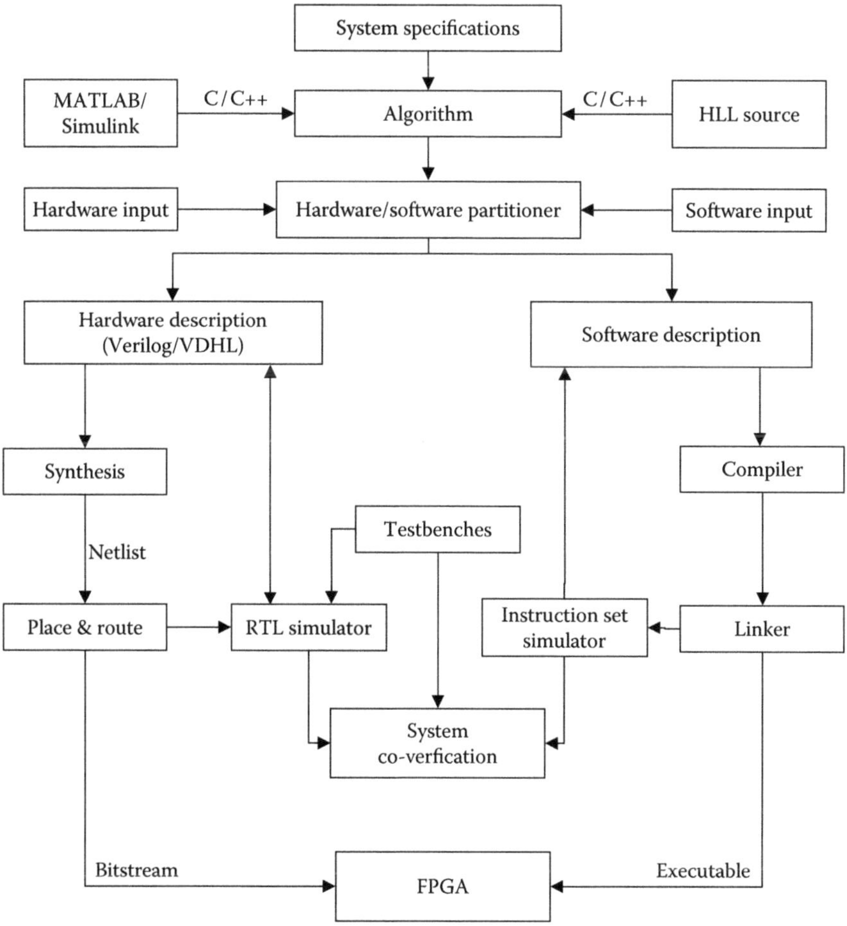

FIGURE 3.4
A typical HLL to hardware design flow.

sources. Xilinx offers a system edition of its ISE design tool that combines a DSP and embedded processing development environment with its more conventional FPGA design environment [14]. Altera provides SoPC Builder with a set of unbundled tools with similar functionality (i.e., Quartus II, Nios II EDK and DSP builder) [15]. Both these tool chains offer seamless integration with the MATLAB and Simulink tools from The Mathworks [31]. In addition, complete system-level tool chains are available from third party sources, such as Mentor Graphics [32] and Synopsys [33].

Even with the advances in EDA tools, good FPGA design still requires a significant amount of domain knowledge. Therefore, to bring FPGA computing into the mainstream, where software programmers still predominantly focus on sequential program design, a number of initiatives looking at synthesis from high-level languages (HLL) have been initiated. Enabling synthesis from a HLL also allows designers to leverage on the large amount of open-source software that is currently available. A number of commercial HLL to HDL tools have been released, most of which target translation from the C-language to hardware. These include Handel-C from Agility (acquired from Celoxica) [34], Impulse C from Impulse Accelerated Technologies [35], Dime-C from Nallatech [24], Catapult C from Mentor Graphics [32], Mitrion-C from Mitrionics [36], and C-to-Hardware from Altera [15]. Tools such as Simulink HDL Coder from The Mathsworks [31] and the Bluespec Compiler from Bluespec [37] allow a more algorithmic approach for generating hardware. A typical HLL to Hardware design flow is shown in Figure 3.4.

3.5 Discussion

There are many documented examples in the scientific literature where FPGA-based computing provides superior performance to CPU-based computing. In the DSP and Image processing domains, these include FPGA implementations of the DES data encryption algorithm providing a speedup of up to two orders of magnitude [38] and video applications with speedups of 20–100 times [39]. In the bioinformatics domain, FPGA implementations of the Smith–Waterman pairwise protein sequence alignment have achieved speedups of 120–200 times [40] that of the corresponding software implementations, while FPGA implementations of the basic local alignment search tool (BLAST) have also achieved impressive speedups [41]. These previous implementations have all been carefully hand crafted with significant manual input to identify the parallelism and the processing precision, thereby achieving the best speedup possible. Solutions based on translating from HLLs to hardware, while producing significant application acceleration [42], still have some way to go before they are able to achieve the performance of these manual designs. Pico Computing has recently announced a 5000X acceleration of a graphical tool for visualization of the comparison of two DNA sequences on an FPGA cluster [43] using C-to-FPGA tools provided by Impulse Accelerated Technologies [35]. This particular algorithm was inherently parallel, thus making it an ideal candidate for acceleration using a large number of FPGA devices.

3.6 References

1. Xilinx XC2064/XC2018 Logic Cell Array, http://www.datasheetarchive.com/XC2064-100PC68C-datasheet.html, last accessed Nov. 2009.

2. R.K. Gupta, *Co-Synthesis of Hardware and Software for Digital Embedded Systems*, Kluwer Academic Publishers, 1995.
3. R. Ernst, J. Henkel, T. Benner, Hardware–software co-synthesis for microcontrollers, *IEEE Design & Test of Computers*, 10(4), 64–75, 1993.
4. F. Balarin et al., *Hardware–Software Co-Design of Embedded Systems: The POLIS Approach*, Kluwer Academic Publishers, 1997.
5. A. DeHon, The density advantage of configurable computing, *IEEE Computer*, 33(4), 41–49, 2000.
6. M.J. Wirthlin and B.L. Hutchings, A dynamic instruction set computer, *IEEE Symposium on FPGAs for Custom Computing Machines* (FCCM '95), pp. 99–107, Napa Valley, CA, USA, Apr. 1995.
7. J.R. Hauser, Augmenting a microprocessor with reconfigurable hardware, PhD dissertation, University of California, Berkeley, CA, 2000.
8. Z.A. Ye, A. Moshovos, S. Hauck, P. Banerjee, CHIMAERA: a high-performance architecture with a tightly-coupled reconfigurable functional unit, *Proceedings of the 27th Annual International Symposium on Computer Architecture*, pp. 225–235, Vancouver, BC, Canada, 2000.
9. Triscend E5 Customizable Microcontroller Platform, http://www.datasheetarchive.com/datasheet-pdf/027/DSA00484394.html, last accessed Nov. 2009.
10. Triscend A7 Configurable System-on-Chip Platform, http://www.datasheetarchive.com/datasheet-pdf/04/DSA0065723.html, last accessed Nov. 2009.
11. QuickSilver Technology, Inc., http://www.qstech.com/default.htm, last accessed Nov. 2009.
12. Stretch, Inc., http://www.stretchinc.com/, last accessed Nov. 2009.
13. Raptor FPGA, http://www.aboundlogic.com/index.html, last accessed Nov. 2009.
14. Xilinx, Inc., http://www.xilinx.com/, last accessed Nov. 2009.
15. Altera Corporation, http://www.altera.com/, last accessed Nov. 2009.
16. Altera Licenses MIPS32 Processor Architecture, http://www.edn.com/blog/980000298/post/1410049541.html, Oct. 2009, last accessed Mar. 2010.
17. ARM Cortex-M1, http://www.arm.com/products/CPUs/ARM_Cortex-M1.html, last accessed Nov. 2009.
18. Freescale ColdFire V1 Core, http://www.freescale.com/, last accessed Nov. 2009.
19. Aeroflex Gaisler AB, LEON3 SPARC V8 Processor core, http://www.gaisler.com/cms/, last accessed Nov. 2009.
20. OpenRISC 1000 architecture, http://www.opencores.org/openrisc, last accessed Nov. 2009.
21. A. Patel, C.A. Madill, M. Saldana, C. Comis, R. Pomes, P. Chow, A scalable FPGA-based multiprocessor, *14th Annual IEEE Symposium on Field-Programmable Custom Computing Machines*, pp. 111–120, Napa Valley, CA, USA, Apr. 2006.
22. G. Pfeiffer, S. Baumgart, J. Schroeder, M. Schimmler, A massively parallel architecture for bioinformatics, *Computational Science—ICCS 2009*, International Conference on Computational Science, LNCS 5544, 994–1003, 2009.
23. Star Bridge Systems, Inc., http://www.starbridgesystems.com/, last accessed Nov. 2009.

24. Nallatech, Inc., http://www.nallatech.com/, last accessed Nov. 2009.
25. SRC Computers, LLC, http://www.srccomputers.com/index.asp, last accessed Nov. 2009.
26. Cray, Inc., http://www.cray.com/, last accessed Nov. 2009.
27. SGI, http://www.sgi.com/, last accessed Nov. 2009.
28. HyperTransport Consortium, http://www.hypertransport.org/, last accessed Nov. 2009.
29. DRC Computer Corporation, http://www.drccomputer.com/, last accessed Nov. 2009.
30. XtremeData, Inc., http://www.xtremedatainc.com/, last accessed Nov. 2009.
31. The MathWorks, Inc., http://www.mathworks.com/, last accessed Nov. 2009.
32. Mentor Graphics, Inc., http://www.mentor.com/, last accessed Nov. 2009.
33. Synopsys, Inc., http://www.synopsys.com/, last accessed Nov. 2009.
34. Agility DK Design Suite, http://agilityds.com/, last accessed Nov. 2009.
35. Impulse Accelerated Technologies, Inc., http://www.impulseaccelerated.com/, last accessed Nov. 2009.
36. A.B. Mitrionics, http://www.mitrionics.com/, last accessed Nov. 2009.
37. Bluespec, Inc., http://www.bluespec.com/, last accessed Nov. 2009.
38. C. Patterson, High performance DES encryption in Virtex FPGAs using JBits, *IEEE Symposium on Field-Programmable Custom Computing Machines*, pp. 113–121, Napa Valley, CA, USA, Apr. 2000.
39. Z. Guo, W. Najjar, F. Vahid, K. Vissers, A quantitative analysis of the speedup factors of FPGAs over processors, *International Symposium on Field Programmable Gate Arrays*, pp. 162–170, Monterey, USA, 2004.
40. T.F. Oliver, B. Schmidt, D.L. Maskell, Reconfigurable architectures for biosequence database scanning on FPGAs, *IEEE Transactions on Circuits and Systems II*, 52(12), 851–855, 2005.
41. P. Krishnamurthy et al., Biosequence similarity search on the mercury system, *J. VLSI Signal Processing Systems*, 49(1), 101–121, 2007.
42. Y.L. Aung, D.L. Maskell, T.F. Oliver, B. Schmidt, W. Bong, C-based design methodology for FPGA implementation of ClustalW MSA, *Pattern Recognition in Bioinformatics 2007*, LNCS, 4774, 11–18, 2007.
43. Pico Computing, Inc., FPGA cluster accelerates bioinformatics application by 5000×, http://www.picocomputing.com/pdf/PR_Pico_Bioinformatics_Nov_9_2009.pdf, last accessed Nov. 2009.

4

Parallel Algorithms for Alignments on the Cell BE

Abhinav Sarje and Srinivas Aluru

Alignment of biological sequences enables discovery of evolutionary and functional relationships among them. Computing alignments, as a means to elucidate different kinds of sequence relationships, is a fundamental tool arising in numerous contexts and applications in computational biology. A number of algorithms for sequence alignments have been developed in the past few decades, the most common being the ones for pairwise global alignments (aligning sequences in their entirety [1]) and local alignments (aligning sequences that each contain a substring that is highly similar [2]). Some applications require more complex types of alignments. One such example is when aligning an mRNA sequence transcribed from a *eukaryotic* gene with the corresponding genomic sequence to infer the gene structure [3]. A gene consists of alternating regions called *exons* and *introns,* while the transcribed mRNA corresponds to a concatenated sequence of the exons. This requires identifying a partition of the mRNA sequence into consecutive substrings (the exons) that align to the same number of ordered, nonoverlapping, nonconsecutive substrings of the gene, a problem known as *spliced alignment.* Another important problem is that of *syntenic alignment* [4], for aligning two sequences that contain conserved substrings that occur in the same order (such as genes with conserved exons from different organisms, or long syntenic regions between genomes of two organisms). An illustration of these four kinds of alignments, showing how the regions of two sequences are aligned, is given in Figure 4.1.

Dynamic programming is the most commonly used method for computing pairwise alignments [1–4], and takes time proportional to the product of the lengths of the two input sequences (although the original Smith–Waterman algorithm for local alignment [2] has cubic complexity, it is widely known that this can be implemented in quadratic time, as is shown in [5]; also, [3] presents an algorithm with cubic complexity, but the spliced alignment problem can be treated as a special case of syntenic alignment and solved in quadratic time, as will be described later in this chapter). Various parallel algorithms have also been developed for these methods. Some of these parallelize the computations within a single processor utilizing its vector processing units and single-instruction multiple-data (SIMD) style instructions [6, 7], while other algorithms deal with parallelization across multiple processors [8–10]. We mainly focus on the latter in this chapter, and present algorithms for computing pairwise alignments in parallel on the Cell Broadband Engine (CBE). We first describe the global/local alignment algorithms using dynamic programming, and a basic parallel computation strategy using the wavefront communication pattern. Using this strategy, alignment scores can be computed in parallel across the different synergistic processing elements of the Cell processor. Though this parallel strategy allows efficient computation of alignment scores in linear space, retrieving the actual optimal alignment requires quadratic space. We then present a linear space parallel algorithm for the Cell processor, which overcomes this limitation and computes an actual optimal alignment. Further, we describe how this algorithm can also be extended to the more specialized spliced and syntenic alignment problems.

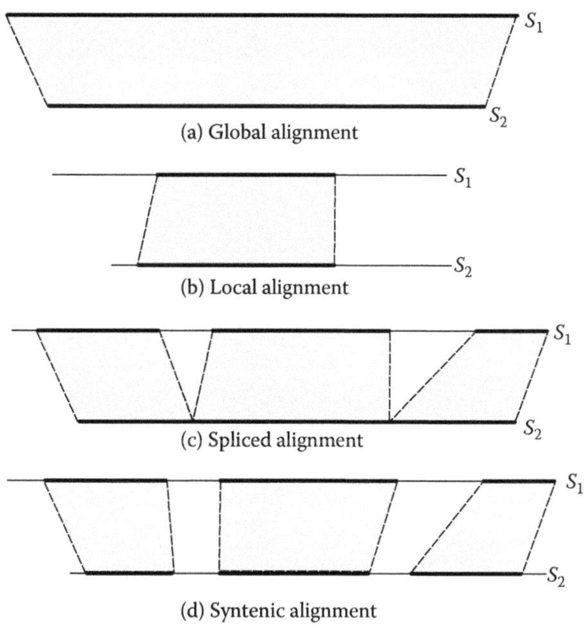

FIGURE 4.1
Genomic alignments—the thick portions of sequences S_1 and S_2 show the segments that are aligned. (a) Global alignment: Both sequences are aligned in their entirety. (b) Local alignment: A substring from each sequence is aligned. (c) Spliced alignment: Ordered series of substrings of one sequence is aligned to the entire second sequence. (d) Syntenic alignment: Ordered series of substrings of one sequence is aligned with ordered series of substrings on the second sequence. For (b), (c), and (d), the goal includes finding the aligning regions such that the score of the resulting alignment, as given by a score function, is maximized. Both the number and boundaries of aligning regions are unknown and need to be inferred by the algorithm. Only the sequences S_1 and S_2 are the input for each alignment problem. (From Sarje, A. and Aluru, S., *IEEE Transactions on Parallel and Distributed Systems*, 20(11):1600–1610, 2009. With permission. © 2009 IEEE.)

4.1 Computing Alignments

We start with a brief description of the sequential dynamic programming algorithm for computing global alignments. Consider two sequences, $S_1 = a_1a_2 \ldots a_m$ and $S_2 = b_1b_2 \ldots b_n$ over an alphabet Σ, and let "–" denote the gap character. A global alignment of the two sequences is a $2 \times N$ matrix, where $N \geq max(m, n)$, such that each row represents one of the sequences with gaps inserted in certain positions and no column contains gaps in both sequences. The alignment is scored as follows: a function, *score*: $\Sigma \times \Sigma \rightarrow \mathbb{R}$, prescribes the score for any column in the alignment that does not contain a gap. We assume the *score* function returns the score for any pair of characters from Σ in constant time. *Affine gap penalty* functions are commonly used to determine the scores of columns involving gaps, so that a sequence of gaps is assigned less penalty than treating them as individual gaps—this is because a mutation affecting a short segment of a genomic sequence is more likely than several individual base mutations. Such a function is defined as follows: for a maximal consecutive sequence of k gaps, a penalty of $h + gk$ is applied. Thus, the first gap in a maximal sequence is charged $h + g$, while the rest of the gaps are charged g each. When $h = 0$, the scoring function is called a *constant gap penalty* function. The score of the alignment is the sum of scores over all the columns.

The global alignment problem with affine gap penalty function can be solved using three $(m + 1) \times (n + 1)$ sized dynamic programming tables, denoted C, D (for deletion), and I (for insertion). An element $[i, j]$ in a table is used to store the optimal score of an alignment between $a_1a_2 \ldots a_i$ and $b_1b_2 \ldots b_j$ with the following restrictions on the last column of the alignment: a_i is matched with b_j in C, a gap is matched with b_j in D, and a_i is matched with a gap in I. The tables can be computed using the following recursive equations, which can be applied row by row, column by column, or antidiagonal by antidiagonal (also called *minor diagonal*):

$$C[i,j] = score(a_i, b_j) + max\begin{cases} C[i-1,j-1] \\ D[i-1,j-1] \\ I[i-1,j-1] \end{cases} \tag{4.1}$$

$$D[i,j] = max\begin{cases} C[i,j-1]-(h+g) \\ D[i,j-1]-g \\ I[i,j-1]-(h+g) \end{cases} \tag{4.2}$$

$$I[i,j] = max\begin{cases} C[i,j-1]-(h+g) \\ D[i,j-1]-(h+g) \\ I[i,j-1]-g \end{cases} \tag{4.3}$$

The first row and column of each table are initialized to $-\infty$ except in the following cases ($1 \le i \le m$; $1 \le j \le n$):

$$C[0, 0] = 0$$
$$D[0, j] = h + gj$$
$$I[i, 0] = h + gi$$

The maximum of the scores among $C[m, n]$, $D[m, n]$, and $I[m, n]$ gives the optimal global alignment score. By keeping track of a pointer from each entry to one of the entries that resulted in the maximum while computing its score, an optimal alignment can be constructed by retracing the pointers from the optimal score at bottom right to the top left corner of the tables. This procedure is known as *trace-back*.

4.2 Sequence Alignments on the Cell Processor

A number of bioinformatics applications dealing with pairwise genomic alignments have been ported to the CBE (e.g., [11–14]). Many such applications employ aligning an input sequence to sequences from a large database and obtaining their alignment scores, for example, BLAST, ClustalW, FASTA, and Ssearch. Porting of these applications to the Cell processor is discussed in [12–14]. These methods for sequence alignments on the CBE are restricted to the basic Smith–Waterman algorithm [2] for local alignments. The basic idea common to them is that all the alignments to be performed between the input sequence and each sequence from the database are independent of each other, and can be computed individually on each of the synergistic processing elements (SPEs) of the CBE. The PowerPC processing element (PPE) assigns sequences for independently computing the alignment scores to the different SPEs, each of which then simultaneously computes the score for the pair of sequences assigned to it and reports the score back to the PPE. Hence, the individual SPEs do not need to synchronize with each other.

Although computation of alignment scores is useful in statistical analyses to assess the alignment quality, or to find a small subset of sequences, which have a high similarity with the query sequence, these implementations do not compute the actual alignment, which is necessary to gain biological insight into the genomic sequences being aligned. Moreover, they work for smaller sequence sizes as only small local stores (256 KB) are available on each SPE to store the whole sequences and the dynamic programming tables to be computed. Therefore, in this chapter we focus on parallel algorithms to compute a single optimal pairwise alignment on the CBE. To perform the alignments in parallel, the input sequences and the dynamic programming table computations need to be distributed among the various SPEs, which also need to synchronize the computations among themselves. In the following section we describe these distribution and communication strategies.

4.3 A Parallel Communication Scheme

In this section, we first describe a parallel communication strategy that is commonly employed by many parallel alignment algorithms on the CBE [15–17]. This scheme, popularly known as the *wavefront communication scheme*, was first proposed by Edmiston et al. [9]. In this method, each table is divided into a $w \times p$ matrix of block, where p is the number of processing elements to be used and w is the number of blocks in one column ($w = p$ in the original algorithm). Therefore, each block contains at most $\lceil n/p \rceil$ columns

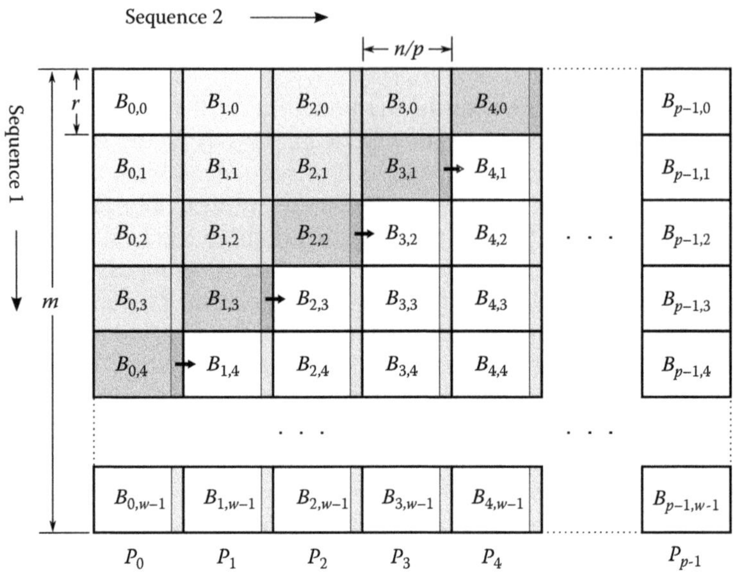

FIGURE 4.2
Block division in the wavefront technique—SPE j is assigned a column of blocks $B_{i,j}$, $0 \le i < w$, as shown by the labels P_j below each column. Block computations follow diagonal wavefront pattern, where for antidiagonal t, blocks $B_{i,j}$ such that $i + j = t$, are computed simultaneously in parallel (shown by blocks in the same shade of gray). SPE P_j $(0 \le j < p)$ sends the rightmost computed column in its assigned block (shown as thin shaded columns) to SPE P_{j+1} for the next iteration. (From Sarje, A. and Aluru, S., *IEEE Transactions on Parallel and Distributed Systems*, 20(11):1600–1610, 2009. With permission. © 2009 IEEE.)

and r rows, where r is a prechosen block size. Let $B_{i,j}$ denote a block, where $0 \le i < w$ and $0 \le j < p$. Each processing element is assigned a unique column of blocks to compute: processor P_j computes the blocks $B_{i,j}$. The blocks are simultaneously computed one antidiagonal at a time. All blocks on an antidiagonal can be computed simultaneously as they depend only on blocks on the previous two antidiagonals—computation of a block $B_{i,j}$ on the antidiagonal t, where $t = i + j$, only depends on blocks $B_{i-1,j}$ and $B_{i,j-1}$ from antidiagonal $t - 1$, and block $B_{i-1,j-1}$ from antidiagonal $t - 2$. Because of the block assignment to processing elements, each of them only needs to receive the last column of a block (plus an additional element) from the previous processing element. An illustration of this wavefront pattern is shown in Figure 4.2. Each SPE receives all of the first sequence (length m) and a distinct $\frac{n}{p}$ length substring of the second sequence. The total number of rounds of block computations in this scheme is equal to the number of block antidiagonals, which is equal to $p + w - 1$, although each SPE computes exactly w blocks. The SPEs need to synchronize with each other after transferring their corresponding rightmost column to the next SPE. For an efficient implementation, this can be achieved using the signal notification registers on the SPEs.

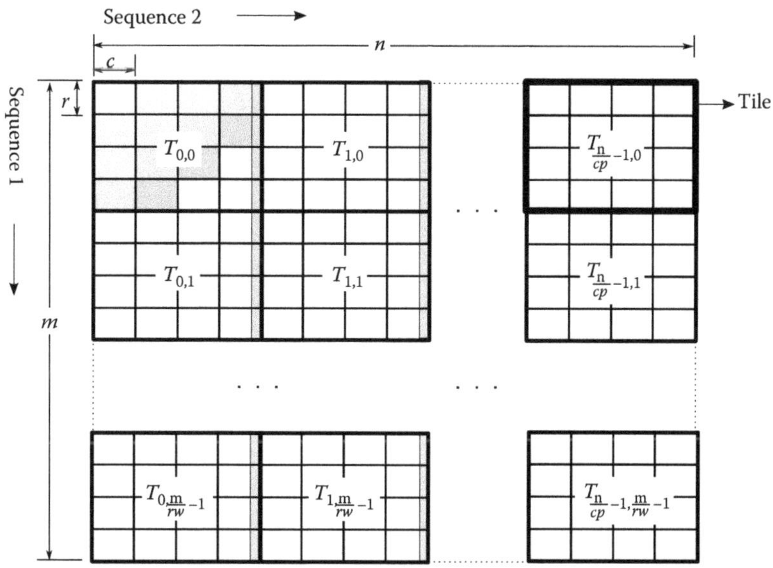

FIGURE 4.3
Tiling scheme to process longer sequences—the bold rectangles correspond to tiles. Shown here is the case when four SPEs are used, resulting in four columns of blocks in each tile. Each tile, denoted by $T_{k,l}$, is processed using the wavefront technique (as in Figure 4.2). The shaded rightmost columns of the tiles need to be explicitly stored in the main memory, when going over the tiles column wise.

The dynamic programming tables can thus be computed on each of the SPEs. A simple extension of the block division enables alignment of longer sequences, and cannot at once fit into the local stores of the SPEs, and we discuss this next.

4.3.1 Tiling Scheme for Aligning Longer Sequences

Because of the small local stores available on the SPEs, to enable optimal alignment score computation of longer sequences, the dynamic programming tables can be first divided into tiles such that the collective memory of all the SPEs is sufficient for computation of a tile at once. Therefore, the two input sequences are partitioned into segments such that each pair of segments, one taken from each input sequence, defines a tile. A single tile consists of $w \times p$ blocks, where each block contains r rows and c columns. The dynamic programming tables are, hence, divided into $\frac{m}{rw} \times \frac{n}{cp}$ tiles. Each tile is computed in parallel among the SPEs using the wavefront scheme described earlier. An illustration of the tiling scheme is presented in Figure 4.3. All the tiles are processed one by one while explicitly storing the last, $(cp - 1)^{\text{th}}$, column of each tile in the main memory (when processing the tiles column-wise). Algorithm 1 gives a pseudocode representation of this tile processing scheme. Wirawan et al. [17] and Aji et al. [15] use this tiling scheme to enable alignment of longer sequences.

Many DMA transfers to/from the main memory are required for the tiling scheme. The PPE initially divides the problem into tiles and instructs the SPEs to process one tile after another. PPE can perform these required communications with the SPEs through mailbox notifications. The SPEs use direct memory access (DMA) transfers to move the corresponding sequence fragments and the last column of the previously computed tile from the main memory to their local stores. Once the processing of a tile is completed, the SPEs transfer the computed scores to the main memory, notify the PPE through mailboxes or signal notifications, and proceed to process the next tile.

Algorithm 1: Tiling scheme. The dynamic programming tables are divided into tiles, and each of them are processed using the wavefront scheme. All the DMA transfers mentioned here are between the main memory and the local stores of the SPEs.

```
for l ← 0 to n/cp - 1 do
    DMA transfer lth segment of S2 to local stores of the SPEs;
    for k ← 0 to m/rw - 1 do
        DMA transfer kth segment of S1 to local stores of SPEs;
        if l = 0 then
            Corresponding SPEs initialize 0th rows and
            columns of the tables;
        else
            DMA transfer (cpl - 1)th column of the tables
            to the local stores of the SPEs;
        end
        Process tile T_{k,l} in parallel using wavefront scheme;
        DMA transfer (cp - 1)th column of computed T_{k,l} to
        main memory;
    end

end
```

4.3.2 Computing the Optimal Alignment Score Using Tiling

To compute the alignment score entry $[i, j]$ of a dynamic programming table, only the entries in previous row $(i - 1)$ and previous column $(j - 1)$ are required. Therefore, to compute only the optimal alignment score of the input sequences, it is sufficient to just store a linear array for the last computed row. In the parallel wavefront scheme on the CBE, this linear space usage can be achieved for score computations by additionally storing a linear sized column array for the last column of the block being computed, which is communicated to the next SPE in charge of computing the next block. Wirawan et al. [17] demonstrate an implementation of this linear space algorithm for computing optimal local alignment scores. They also use this scheme in conjunction with the tiling scheme described earlier to compute optimal alignment scores of longer sequences. Therefore, when only the optimal score needs to be computed, this linear space method is helpful in light of the small local stores available on the SPEs.

4.3.3 Computing an Optimal Alignment Using Tiling

Using the aforementioned linear space technique in conjunction with tiling does not allow retrieval of an actual alignment. To obtain an optimal alignment along with the score, quadratic space is needed when using the above-mentioned tiling scheme. Each tile is computed in parallel among the SPEs, while storing all the rows of the dynamic programming tables. On completion of a tile, the SPEs initiate DMA transfers to move the computed tile (not just the $(cp - 1)^{th}$ column of each tile, as given in Algorithm 1) to the main memory, and then proceed to the next tile. For further parallel efficiency, the SPEs also keep track of their local maximum score when performing local alignments, which are transferred to the main memory at the end, to be used by the PPE to pick the optimal one. This information can then be used by the PPE to perform a trace-back sequentially on the fully stored dynamic programming tables residing in the main memory to retrieve an optimal alignment. This strategy is demonstrated for the Smith–Waterman local alignment algorithm in [15]. This method enables obtaining an optimal alignment of longer sequences, though it is still restricted by the size of the main memory because of its quadratic space usage. It will also be slower due to large memory (quadratic sized) transfers taking place from the SPEs to the main memory after computation of each tile.

Although the tiling scheme for alignment on the Cell processor is useful in certain cases, it has the following drawbacks: (1) Linear space usage can provide an efficient implementation, but this yields only the optimal alignment score. (2) When an actual alignment is required, this scheme can align longer sequence than what can fit in the local stores of the SPEs, but the main memory usage is still quadratic, which can become a limiting factor; moreover, it performs slower. In the following section, we describe a parallel approach as an extension of the wavefront scheme, for computing an optimal alignment using only linear amount of space on the SPEs and the main memory. The alignments are also obtained in parallel during the computations on the SPEs. Therefore, this approach delivers much faster performance and avoids DMA transfers to the main memory until the whole alignment computation is finished. With only linear space usage, longer sequences can be aligned at once.

4.4 A Hybrid Parallel Algorithm

In the rest of the chapter, we focus on a parallel approach incorporating a linear space strategy to increase the size of problems that can be solved using the collective SPE memory and also infer an optimal alignment. Hirschberg [18] presented a divide-and-conquer algorithm to obtain an optimal alignment while using linear space, and we incorporate this strategy in the parallel algorithm discussed later. This scheme should be sufficient for most global/local/spliced alignment problems as the sequences are unlikely to exceed several thousand bases. This alignment method for the CBE is based on the parallel algorithm by Aluru et al. [8], which we describe subsequently.

4.4.1 Parallel Alignment Scheme Using Prefix Computations

The parallel algorithm for computing global alignments in linear space given in [8] consists of two phases: (1) *problem decomposition phase* and (2) *subproblem alignment phase.* In the first phase, the alignment problem is divided into p nonoverlapping subproblems, where p is the number of processing elements. Once the problem is decomposed, each processing element performs a linear space alignment algorithm, computing an optimal alignment for the corresponding subproblem. The result from each processing element is then simply concatenated to obtain an optimal alignment of the actual problem. We describe the problem decomposition phase first.

Initially, the sequence S_1 is provided to all the processors and S_2 is equally divided among them—each processor receives a distinct block of $\frac{n}{p}$ consecutive columns to compute. Define p *special columns*, $C_k = (k+1)\times\frac{n}{p}$, $0 \leq k \leq p-1$, of a table to be the last columns of the blocks allocated to each processing element, except for the last one. The intersections of an optimal alignment path with these special columns define the segment of the first sequence to be used within a particular processing element independently of other blocks, thereby splitting the problem into p subproblems.

To compute the intersections of an optimal path with the special columns, the information on the special columns is explicitly stored. In addition to the score values, for each entry of a table a *pointer* is also computed. This represents the table and row number of the entry in the closest special column to the left that lies on an optimal path from C[0, 0] to the entry. The pointer information is also explicitly stored for the special columns. Conceptually, these pointers give the ability to perform a trace-back through special columns without considering other columns. The entries of the dynamic programming tables are computed row by row using parallel prefix operation as described later in linear space (storing only the last computed row, and the special columns, thereby using $O(m + \frac{n}{p})$ space).

Parallel prefix is a basic operation in parallel computing to compute prefix sums. Given N data items $x_0, x_1, \ldots, x_{N-1}$, and a binary associative operator $\otimes$ that operates on these data items and produces a result of the same type, the parallel prefix operation is to compute the N partial sums $s_0, s_1, \ldots, s_{N-1}$, where $s_i = x_0 \otimes x_1 \otimes x_2 \otimes \ldots x_i$ in parallel. This operation is used to compute the table entries. Consider computing row i of the tables C, D, and I after the $(i-1)^{\text{th}}$ rows are already computed. The ith rows of C and I can be computed directly as they depend only on $(i-1)^{\text{th}}$ rows (see Equations 4.1–4.3). After computing them, the ith row of D can be computed using parallel prefix. Separating the terms that are already computed, let

$$W[j] = max \begin{cases} C[i,j-1] - (g+h) \\ I[i,j-1] - (g+h) \end{cases}$$

Then,

$$D[i,j] = max \begin{cases} W\,[j] \\ D\,[i,j-1] - g \end{cases}$$

Let

$$\begin{aligned} X[j] &= D[i,j] + jg \\ &= max \begin{cases} W\,[j] + jg \\ D\,[i,j-1] + (j-1)g \end{cases} \\ &= max \begin{cases} W[j] + jg \\ X[j-1] \end{cases} \end{aligned}$$

As $W[j] + jg$ is known for all j, $X[j]$'s can be computed using parallel prefix with *max* as the binary associative operator. Then, $D[i, j]$ $(1 \le j \le n)$ can be derived using

$$D[i, j] = X[j] - jg.$$

On completion, a trace-back procedure along the special columns can be used to split the problem into p subproblems in $O(p)$ time. The problem decomposition phase is visualized in Figure 4.4. Once the problem is divided among the processors, in the second phase each processing element performs an alignment on its corresponding segments of sequences while adopting Hirschberg's technique [18, 19] to use linear space.

The hybrid parallel alignment algorithm on the CBE presented here is a combination of this special–columns-based parallel alignment algorithm with Edmiston's wavefront communication pattern described previously in Section 4.3. In the wavefront alignment scheme, each processing element works on a block of the tables independently, communicating the last column

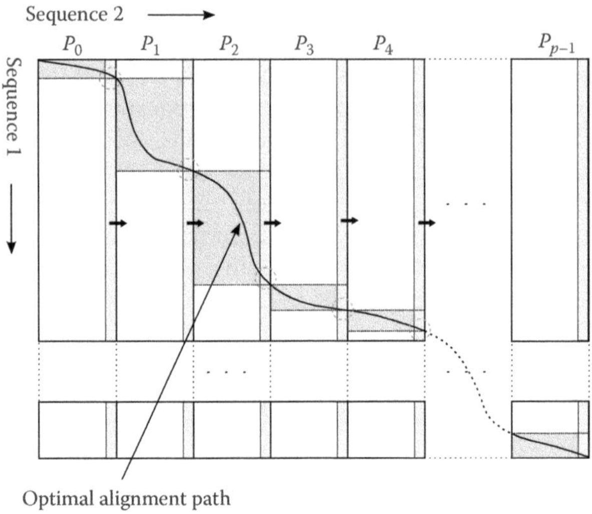

FIGURE 4.4
Block division in parallel-prefix based special columns technique—the second sequence is divided into vertical blocks, which are assigned to different processors P_i. Special columns constitute the shaded rightmost column of each vertical block and the dotted circles show intersection of an optimal alignment path with the special columns, which are used for problem division. The shaded rectangles around the optimal alignment path represent the subdivisions of the problem for each processor. (From Sarje, A. and Aluru, S., *IEEE Transactions on Parallel and Distributed Systems*, 20(11):1600–1610, 2009. With permission. © 2009 IEEE.)

to the next processing element when done and then starts computation on its next block; the parallel prefix approach requires the processing elements to communicate a single element when computing each row. If implemented as such on the CBE, these short but frequent communications for each row increase channel stalls in the SPEs, which is reduced to one bulk communication per block of size r in the wavefront scheme. Each communication leads to a synchronization event among the SPEs. To make most use of parallelism on the Cell processor, such events should be minimized. Moreover, the block size can be optimized for DMA transfer in the wavefront communication scheme, which makes it a better choice for the CBE. Furthermore, adopting the space-saving method is particularly important for the CBE because of the small local store on each SPE.

4.4.2 Problem Decomposition Using Wavefront Scheme

As described earlier in Section 4.3, each dynamic programming table is partitioned into a $w \times p$ matrix of blocks, the size of each block being $r \times \frac{n}{p}$, where $r = \frac{m}{w}$ is the number of rows in a block. Each row of blocks contains as many blocks as SPEs (=p). Each column of blocks is assigned to a single SPE. The parallel decomposition phase of the special–columns-based algorithm [8] is modified to incorporate the wavefront communication scheme and store only the last column (special column) for each SPE block. This also enables use of double buffering in moving input column sequence, and overlapping of DMA transfers with block computations. Each SPE transfers portions of the second sequence allotted to it by the PPE from the main memory to its local store. For each computation block, it transfers blocks of the first sequence using double buffering and performs the table computations in linear space, while storing all of the last column. Once done, it transfers the recently computed block of last column data to the next SPE and continues computation on the next block. This scheme for SPE P_j, $0 \leq j < p$ with a total of p SPEs, is shown as pseudocode in Algorithm 2. Once the special columns containing pointers to the previous special columns are computed, the segments of the first sequence are found, which are to be aligned to the segments of the second sequence on the corresponding SPEs, thereby decomposing the problem into p independent subproblems. This is followed by the sequential alignment phase.

Algorithm 2: Problem decomposition phase of the parallel space-saving algorithm for SPE P_j.

```
Start DMA transfer of S'_2, the allocated S_2 segment for SPE P_j
l_2 ← length(S'_2);
w ← length(S_1)/r;
Start DMA transfer of currentBlock, the first r characters of S_1;
for r ← 1 to w - 1 do
    if j ≠ 0 then
        Receive signal from SPE P_{j-1}
    end
    Start DMA transfer of nextBlock, the next r characters of S_1;
    Wait for completion of currentBlock transfer;
    for i ← 1 to r do
        Compute entries of row i for three tables re-using
        single row buffer;
        specialColumn[i] ← last entry in row i;
    end
    if j ≠ p - 1 then
        DMA transfer last computed block of specialColumns
        to next SPE P_{j+1};
        Signal the next SPE P_{j+1} that its first column has
        been written;
    end
    currentBlock ← nextBlock;
end
Wait for completion of currentBlock transfer;
Perform linear space table computation on currentBlock
storing the last column in specialColumns array;
DMA transfer last block of computed specialColumns to next
SPE;
```

4.4.3 Subproblem Alignment Phase Using Hirschberg's Technique

Once the alignment problem is decomposed into subproblems among all the SPEs, each SPE simultaneously computes optimal alignments for its local subproblem making use of Hirschberg's space-saving technique [18, 19], which reduces the space usage from $O(mn)$ to $O(m + n)$ while enabling retrieval of an optimal alignment. This method is a divide-and-conquer technique where the problem is recursively divided into subproblems, the results of which are combined to obtain an optimal alignment of the original problem [20]. In this scheme, one of the input sequences is divided into two halves, and tables are computed for each half aligned with the other input sequence. This is done in the normal top-down and left-to-right fashion for the upper half and in a reverse bottom-up and right-to-left manner (aligning the reverses of the input sequences) for the lower half. For these computations, it is sufficient to store a linear array for the last computed row. Once the middle two rows are obtained from the corresponding two halves, they are combined to obtain the optimal alignment score, dividing the second sequence at the appropriate place where the optimal alignment path crosses these middle rows. Care needs to be taken to handle the gap continuations across the division, and the possibility of multiple optimal alignment paths. The problem is subsequently divided into two subproblems, and this is repeated recursively for each subproblem. An illustration of the recursion using this scheme is shown in Figure 4.5.

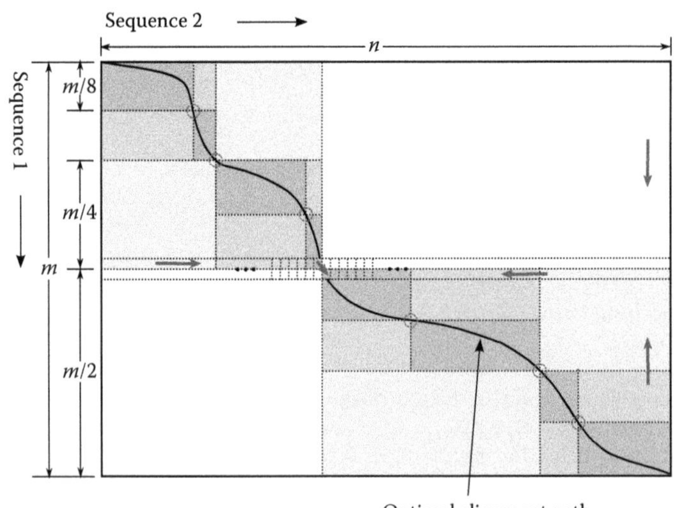

FIGURE 4.5
The sequential recursive space-saving scheme—in Hirschberg's technique, the problem is recursively divided into subproblems around an optimal alignment path, while using linear space. The middle two rows are enlarged for the first recursion showing an example of an optimal alignment path crossing them (not shown for subsequent divisions). The four bold arrows show the direction of computations for the two halves. (From Sarje, A. and Aluru, S., *IEEE Transactions on Parallel and Distributed Systems*, 20(11):1600–1610, 2009. With permission. © 2009 IEEE.)

On completion, each SPE contains an optimal alignment of its subproblem, and writes it to the main memory through DMA transfers. A concatenation of these alignments gives an overall optimal alignment.

Huang [21] describes how to perform space-saving local alignment by using space-saving global alignment as a building block. This technique can be used in conjunction with this hybrid algorithm to derive a space-saving local alignment on the CBE that produces an optimal alignment.

4.4.4 Further Optimizations: Vectorization and Memory Management

SPEs are vector processing units with 128-bit vectors. We present here a basic vectorization scheme for this algorithm, to take advantage of this level of parallelism on the CBE. In the problem decomposition phase, each of the table entries contains a score value along with a pointer to previous special column comprising the table number and row number. Using arrays of vectors, all these data for one table entry are grouped together into one vector. To minimize space usage, single-dimensional arrays *vecEntry*[], representing a single row of the tables, are reused while computing each row within a block. The table entry for column j during computation of a particular row i is represented as the vector:

$$vecEntry\,[j] = \left\langle score_{(i,j)}, tableNum_{(i,j)}, rowNum_{(i,j)} \right\rangle \tag{4.4}$$

Here, $score_{(i,j)}$ represents the alignment score corresponding to the table entry $[i,j]$, and the two entries $tableNum_{(i,j)}$ and $rowNum_{(i,j)}$ are, respectively, the table number and the row number, representing the pointer to the previous special column. Hence, there are three such vector arrays, corresponding to the tables C, D, and I for global alignment. Subsequent to problem decomposition phase, each processor runs Hirschberg's space-saving technique–based algorithm sequentially on its assigned local segments of the input sequences. During this phase, only the scores in each entry of the tables need to be stored. Hence, the vector construction is different—each entry in a particular vector corresponds to the score in each of the three different tables. Such a vector for column j, during computation of row $[i]$, is defined as.

$$vecTables[j] = \left\langle C[i,j], D[i,j], I[i,j] \right\rangle \tag{4.5}$$

where $C[i, j]$, $D[i, j]$, and $I[i, j]$ represent the alignment scores for the respective tables. To use linear space, a single row is stored at a time in the vector array *vecTables* and is then reused for each row. This results in a single array for all the three tables. This way of vectorizing also helps in using various efficient SPE *intrinsics* for computation of the entries.

As the vector buffers for table computation used in the two phases are constructed differently, dynamic memory management can be used to minimize memory usage when integrating the two phases. The linear space sequential algorithm used in the second phase is a recursive algorithm. Owing to small local storage on each SPE, recursive implementations on the Cell are not recommended, but in this case the same row buffer can be reused for table calculations during the recursion, limiting the extra memory used within each step of recursion so that the stack does not grow rapidly. As mentioned previously, the lengths of sequences are split into two in each recursive call, which makes the number of recursive calls linear in the order of sequence lengths. Actual alignments are obtained in parallel on all the SPEs during the recursion [19].

4.4.5 Space Usage

The local store space usage for table computations (apart from space needed for input sequences and output alignment) on a single SPE during the problem decomposition phase is $(m+\frac{n}{p})sy$ bytes, where s is the number of dynamic programming tables used (three in the case of global or local alignment), and y is the number of bytes needed to represent a single element of a single table (this comprises *score, tableNum,* and *rowNum*). The computation space usage during second phase is lower: $(\frac{n}{p}sy')$ bytes, where y' is number of bytes required to store a single table entry (here it is just the *score*). Owing to small local store of 256 KB, a limit is put on the maximum input sequence lengths.

4.4.6 Performance of the Hybrid Algorithm

Here we present some basic performance graphs for the hybrid parallel algorithm for global alignment on the CBE. More detailed results can be found in [16, 22]. The implementation used for these results was developed on the IBM Cell SDK 3.0, compiled with 03 optimization level, and run on a QS20 Cell Blade. (The *CellBuzz* cluster located at the Sony-Toshiba-IBM Center of Competence in Georgia Institute of Technology, Atlanta, U.S.A., was used for this purpose.) A QS20 Cell blade contains two Cell processors connected by an extension of the EIB through a coherent interface, providing a total of 16 SPEs. For these tests, the block size r was chosen to be 128 to optimize the DMA transfers.

The runtimes for varying number of SPEs are shown in Figure 4.6 along with the speedups. The speedups shown are obtained by comparing the parallel Cell implementation with (A1) the parallel implementation running on a single SPE on the Cell processor, (A2) a sequential implementation of the Hirschberg's space-saving technique-based global alignment algorithm for a single SPE on the Cell processor (to completely eliminate the parallel decomposition phase), and (A3) a generic sequential implementation run on a desktop with a 3.2 GHz Pentium 4 processor. On one SPE, the parallel implementation obviously performs worse than the serial implementation, as it includes the additional problem decomposition phase, which computes the whole table to merely return the entire problem as the subproblem to solve sequentially. This is used to study the scaling of the algorithm, and a

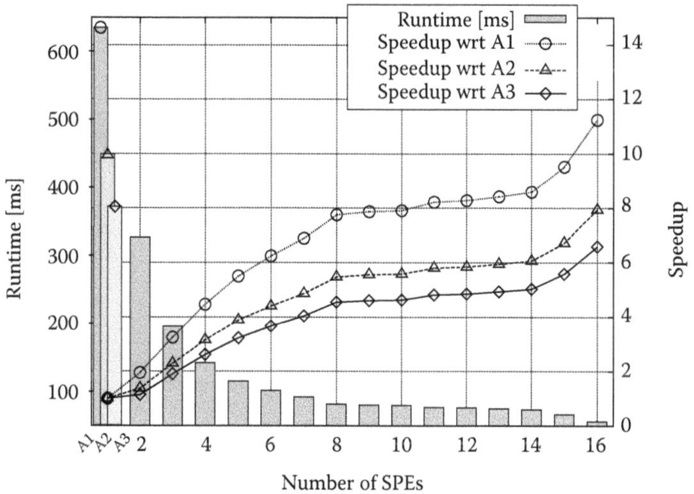

FIGURE 4.6
Runtimes and speedup of global alignment implementation for an input of size 2048 × 2048. A1 is the parallel implementation running on single SPE, A2 is a sequential implementation on one SPE and A3 is a sequential implementation running on Pentium 4 processor. Both these sequential implementations do not contain the problem decomposition phase. (From Sarje, A. and Aluru, S., *IEEE Transactions on Parallel and Distributed Systems*, 20(11):1600–1610, 2009. With permission. © 2009 IEEE.)

speedup of 11.25 on 16 SPEs is obtained. When compared with the sequential implementations, a speedup of almost 8 over a single SPE, and a speedup of more than 6.5 over the Pentium 4 processor are obtained.

It can be seen in the runtime/speedup graph (Figure 4.6) that the runtimes only show a marginal improvement as the number of SPEs is increased from 8 to 12, as opposed to the near linear scaling exhibited below 8 and beyond 12. The latency for data transfer from one Cell processor to the other Cell processor on the blade (*off-chip* communication) is much higher than any data transfer between components on a single processor (*on-chip* communication), and these communication times are significant compared to the computational running time of the implementation. On using more than 8 SPEs, both the processors on the Cell blade are used and data needs to be transferred from one processor to the other. Owing to the higher off-chip communication latency, the runtime using 9 SPEs is similar (or even worse in case of other alignment problems discussed in later sections) to the runtime using 8 SPEs. A tradeoff is created with the off-chip communication time and computation time on the two processors. When amount of computation exceeds the communication time, the runtime further starts to decrease, thereby increasing the speedups as seen in Figure 4.6 for more than 12 SPEs.

To assess the absolute performance of the Cell implementation, the metric of number of cells in the dynamic programming tables updated per second

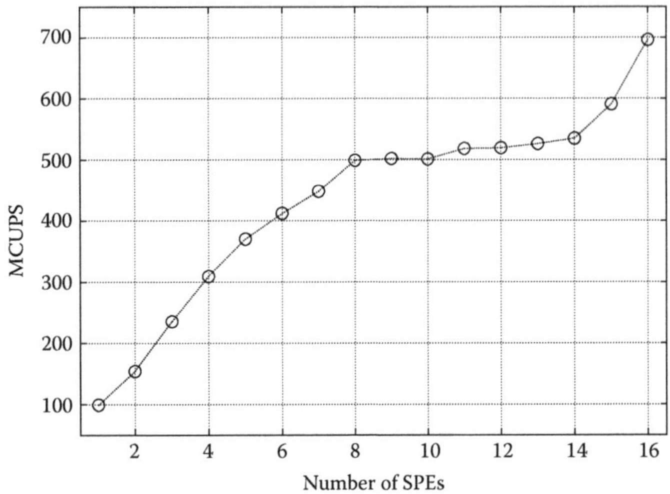

FIGURE 4.7
Cell updates per second for the global alignment on input size of 2048 × 2048 is shown in this graph for increasing number of SPEs and is given in MCUPS (10^6 CUPS). CUPS for 1 SPE is shown for the parallel implementation running on a single SPE. (From Sarje, A. and Aluru, S., *IEEE Transactions on Parallel and Distributed Systems*, 20(11):1600–1610, 2009. With permission. © 2009 IEEE.)

(CUPS) is used and the results are shown in Figure 4.7, with varying number of SPEs. For an implementation where only the alignment scores are computed, the absolute performance obtained would be higher because of algorithmic differences [16].

4.5 Algorithms for Specialized Alignments

We next describe how to extend the parallel global alignment algorithm to more specialized alignment problems of spliced alignments and syntenic alignments. Both these alignment problems involve identification of an ordered set of subregions from one (spliced alignment) or both (syntenic alignment) input sequences, which form a part of the optimal alignment, while the remaining subregions are unaligned.

4.5.1 Spliced Alignments

During the synthesis of a protein, mRNA is formed by transcription from the corresponding gene, followed by removal of the introns and splicing together of the exons. To identify genes on a genomic sequence, or to infer gene structure, one can align processed products (mRNA, EST, cDNA, etc.) to the genomic sequence. To solve this spliced alignment problem, a solution similar to the one for global alignments is described here. While Gelfand et al.'s algorithm [3] has an $O(m^2n + mn^2)$ runtime complexity, an $O(mn)$ algorithm can be easily derived as a special case of Huang's $O(mn)$ time syntenic alignment algorithm [4] by disallowing unaligned regions in one of the sequences. This algorithm uses the three tables as before along with a fourth table H, which represents those regions of the gene sequence that are excluded from the aligned regions (i.e., they correspond to introns or other unaligned regions). A large penalty d is used in table H to prevent short spurious substrings in the larger sequence from aligning with the other sequence. Intuitively, a sequence of contiguous gaps with penalty greater than the threshold d is replaced by a path in the table H representing this region to be unaligned. The four tables are computed as follows:

$$C[i,j] = score\left(a_i, b_j\right) + max\begin{cases} C[i-1,j-1] \\ D[i-1,j-1] \\ I\,[i-1,j-1] \\ H\,[i-1,j-1] \end{cases} \tag{4.6}$$

$$D[i,j] = max\begin{cases} C[i,j-1] - (h+g) \\ D[i,j-1] - g \\ I[i,j-1] - (h+g) \\ H[i,j-1] - (h+g) \end{cases} \tag{4.7}$$

$$I\,[i,j] = max\begin{cases} C[i-1,j]-(h+g) \\ D[i-1,j]-(h+g) \\ I[i-1,j] - g \\ H[i-1,j] - (h+g) \end{cases} \tag{4.8}$$

$$H[i,j] = max\begin{cases} C[i-1,j]-d \\ D\,[i-1,j]-d \\ H[i-1,j] \end{cases} \tag{4.9}$$

For a parallel algorithm for spliced alignments on the CBE, the same techniques as described for parallel global alignment can be followed. Algorithm 2 is used to compute the special columns for the four tables in this case. The vectorization also needs to incorporate the additional table H. For the problem decomposition phase, the vectorization used on the SPE for each of the tables is the same as that given by Equation 4.4. There will be four such arrays, one for each of the tables. Owing to the presence of the fourth array, memory usage for this problem is higher than that for the global alignment problem. The vectorization for the second phase includes

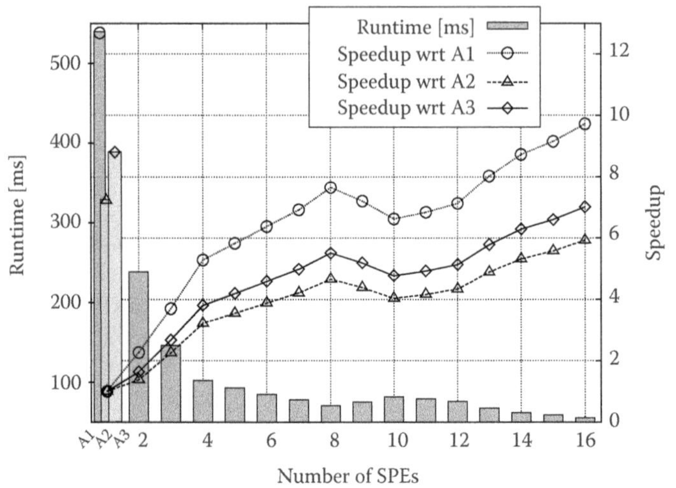

FIGURE 4.8
The runtimes of the spliced alignment implementation and the respective speedups on various number of SPEs for a synthetic input of size 1408 × 1408 is shown in this graph. The speedups are obtained by comparison with (A1) parallel implementation running on one SPE, (A2) sequential implementation for a single SPE, and (A3) sequential implementation on a Pentium 4 desktop. (From Sarje, A. and Aluru, S., *IEEE Transactions on Parallel and Distributed Systems,* 20(11):1600–1610, 2009. With permission. © 2009 IEEE.)

an entry for the fourth table, using the following structure for column [j] and a particular row i:

$$vecTables[j] = \langle C[i,j], D[i,j], I[i,j], H[i,j] \rangle \tag{4.10}$$

4.5.2 Performance of Parallel Spliced Alignment Algorithm

An implementation of the spliced alignment algorithm, as an extension and modification of the global alignment implementation, is used to obtain the following performance graphs. Figure 4.8 shows the runtimes and speedups obtained from a synthetic dataset on varying number of SPEs.

To demonstrate the performance on real biological data, Figure 4.9 shows the runtimes and speedups for aligning the *phytoene synthase* gene from *Lycopersicum* (tomato) with the messenger ribonucleic acid (mRNA) corresponding to this gene's transcription. The scaling obtained is similar to that obtained for global alignment implementation. The difference in speedups in Figures 4.8 and 4.9 is mainly attributable to the different input sizes (the artificial data size is larger than the biological data). Moreover, the synthetic dataset is random, which results in a more uniform problem decomposition among the SPEs, while in the actual biological data the problem sizes for each SPEs may be quite different because of the presence of larger unaligned regions.

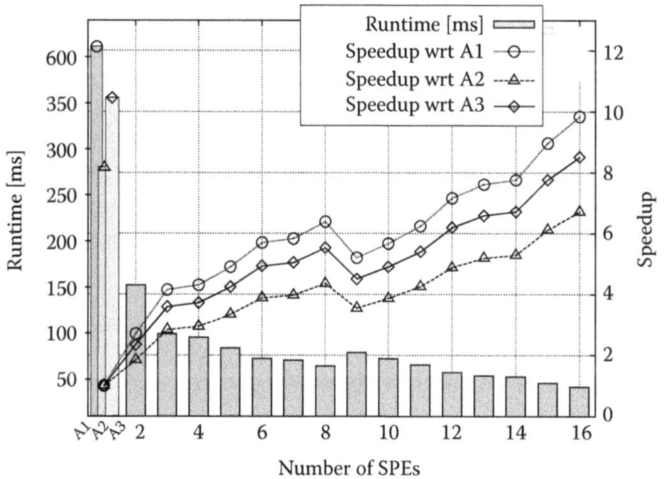

FIGURE 4.9
This graph shows the runtimes and speedups of spliced alignment implementa- tion on various number of SPEs on the Cell blade for phytoene synthase gene from *Lycopersicum* with its mRNA sequence (1792 × 872). (From Sarje, A. and Aluru, S., *IEEE Transactions on Parallel and Distributed Systems*, 20(11):1600–1610, 2009. With permission. © 2009 IEEE.)

4.5.3 Syntenic Alignments

Syntenic alignment, used to compare sequences with intermittent similarities, is a generalization of spliced alignment allowing unaligned regions in both the sequences. This is used to discover an ordered list of similar regions separated by dissimilar regions that do not form part of the final alignments. This technique is applicable to comparison of two genes with conserved exons, such as counterpart genes from different organisms. A dynamic programming algorithm for this has been developed by Huang [4]. Similar to spliced alignment, a large penalty d is used to prevent alignment of short substrings. This dynamic programming algorithm also has four tables, but with an extension in the table H that both sequences can have substrings excluded from aligning. Table definitions for C, D, and I remain the same as Equations 4.6–4.8 for spliced alignment. Definition of table H is modified as follows:

$$H[i,j] = max\begin{cases} C[i-1,j]-d \\ D[i-1,j]-d \\ C[i,j-1]-d \\ I[i,j-1]-d \\ H[i-1,j] \\ H[i,j-1] \end{cases} \tag{4.11}$$

A parallel algorithm for solving the syntenic alignment problem is described in [10] that is similar to the parallel global alignment algorithm described earlier in this chapter. To develop a parallel algorithm for the

CBE, Algorithm 2 is used with adaptation to compute the modified table H. Table H can derive scores from either an entry in previous row or a previous column. This directionality information is important to retrieve the alignment and needs to be stored explicitly. Another way to view this extra information is to split the table H into two, H_h and H_v, where they have the restrictions of alignment paths going only horizontally or only vertically, respectively. Because of this overhead, space requirement in syntenic alignment implementation is even higher. The same scheme for vectorization can be followed for construction of the table entries as in Equations 4.4 and 4.10.

4.5.4 Performance of Parallel Syntenic Alignment Algorithm

The results shown here for syntenic alignment implementation on the CBE have been obtained using both synthetic data and alignment of a copy of the *phytoene synthase* gene from *Lycopersicum* (tomato) and *Zea mays* (maize). The runtimes and speedups of the syntenic alignment implementation run on QS20 Cell blade are shown in Figure 4.10. The performance results for syntenic alignment of a copy of the *phytoene synthase* gene from *Lycopersicum* (tomato) and *Zea mays* (maize) are shown in Figure 4.11. The speedup is better for the biological data mainly because of its larger size than the synthetic dataset.

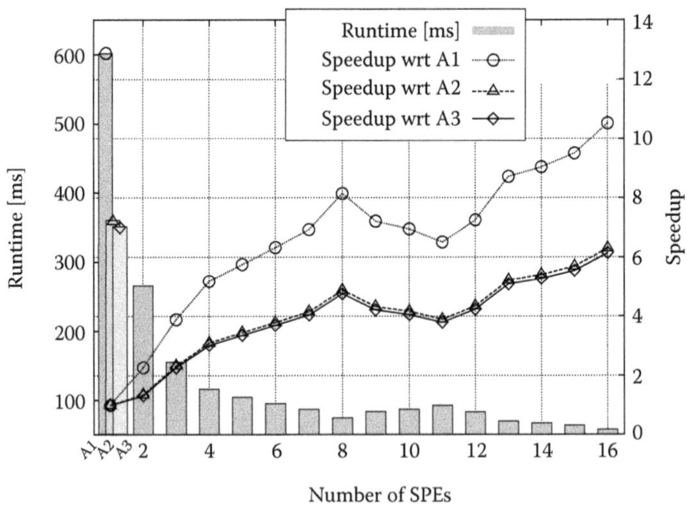

FIGURE 4.10
The runtimes (in milliseconds) and speedups of syntenic alignment running on the Cell blade for a synthetic input data size of 1408 × 1408. *A*1 is the parallel algorithm running on one SPE, *A*2 is sequential algorithm on single SPE, and *A*3 is the sequential algorithm on a Pentium 4 desktop. (From Sarje, A. and Aluru, S., *IEEE Transactions on Parallel and Distributed Systems*, 20(11):1600–1610, 2009. With permission. © 2009 IEEE.)

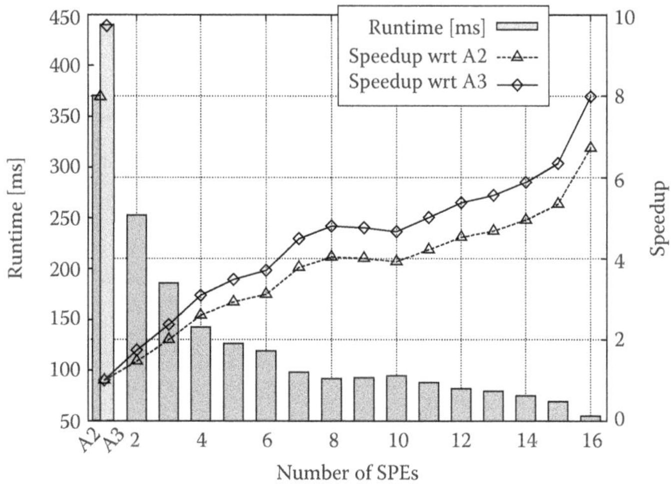

FIGURE 4.11
The runtimes (in milliseconds) and speedup of syntenic alignment implemen- tation for the phytoene synthase gene from *Lycopersicum* (tomato) and *Zea mays* (maize) (1792 x 1580) on the Cell blade. Speedup comparison is done against *A*2 and A3. (From Sarje, A. and Aluru, S., *IEEE Transactions on Parallel and Distributed Systems*, 20(11):1600–1610, 2009. With permission. © 2009 IEEE.)

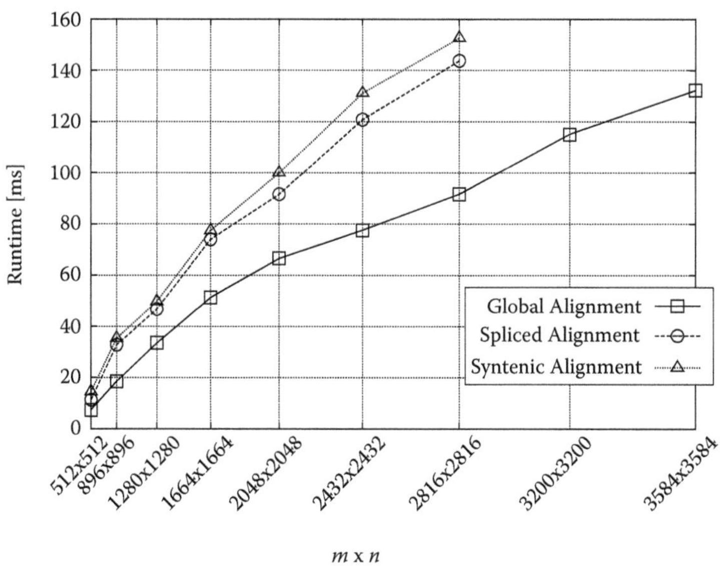

FIGURE 4.12
Scaling of the three alignment implementations with increase in input data size. x-axis is the product of lengths m and n of the two input sequences. This shows that runtime of our implementations scales linearly with $m \times n$ as expected. (From Sarje, A. and Aluru, S., *IEEE Transactions on Parallel and Distributed Systems*, 20(11):1600–1610, 2009. With permission. © 2009 IEEE.)

4.6 Ending Notes

All the three hybrid parallel alignment algorithms run in time proportional to the product of the lengths of the input sequences. This fact is supported by the scaling graphs shown in Figure 4.12. It shows a linear scaling of runtimes of the algorithms with varying product of input sequence sizes.

The performance of the hybrid parallel algorithms for pairwise global/local, spliced, and syntenic alignments for biological sequences show that the Cell processor is a promising platform for developing high-performing applications in bioinformatics that use sequence alignments as a fundamental tool. Depending on the type of the application, the various parallel algorithms presented in this chapter would be helpful to develop an efficient implementation. The alignment algorithms provide an easy overlapping of computations with DMA transfers, a key to achieve high parallel efficiency on multicore architectures. These algorithms clearly demonstrate the use of parallelism at the level of SPEs through the problem decomposition and data distribution, and within the SPEs through vectorization.

Acknowledgment

Portions of text in this chapter have been taken from the article [16]. © 2009 IEEE.

4.7 References

1. S. B. Needleman and C. D. Wunsch. A general method applicable to the search for similarities in amino acid sequence of two proteins. *Journal of Molecular Biology*, 48:443–453, 1970.
2. T. Smith and M. Waterman. Identification of common molecular subsequences. *Journal of Molecular Biology*, 147:195–197, 1981.
3. M. S. Gelfand, A. A. Mironov, and P. A. Pevzner. Gene recognition via spliced sequence alignment. *Proceedings of the National Academy of Sciences USA*, 93(17):9061–9066, 1996.
4. X. Huang and K. Chao. A generalized global alignment algorithm. *Bioinformatics*, 19:228–233, 2003.
5. O. Gotoh. An improved algorithm for matching biological sequences. *Journal of Molecular Biology*, 162(3):705–708, December 1982.

6. M. Farrar. Striped Smith–Waterman speeds database searches six times over other SIMD implementations. *Bioinformatics*, 23(2):156–161, 2007.
7. T. Rogens and E. Seeberg. Six-fold speed-up of Smith–Waterman sequence database searches using parallel processing on common microprocessors. *Bioinformatics*, 16(8):699–706, 2000.
8. S. Aluru, N. Futamura, and K. Mehrotra. Parallel biological sequence comparison using prefix computations. *Journal of Parallel and Distributed Computing*, 63:264–272, 2003.
9. E. W. Edmiston, N. G. Core, J. H. Saltz, and R. M. Smith. Parallel processing of biological sequence comparison algorithms. *International Journal of Parallel Programming*, 17(3):259–275, 1988.
10. N. Futamura, S. Aluru, and X. Huang. Parallel syntenic alignments. *Parallel Processing Letters*, 63(3):264–272, 2003.
11. A. M. Aji and W. Feng. Optimizing performance, cost, and sensitivity in pairwise sequence search on a cluster of playStations. In *8th IEEE International Conference on BioInformatics and BioEngineering, 2008. BIBE 2008*, pages 1–6, 2008.
12. S. Isaza, F. Sanchez, G. Gaydadjiev, A. Ramirez, and M. Valero. Preliminary analysis of the cell BE processor limitations for sequence alignment applications. In *SAMOS '08: Proceedings of the 8th International Workshop on Embedded Computer Systems*, pages 53–64, Springer-Verlag, Berlin, 2008.
13. V. Sachdeva, M. Kistler, E. Speight, and T. K. Tzeng. Exploring the viability of the cell broadband engine for bioinformatics applications. *Parallel Computing*, 34(11):616–626, 2008.
14. H. Vandierendonck, S. Rul, M. Questier, and K. Bosschere. Experiences with parallelizing a bio-informatics program on the cell BE. In *High Performance Embedded Architectures and Compilers (HiPEAC'08)*, volume 4917/2008, pages 161–175. Springer, Berlin, January 2008.
15. A. M. Aji, W. Feng, F. Blagojevic, and D. S. Nikolopoulos. Cell-SWat: modeling and scheduling wavefront computations on the Cell broadband engine. In *Proc. of the 2008 Conference on Computing Frontiers (CF'08)*, pages 13–22, New York, NY, USA, 2008. ACM.
16. A. Sarje and S. Aluru. Parallel genomic alignments on the Cell Broadband Engine. *IEEE Transactions on Parallel and Distributed Systems*, 20(11):1600–1610, 2009.
17. A. Wirawan, K. C. Keong, and B. Schmidt. Parallel DNA sequence alignment on the cell broadband engine. *Workshop on Parallel Computational Biology (PBC 2007)*, LNCS, 4967:1249–1256, 2008.
18. D. S. Hirschberg. A linear space algorithm for computing maximal common subsequences. *Communications of the ACM*, 18(6):341–343, 1975.
19. E. W. Myers and W. Miller. Optimal alignments in linear space. *Computer Applications in Biosciences*, 4(1):11–17, 1988.
20. S. Aluru. *Handbook of Computational Molecular Biology* (Chapman & Hall/CRC Computer and Information Science Series). Chapman & Hall/CRC, Boca Raton, FL, 2005.
21. X. Huang. A space-efficient algorithm for local similarities. *Computer Applications in the Biosciences*, 6(4):373–381, 1990.
22. A. Sarje and S. Aluru. Parallel biological sequence alignments on the cell broadband engine. In *IEEE International Symposium on Parallel and Distributed Processing, 2008. IPDPS 2008*, pages 1–11, 2008.

5

Orchestrating the Phylogenetic Likelihood Function on Emerging Parallel Architectures

Alexandros Stamatakis

The reconstruction of phylogenetic (evolutionary) trees from molecular sequence data is a comparatively old problem in bioinformatics, given that Joe Felsenstein's seminal paper [1] on computing the maximum likelihood score on trees was already published in 1981, that is, almost 3 decades ago. However, significant advances in wet-lab molecular sequencing techniques with the introduction of, for example, the 454 sequencers [2], are generating a highly challenging, unprecedented molecular data flood. In addition, recent years have witnessed the emergence of multicore and other parallel architectures such as graphics processing units (GPUs) or the IBM Cell that pose new challenges to the field of phylogenetic or phylogenomic analysis (reconstruction of phylogenies at the genome scale), in particular with respect to orchestrating the phylogenetic likelihood function (PLF). In fact, the phyloinformatics community faces a continuous struggle to keep up with the rapid speed of data accumulation and provide ever more scalable and powerful analysis tools; that is, we just try to keep pace with data accumulation.

Memory *footprints* of more than 50 GB, just to compute the likelihood, on a single, fixed tree as well as resource requirements exceeding 2 million central processing unit (CPU) hours to simply conduct a comprehensive and thorough real-world ML analysis on one large phylogenomic dataset are becoming the norm, rather than the exception.

In the present chapter I will attempt to review the underlying concepts, current developments, and advances in orchestrating the PLF on parallel computer architectures ranging from field-programmable gate arrays (FPGAs) up to the IBM BlueGene/L supercomputer. The PLF typically consumes more than 95% of total execution time in current state-of-the-art maximum likelihood and Bayesian tools for phylogenetic tree reconstruction. The acceleration and parallelization of the PLF is thus—apart from algorithmic innovations—the key component to handle the data flood to improve scalability of the respective tools. I will also outline potential future developments and challenges.

This chapter is organized as follows: In Section 5.1 I will briefly introduce the field of phylogenetic inference and its applications to medical and biological research. In the following Section 5.2, I will outline the PLF and respective important numerical issues and sequential optimization strategies. In the subsequent Section 5.3, I will discuss the basic fine-grain parallelization strategies for the PLF. Thereafter (Section 5.4), I will review recent adaptations to accelerators and supercomputers. I will conclude in Section 5.5 with a summary of potential future challenges with respect to the PLF per se, parallelization strategies, and necessary adaptations to different input dataset shapes.

5.1 Phylogenetic Inference

The goal of phylogenetic inference consists in reconstructing the evolutionary history of a set of n present-day organisms from their respective molecular sequence data. Those n organisms, also called *taxa*, may also be represented by a concatenation of molecular data from various genes or even the whole genome. Thus, the molecular sequence representing one taxon may consist of a mixture of DNA, protein, and even morphological or binary characters. A phylogenetic tree, or simply a phylogeny, is usually represented as unrooted binary tree, where the n present-day taxa are located at the leaves of the tree and the inner nodes represent extinct common ancestors.

The input data for a phylogenetic analysis under maximum likelihood consists of a "good" multiple sequence alignment of the *n taxa*; that is, by insertion of gaps, or essentially nucleotide insertion and deletion events, all sequences in the input data will have the same length *m* after the alignment step. While there also exist simpler methods for alignment-free tree reconstruction, they have been shown to be generally less accurate [3, 4] than alignment-based methods. Alignments that comprise sequence data from several genes are called *multigene* or *phylogenomic alignments*. A simple example for a multiple sequence alignment of DNA data for human, mouse, cow, and chicken is provided below.

```
Cow        ATGGCATATCCCA-ACAACTAGGATTCCAAGA----
Chicken    ATGGCCAACCACTCCCAACTAGGCTTTC-AGACGCC
Human      ATGGCACAT---GCGCAAGTAGGTCTAC-AGACGCT
Mouse      ATGGCCCATTCCAACTTGGTCTACAAGACGCCACAT
```

An open issue, especially within the context of real-world analyses, is the definition of what a "good" multiple sequence alignment actually is, since no objective criterion is available to judge alignment quality. In a recent *Science* paper Loytynoja and Goldman [5] challenged the established opinion on how a "good" alignment should look like by arguing in favor of a phylogeny-aware view of the alignment problem. In addition, the multiple sequence alignment problem is a computationally hard problem by itself. Since, this chapter mainly focuses on the computational aspects of phylogenetic inference, we will just assume that the alignment is given, though the problems of phylogenetic inference and multiple sequence alignment should be solved simultaneously in an ideal world, but current approaches [6–8] are still too slow and too resource intensive for practical purposes, especially when we consider current input alignment growth.

It is important to note that provided any biologically meaningful optimality criterion to score a given tree topology, such as ML or Maximum Parsimony [9], the underlying optimization problem for finding the optimal tree is NP-hard [10, 11]. The number of distinct alternative unrooted tree topologies for n taxa is $\Pi_{i=3}^{n}(2i-5)$ [12]. While there exists a vast amount of literature covering heuristic search algorithms for the ML optimization problem (see [13] for an overview), they all rely on repeatedly executing the likelihood function to explore the tree space, which represents the main computational and memory bottleneck.

Phylogenetic trees have many important applications in medical and biological research; current state-of-the-art ML phylogeny programs such as PAML [14], PHYML [15], PAUP [16], GARLI [17], RAxML [18], IQPNNI [19], TEEFINDER [20], or likelihood-based Bayesian programs such as MrBayes [21], PhyloBayes [22], or BEAST [23] have accumulated well above 20,000 citations to date. Phylogenies can be used, for example, to infer the evolutionary history of pappilomaviruses that are associated with cervical cancer [24], to

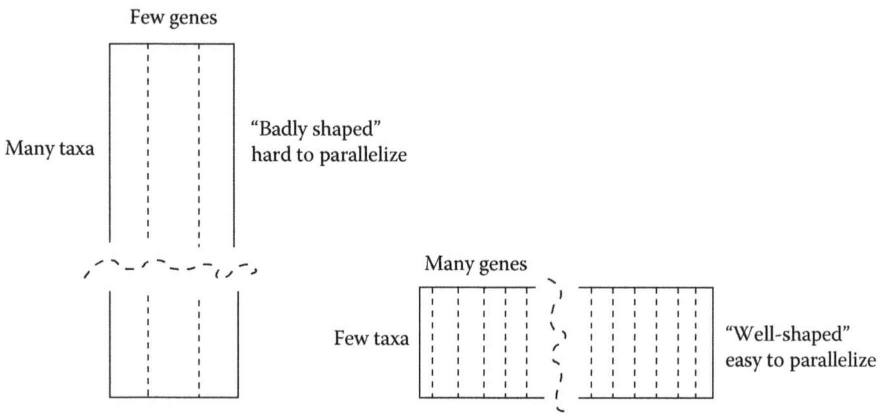

FIGURE 5.1
Badly shaped and well-shaped alignments.

disentangle the evolutionary history of Acer [25] (maple trees), or to analyze bacterial communities in permafrost soils [26]. A recent phylogenomic study in *Nature* improved the accuracy of the animal tree of life [27] while another recent study in *Science* assessed the rates of evolution (essentially the speed of evolution) and their association to life history in flowering plants [28]. Those two papers also point toward a fundamental problem that will need to be tackled in the future: the phylogenomic study [27] contains less than 100 taxa but a large number of 150 genes—an on-going follow-up study [29] even comprises about 1,000 genes. The study by Smith and Donoghue [28] is based on two datasets with less than ten genes, but more than 4,000 and 13,000 taxa respectively. The datasets used have significantly distinct shapes that have an impact on algorithm design, scalability, and future parallelization strategies that need to be deployed. Here I introduce the term "well-shaped" alignments for few-taxa/many-gene input datasets and "badly shaped" for many-taxa/few-gene datasets (see Figure 5.1). Evidently, badly shaped datasets are harder to analyze algorithmically and also more difficult to parallelize.

5.2 The Phylogenetic Likelihood Function

As already mentioned, the input for a phylogenetic analysis under ML consists of a multiple sequence alignment with n sequences (also denoted as taxa or tips) and m alignment columns. The branch length values on the tree that are returned by ML essentially represent the relative time of evolution between nodes in the tree. Here we will initially consider only how to compute the likelihood on a fixed, given, tree topology. Apart from the tree topology one also needs several ML model parameters. One important parameter is the instantaneous nucleotide substitution matrix Q, which contains the transition probabilities for time dt between binary (2 x 2 matrix, states: `0` or `1`), nucleotide (4 x 4 matrix, states: `A`, `C`, `G`, `T`), or for instance, amino acid (20 x 20 matrix) characters. The transition probability matrix for time (branch length) t is then computed as $P(t) = e^{Qt}$ and can be computed via a respective eigenvector/eigenvalue decomposition. Note that there also exist various models to accommodate ribonucleic acid (RNA) secondary structure information, that is, models that allow to group together columns and hence let them evolve together in the respective DNA/RNA alignment. For secondary structure there exist 6-state (6 x 6 Q-matrix), 7-state, and 16-state models (for a summary see [30]). Recently, 61-state codon models (see, e.g., [31]) for protein-coding genes, which group together triplets of DNA characters, have also received considerable attention. As we will see below, the computational complexity for computing the likelihood of a single column is directly proportional to the square of the number of states. Using DNA as an example, in addition to the Q matrix we also need the prior probabilities of observing the nucleotides, for example, π_A, π_C, π_G, π_T for DNA data, which can be determined empirically from the alignment or obtained via an ML estimate. We also need the α shape parameter that forms part of the Γ model [32] of rate heterogeneity. The Γ model accounts for the biological fact that different columns in the alignment evolve at different speeds. While the Γ model is well-established and the de facto standard, there exist computationally much more efficient ways to incorporate rate heterogeneity, such as the CAT (category) approximation of rate heterogeneity [33], which represents perhaps the most underestimated paper of the author, despite its huge computational advantages, especially in the phylogenomic era. Finally, one also requires the $2n - 3$ branch lengths in the unrooted tree topology.

Given all these parameters to compute the likelihood of a fixed *unrooted* binary tree topology, initially one needs to compute the entries for all internal probability vectors (located at the inner nodes) that contain the probabilities $P(A)$, $P(C)$, $P(G)$, $P(T)$ of observing an `A`,`C`,`G`, or `T` at each site/column c, where $c = 1 \ldots m$ of the input alignment at the specific inner node. Those probability vectors are computed bottom-up from the tips toward a virtual root that can be placed into any branch of the tree. This important property of ML holds as long as the nucleotide substitution model is time reversible; that is, evolution occurred in the same way if followed forward or backward in time. The most commonly used and general model for DNA substitution is the general time reversible (GTR) model [34] of nucleotide substitution. However, there also exist proposals for comparatively efficient and more realistic nonreversible substitution models [35]. The procedure described earlier for computing the likelihood is also know as the Felsenstein pruning algorithm [1]. Under certain standard model restrictions (time reversibility of the model) the overall likelihood score will be the same regardless of the placement of the virtual root.

As already mentioned, every probability vector entry $\vec{L}(c)$ at position c ($c = 1 \ldots m$) at the tips and at the inner nodes of the tree topology contains the four probabilities P(A), P(C), P(G), P(T) of observing a nucleotide `A`, `C`, `G`, `T` at a specific column c of the input alignment. The probabilities at the tips (leaves) of the tree for which observed data (e.g., the DNA sequences of the currently living organisms under study) *is* available are set to 1.0 for the observed nucleotide character at the respective position c, for example, for the nucleotide `A`: $\vec{L}(c) = [1.0,\ 0.0,\ 0.0,\ 0.0]$. Given a parent node k, and two child nodes i and j (with respect to the virtual root), their probability vectors $\vec{L}^{(i)}$ and $\vec{L}^{(j)}$, the respective branch lengths leading to the children b_i and b_j, and the transition probability matrices $P(b_i)$, $P(b_j)$, the probability of observing an `A` at position c of the ancestral (parent) vector $\vec{L}_A^{(k)}(c)$ is computed as follows:

$$\vec{L}_A^{(k)}(c) = \left(\sum_{S=A}^{T} P_{AS}(b_i)\vec{L}_S^{(i)}(c) \right)\left(\sum_{S=A}^{T} P_{AS}(b_j)\vec{L}_S^{(j)}(c) \right) \tag{5.1}$$

As already mentioned, the transition probability matrix $P(b)$ for a given branch length b is obtained from Q via $P(b) = e^{Qb}$. Once the two probability vectors $\vec{L}^{(i)}$ and $\vec{L}^{(j)}$ to the left and right of the virtual root (vr) have been computed, the likelihood score $l(c)$ for an alignment column c ($c = 1 \ldots m$) can be calculated as follows, given the branch length b_{vr} between nodes i and j:

$$l(c) = \sum_{R=A}^{T} (\pi_R \vec{L}_R^{(i)}(c) \sum_{S=A}^{T} P_{RS}(b_{vr})\vec{L}_S^{(j)}(c)) \tag{5.2}$$

The overall score is then computed by summing over the per-column log likelihood scores as indicated in Equation 5.3.

$$LnL = \sum_{c=1}^{m} log(l(c)) \tag{5.3}$$

An important property of the likelihood function is the assumption that sites evolve independently; that is, all entries c of the probability vectors $\vec{L}$ can be computed independently. This property represents the main source of fine-grain parallelism in the PLF. Therefore, for a *full* tree traversal, only one single reduction operation and hence synchronization point is required when the virtual root is reached (see Equation 5.3).

When the Γ model of rate heterogeneity is used, the computation is slightly more complex, since initially the Γ function is approximated by usually four discrete rates r_0, r_1, r_2, r_3 using standard numerical techniques (see, e.g., [36]). Then, for each branch we need to compute four transition probability matrices: $P(t) = e^{Qtr_0}, \ldots, P(t) = e^{Qtr_3}$ and also need to calculate a separate probability vector $\vec{L}$ for every discrete rate, which results in a four-fold increase in floating point operations and memory consumption of the inner probability vectors. The log likelihood at the root is then calculated as.

$$LnL = \sum_{c=1}^{m} log(0.25 \times (l_0(c) + l_1(c) + l_2(c) + l_3(c))) \tag{5.4}$$

where $l_0(c), \ldots, l_3(c)$ are the per-site likelihoods at column c of the alignment for the 4 discrete Γ rates $r_0, \ldots, r_3$.

To compute the maximum likelihood value for such a fixed tree topology all individual branch lengths, as well as the substitution rates in the Q matrix and the α shape parameter of the Γ distribution, must also be optimized via an ML estimate. For the Q matrix and the α shape parameter the most common approach in state-of-the-art ML implementations consists in using Brent's algorithm [37]. A key computational issue is that to evaluate changes in Q or α the entire tree needs to be retraversed, that is, a *full* tree traversal needs to be conducted from the leaves toward the virtual root to correctly recompute the likelihood. For the optimization of branch lengths the Newton–Raphson method is commonly used. To optimize the branches of a tree, the branches are repeatedly visited and optimized one by one until the achieved likelihood improvement (or branch length change) is smaller than some predefined ϵ. Since the branch length is optimized with respect to the likelihood score, the Newton–Raphson method operates only on a single pair of probability vectors $\vec{L}^{(i)}, \vec{L}^{(j)}$ that are located at either ends of the branch to be optimized. The Newton–Raphson method requires the computation of the first and second derivative of the likelihood function. Because we intend to maximize the likelihood function, we need to determine the root of the first derivative of the likelihood function. Note that a reduction operation to compute the overall value (accumulate over all columns) for the first and second derivative is required at *every* iteration of the Newton–Raphson procedure. Evidently, when a branch of the tree is updated this means that a large number of probability vectors $\vec{L}$ in the tree are affected by this change and hence need to be recomputed to maintain a state that is consistent with the new branch length configuration.

An important implementation issue is the assignment of memory space for the probability vectors to inner nodes of the tree. There exist two alternative approaches: a separate vector can be assigned to each of the three outgoing branches of an inner node (PHYML uses this approach), or only one vector can be assigned to each inner node (GARLI, RAxML, and MrBayes, among others deploy this technique). In the latter case, which evidently is

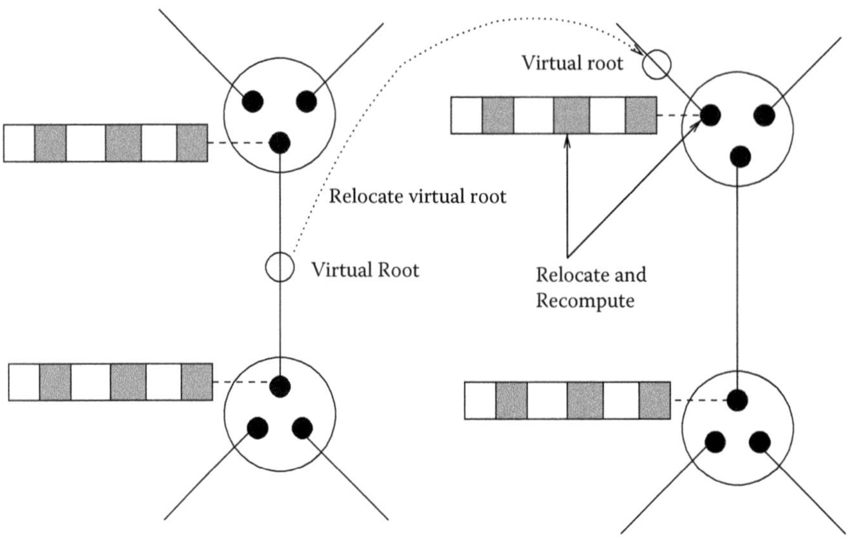

FIGURE 5.2
Rooted organization of the probability vectors at inner nodes. This figure also shows the cyclic distribution of probability vector entries to two threads.

significantly more memory efficient, the probability vectors always maintain a rooted view of the tree; that is, they are oriented toward the current virtual root of the tree. In the case that the virtual root is then relocated to a different branch (e.g., to optimize the respective branch length), a certain number of vectors, for which the orientation to the virtual root has changed, need to be recomputed. If the tree is traversed in an intelligent way, for example, for branch length optimization, the number of probability vectors that will need to be recomputed after relocations of the virtual root can be minimized. An example for this type of data organization is provided in Figure 5.2. RAxML also uses this type of rooted probability vector organization to handle large-scale alignments, since current phylogenomic datasets can require up to 89 GB of main memory under the Γ model, even when using this efficient organization of the inner (ancestral) vectors.

5.2.1 Avoiding Numerical Underflow

The methods deployed to avoid numerical underflow via appropriate scaling mechanisms represent an important implementation and performance issue. As can be derived from Equation 5.1 the values in the probability vectors $\vec{L}$ at the inner nodes of the tree will progressively become smaller as we approach the virtual root in the tree, since we are always conducting multiplications with probability values in the transition probability matrix $P(t)$. Therefore, for trees with many taxa, measures need to be taken to avoid numerical underflow in the probability vectors. A detailed analysis of numerical issues regarding large-scale phylogenetic analyses is provided in [38].

Scaling of the probability vector entries may be conducted as follows: at a column c of an ancestral probability vector for DNA data $\vec{L}$ we scale the entries if $\vec{L}_A(c) < \epsilon \wedge \vec{L}_C(c) < \epsilon \wedge \vec{L}_G(c) < \epsilon \wedge \vec{L}_T(c) < \epsilon$, where $\epsilon = 1/2^{256}$ for double precision (DP) and $\epsilon = 1/2^{32}$ for single precision (SP) arithmetics. Note that the decision to scale only a probability vector entry when *all* values in that column c are smaller than ε may potentially be dangerous if the differences between the individual values become too large. So far, we have not observed any numerical instability because of this strategy in RAxML, but this potential problem may become prevalent for 61-state codon models.

If probability vector column c at vector $\vec{L}$ needs to be scaled, we simply multiply all entries $\vec{L}_A(c), \vec{L}_C(c), \vec{L}_G(c), \vec{L}_T(c)$ by 2^{256} under DP or 2^{32} under SP, respectively (see Section 5.2.3 for a discussion of single vs. double precision arithmetics trade-offs).

To correct for the scaling multiplications once the virtual root is reached, we need to keep track of the total number of scaling events conducted per column. We use integer vectors $\vec{U}$ that maintain the scaling events and correspond to the respective probability vectors at inner nodes. As we traverse the tree to compute an ancestral vector $\vec{L}^{(k)}$ from two child vectors $\vec{L}^{(i)}$ and $\vec{L}^{(j)}$ the scaling vector is initially updated as follows $\vec{U}^{(k)}(c) := \vec{U}^{(i)}(c) + \vec{U}^{(j)}(c)$. Then, if an entry of $\vec{L}^{(k)}$ needs to be scaled at position c we increment $\vec{U}^{(k)}(c) := \vec{U}^{(k)}(c) + 1$. The scaling vectors at the tips of the tree are not allocated, but implicitly initialized with 0.

At the virtual root, given $\vec{L}^{(i)}$, $\vec{L}^{(j)}$ and the corresponding scaling vectors $\vec{U}^{(i)}$, $\vec{U}^{(j)}$ we can compute the likelihood as follows:

$$l(c) = \frac{1}{2^{256}}^{U^{(i)}(c)+U^{(j)}(c)} \left(\sum_{R=A}^{T} (\pi_R \vec{L}_R^{(i)}(c) \sum_{S=A}^{T} P_{RS}(b_{vr}) \vec{L}_S^{(j)}(c)) \right) \tag{5.5}$$

If we take the logarithm of $l(c)$ and $\epsilon = 1/2^{256}$ this can be rewritten as:

$$log(l(c)) = (U^{(i)}(c) + U^{(j)}(c) \log(\epsilon)) + \log\left(\sum_{R=A}^{T} (\pi_R \vec{L}_R^{(i)}(c) \sum_{S=A}^{T} P_{RS}(b_{vr}) \vec{L}_S^{(j)}(c)) \right) \tag{5.6}$$

This does not appear to be the most efficient method for likelihood scaling, in particular because of the conjunction of comparisons on floating point numbers that we use to decide if scaling is required at every single iteration of the `for` loop over a vector $\vec{L}$. However, our experiments with various alternative scaling methods indicate that this method is indeed very efficient.

An alternative to the aforementioned method consists of keeping track of scaling events across all entries $c = 1 \ldots m$ of a probability vector $\vec{L}$. In this case every inner node of the tree will store only one integer value u instead of an integer array $\vec{U}$. As we traverse the tree to compute an ancestral vector $\vec{L}^{(k)}$ from two child vectors $\vec{L}^{(i)}$ and $\vec{L}^{(j)}$ the single scaling entry at node k, u_k, is computed as $u_k := u_i + u_j$. During the computation of vector $\vec{L}^{(k)}$ we simply count the total number s of scaling events that occurred along this vector, where $0 \leq s \leq m - 1$ and add it to the node scaling value $u_k := u_k + s$. When we reach the root we compute the likelihood across all sites as follows:

$$LnL = (u_i + u_j)\log(\epsilon) + \sum_{c=1}^{m} \log(l(c)) \tag{5.7}$$

The method outlined in Equation 5.7, which was suggested to us by Minh Bui from the University of Vienna, clearly requires less arithmetic operations and less memory for scaling. On a large phylogenomic DNA dataset with 404 taxa and 11 genes we measured a performance improvement of 7% for RAxML with the aforementioned scaling method. While all of this may appear relatively simple, in the real world and for a widely used tool such as RAxML it is not. The problem is that the aforementioned method does not allow to obtain correct per-site (per-column) log likelihood values $log(l(c))$ that are required, for example, to conduct some of the standard statistical tests (Goldman et al. provide an excellent summary of statistical tests in [39]) to assess if two trees have significantly different likelihood scores or not. In reality we face difficult software engineering issues that are outlined by the aforementioned example, and constantly have to strive for a balance between efficiency and code complexity. In the aforementioned example we decided to integrate both approaches, for example, scaling with scaling vectors and with per-node scaling counters in RAxML to offer faster scaling and faster likelihood computations when the per-site log likelihood scores are not required, which is the case for tree searches, but at the same time maintain all functionalities of RAxML, for example, the computation of the extended likelihood weights ELW statistics [40] or the option to print per-site log likelihood values to a file for usage with the CONSEL package [41].

Overall, more research is required to devise better scaling procedures, which as we demonstrate here can have a huge impact on program performance, especially under SP, which in turn is important for GPU implementations of the PLF.

5.2.2 Memory Requirements

The memory requirements for ML-based phylogeny programs are dominated by the space required for the inner probability vectors $\vec{L}$ and, to a lesser extent, the inner scaling vectors $\vec{U}$ or per-node scaling numbers u. Depending on the memory organization and data structures used, we need to assign at least one probability vector and one scaling vector to each of the n–2 inner (ancestral) nodes of the tree. Since for the values at the leaves we only have, for example, 15 alternative probability vector entries using ambiguous DNA character encoding, we need to store only one vector $\vec{L}$ of length 15 that can then be accessed using the raw input sequences as an index. The input sequences can be stored as simple character arrays and the respective small table may be viewed as a lookup table. Hence, the memory requirements for computing the likelihood on a DNA alignment (without accommodating for rate heterogeneity) with n taxa and m columns are $n \cdot m \cdot 1$ bytes for the input sequences, $(n - 2) \cdot m \cdot 4 \cdot 8$ bytes for the probability vectors, and $(n - 2) \cdot m \cdot 4$ bytes for the scaling vectors. If we use the standard Γ model of rate heterogeneity with 4 discrete rates $r_0, \ldots, r_3$ the space requirements for the probability vectors increase to $(n - 2) \cdot m \cdot 16 \cdot 8$ bytes, while the remaining values remain unchanged. Hence, the memory requirements are dominated by the space required for the probability vectors located at the inner nodes of the tree and can be reduced by almost factor 2 using single precision arithmetics. In addition, when the CAT approximation of rate heterogeneity [33] is being used, memory requirements can be reduced by a factor of 4 compared to the standard Γ model. This is an important issue, since we are receiving an increasing number of reports from RAxML users that they are encountering memory shortages. Memory requirements can be reduced dramatically, especially for large phylogenomic alignments by using SP arithmetics and the CAT approximation. On a protein dataset with 232 taxa and 349,718 alignment sites, the Γ model under double precision requires 44GB of main memory compared to only 5GB of memory for CAT and single precision. Such a reduction in memory requirements, which can be further improved (see [42]) by taking into account the large number of missing sequence data in phylogenomic alignments, will help many typical users who do not have access to high-performance computing (HPC) resources to conduct large-scale analyses on their desktop.

5.2.3 Single or Double Precision?

One of the key design decision when implementing the PLF is whether to use single or double precision floating point arithmetics for the implementation. As outlined in the previous section, a single-precision implementation can yield significant computational advantages and memory savings. Memory savings are the main reason why MrBayes uses single precision. Note that memory consumption is an even more critical issue for Bayesian approaches that typically deploy a Metropolis-Coupled Markov-Chain Monte-Carlo [43] approach with several heated and one cold chain. The usage of multiple chains means that as many trees and associated data structures as there are chains need to be kept in memory and hence the memory footprint is proportional to the number of chains as well. The key dilemma here is that more chains will yield better results, or at least higher confidence for convergence assessment, but that they will also require more memory. A reduction of the number of chains because of memory shortage may have fatal effects on the quality of the trees produced by Bayesian inferences (see [44] for a discussion of potential pitfalls of Bayesian phylogenetic inference).

Another key driving factor to work on SP implementations is the current performance gap between SP and DP arithmetics on GPUs that amounts to one order of magnitude. At present, it is hard to predict if, as has happened with the IBM Cell, DP performance will dramatically improve on GPUs in the future. The SP-DP performance gap on GPUs is also one of the main reasons why GARLI (Derrick Zwickl, personal communication) and RAxML [45] have recently been ported to SP arithmetics.

The standard approach currently consists in conducting the numerically sensitive operations, such as Eigenvector/Eigenvalue decomposition, computation of the $P(t)$ matrix under DP, then cast $P(t)$ to SP and conduct the computation of the probability vector entries in $\vec{L}$, that is, the main computational bulk, under SP. The "classic" ML programs RAxML and GARLI also allow for a more straightforward assessment of the impact of the induced loss of precision on topological accuracy, because every search under SP will just return one single tree that can then be compared to the tree returned by the DP likelihood kernel.

Following an analysis in [45] and a personal communication with Derrick Zwickl, it is not necessary to adopt a mixed-precision approach as proposed, for example, for systems of linear equations in [46], but it suffices to use SP all the way, since the tree topology has the by far largest influence on the likelihood score and small deviations in likelihood scores between SP and DP are negligible. The observation that modifications of the tree topology yield the largest improvements in likelihood scores is also used to devise fast heuristic search algorithms (see [15, 17, 18, 47–49] for examples). As outlined in [45] SP also allows for better exploitation of general-purpose CPUs by SSE3 vector instructions than DP, because four, instead of two, operations can be executed per cycle. Surprisingly, current commercial compilers such as the Intel `icc` compiler (versions 10.x and 11.x) are not able to automatically vectorize the RAxML code despite the fact that the loops appear to be relatively straightforward to vectorize to the human eye.

However, there are two major drawbacks to the usage of SP. As described in Section 5.2.1 the scaling threshold ε for SP is significantly smaller than for DP, which means that a *significantly* higher number of scalings, that is, multiplications by 2^{32}, are required. We find that the number of such multiplications increases by one order of magnitude in the SP version of RAxML. This actually led to a performance decrease, despite using SSE3 intrinsics under SP, compared to DP. However, in the experiments described in [45], we used only single-gene alignments with relatively small memory footprints. A follow-up analysis on the aforementioned phylogenomic alignment with 232 taxa and 349,718 columns showed that the SP version is actually faster than the DP version, both under Γ and under the CAT approximation. Under the WAG+Γ [50] the DP version (without SSE3) required 37.7 hours, compared to the SP version that required 22 hours and the SSE3 vectorized SP version that required only 13.4 hours for the first six iterations of the search algorithm on a SUN x4600 multicore system with 32 CPUs and 64GB of main memory. We assume that this impressive speedup is partially due to cache effects, since the memory footprint is reduced from 44 GB to about 22 GB, but this hypothesis requires further analysis.

While these results are promising, the *second* major drawback of SP is a loss of accuracy for datasets with more than 1,000 taxa. While we have demonstrated in [45] that the SP implementation in RAxML is reliable up to about 2,000 taxa, further tests have revealed that apparently for alignments with more than 2,000 taxa the loss in precision induced by SP arithmetics is too great in order to achieve numerical stability. We have found that at least the RAxML SP implementation is not able to optimize branch lengths and model parameters on trees with 4,000, 6,000, and 7,000 taxa. This may simply be a problem that is associated with the specific implementation in RAxML, but we are not aware of any other study that assesses SP precision issues in the ML function on such large datasets.

To this end, SP seems to be a feasible solution for handling memory-intensive phylogenomic datasets up to 500 or 1,000 taxa, but SP implementations of the ML function should be handled with extreme care when many-taxon trees are analyzed. Although the SP implementation in MrBayes may be considerably more stable, these problems may nonetheless occur with Bayesian inferences and it would be important to conduct comparative studies on many-taxon trees using SP and DP arithmetics for the most widely used Bayesian inference programs.

5.3 Parallelization Strategies

As outlined in the previous section and as can be derived from Equations 5.1, 5.2, and 5.3 the bulk of all PLF computations consists of `for`-loops over the length m of the vectors $\vec{L}$. These `for`-loops require about 95% of total execution time in all standard likelihood-based phylogenetic tools and are thus the candidate functions for a fine-grain parallelization. An important property of the likelihood function that actually enables such a fine-grain parallelization is the assumption that sites evolve independently; that is, all entries c of the probability vectors $\vec{L}$ can be computed simultaneously. This property represents the main source of fine-grain parallelism in the PLF (see, e.g., [51]).

5.3.1 Parallel Programming Paradigms

The most straightforward approach is to use OpenMP for parallelization [52]. However, OpenMP has some major drawbacks: for nonexpert users it will be difficult to install an OpenMP-based compiler. Moreover, a commercial compiler is required, since the current `gcc` version implements a strict fork-join paradigm for the threads, in contrast to a synchronization via barriers that avoids thread initialization and termination for every parallel region that is traversed as implemented in the `icc` copiler. This is a performance critical issue for OpenMP-based PLF implementations, since a prohibitively large number of parallel regions will be entered and exited during a tree search. The advantage of using Pthreads instead is that the program compiles out of the box and hence a significantly larger number of users are able to fully exploit the capabilities of their multicore machines. Another reason for using Pthreads is the nondeterminism in OpenMP-based reduction operations; that is, the reduction operation conducted at the virtual root of the tree (see Equation 5.3) may yield different results when invoked repeatedly on exactly the same tree. Since this will generate, and has generated, extremely hard-to-detect numerical bugs we have decided to completely abandon OpenMP in favor of Pthreads to have full control over such issues. A Pthreads-based implementation also allows more easily to separate the address spaces of the threads such that it is easier to maintain an MPI [41] as well as Pthreads-based version of the code. Some additional software engineering issues regarding design choices for parallel programming paradigms in the PLF are covered in [42] and [53].

5.3.2 General Fine-Grain Parallelization

The general fine-grain parallelization scheme for the PLF is outlined by the example of the Pthreads-based version of RAxML. An example for the overall parallelization strategy is outlined in Figure 5.3 for a *full* tree traversal. The basic underlying concept is based upon a clear separation of tasks between the master and worker threads/processes. The only thread that actually needs to "understand" tree topologies and conduct the heuristic tree search is the master thread. The worker threads essentially only allocate and operate on their private fraction of the probability vectors $\vec{L}$ and are used to perform the floating-point intensive likelihood computations.

The probability vectors are enumerated consistently across all workers and the master thread. Thereby, an operation or a series of operations on the probability vectors can be described and communicated as a sequence of operations on those vectors. We call such a data structure that contains the number and order of vectors as well as the respective branch lengths a

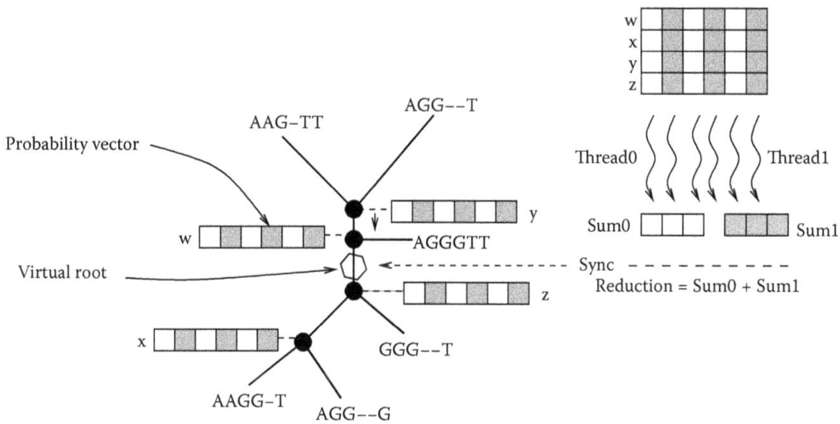

FIGURE 5.3
Parallel fine-grain computation of the likelihood score on a given tree with given branch lengths. Probability vectors are enumerated consistently by w, x, y, z.

traversal descriptor. In the case of a full tree traversal such a traversal descriptor will hence contain all numbers that correspond to inner probability vectors and the order in which they are combined represents the tree structure. One important property of current search algorithms is that only a part of the tree will usually be traversed when a topological change has been applied to the tree. This type of traversal, which frequently entails only the recomputation of just a few probability vectors is called *partial traversal.* In addition to the tree traversals that reflect Equation 5.1, other function types, specifically the computation of the likelihood score at the root (Equation 5.3) or the optimization of a specific branch length, also require reduction operations for computing the log likelihood score across sites or the 1st and 2nd derivative across sites, respectively.

For the sake of completeness we provide the traversal data type definition as used in RAxML below:

```
typedef struct
{
  int tipCase; /* tip case */
  int pNumber; /* ancestral vector number */
  int qNumber; /* left child number */
  int rNumber; /* right child number */
  double qz[NUM_BRANCHES]; /* branch length(s) for p <-> q */
  double rz[NUM_BRANCHES]; /* branch length(s) for p <-> r */
} traversalInfo;
typedef struct
{
  traversallnfo *ti; /* array of traversallnfo entries */
  int count;         /* number of nodes to traverse */
} traversalData;
```

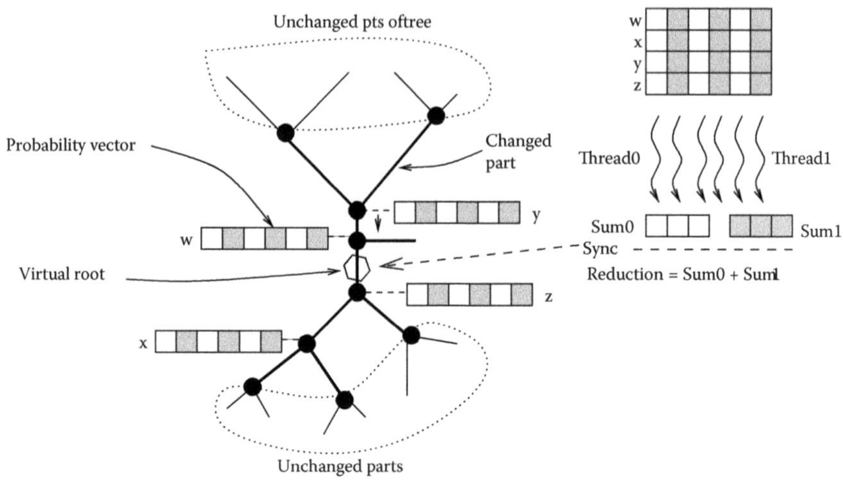

FIGURE 5.4
Example for a parallel partial tree traversal. Not all probability vectors are re-computed in this partial reevaluation of the tree. Probability vectors are enumerated consistently by w, x, y, z.

The variable `tipCase` is used to determine if the ancestral vector has two children that are tips/leaves, one child that is a leaf, or two children that are also ancestral vectors. The respective branch lengths between nodes $p \leftrightarrow q$ and $p \leftrightarrow r$ are actually arrays of double values to accommodate partitioned models with a per-partition estimate of branch lengths (see Section 5.3.3 for a more detailed discussion). Partial tree traversals, that is, relatively short `traversalData` arrays, with `count` typically ≤ 10 will dominate the communication between master and workers. An example for a parallel partial tree traversal is provided in Figure 5.4. Note that the definition of `traversalInfo` lacks an additional field `operationType`, that is, computation of an ancestral vector, computation of the likelihood at the root, or branch length optimization. This field is missing because the data structure has evolved over various redesign cycles, but a future RAxML version will contain a cleaner interface definition that will allow for a better separation of the PLF implementation from the search algorithm.

The master thread steers the tree search and orchestrates the optimization of the branch lengths and model parameters. During the model parameter optimization phase the tree needs to be fully traversed to optimize the rates in the Q matrix and the α shape parameter of the Γ distribution. In this case the master thread generates a full tree traversal list that remains fixed during the model parameter optimization process because the tree topology and the traversal order will not be changed during model parameter optimization. When a model parameter (Q or α) has been changed, every worker thread can independently update its fraction of the likelihood array entries for the full tree traversals and the threads only need to be synchronized when the virtual root is reached and the likelihood score is computed (see Figure 5.3). Therefore, every thread can conduct a relatively large fraction of independent work per alignment column during the model parameter optimization phase.

In contrast to this, if we consider the branch length optimization process, it requires several Newton–Raphson iterations and therefore synchronizations coupled with reduction operations at every individual branch. The synchronization-to-computation ratio for branch length optimization is thus significantly less favorable than for optimization of the model parameters (Q, α) and partial tree traversals. Another important issue is that distinct branches in the tree cannot be optimized simultaneously, since branch length alterations are not independent from each other; that is, we need to repeatedly traverse the tree and optimize one branch at a time. In addition, a subset of branches will also need to be reoptimized after topological changes during the tree search. The considerations regarding branch length optimization are important for the load balance issues discussed in Section 5.3.3.

5.3.2.1 A Library for the PLF

The organization and clear separation of tasks between master and workers described earlier is a generally applicable concept that is not RAxML specific. In the course of several code reorganizations, the calls to the likelihood function have become transparent; that is, the caller does not need to know if they will be executed sequentially or in parallel. The logical next step would thus be to develop a library that implements the PLF and that can use several parallel architectures in a way that is not visible to the actual tree search algorithm. Ideally, the community would require a highly optimized BLAS-like (basic linear algebra subroutines) kernel for the PLF, which represents one of the most important functions in bioinformatics (Felsenstein's seminal paper that introduces the likelihood function [1] for trees has been cited 3,741 times according to Google Scholar).

In Figure 5.5 we provide an abstract description of the architecture of such a library. On the basis of the prolegomena, one would mainly require a low latency interconnect to the hardware platform on which the actual computations are performed, to carry out the frequent reduction and synchronization operations. Bandwidth is not a major issue, since as mentioned before, traversal descriptors will usually be short; that is, comprise only a few nodes on average. As mentioned before, for full tree traversal that is required for model parameter optimizations it suffices to broadcast the traversal order once at the start of a model parameter optimization phase and then reuse it for all successive iterations that optimize the Q and α parameters. As outlined in Figure 5.5 and implemented in RAxML the memory needed to hold the probability vectors should exclusively be allocated by the worker threads and be hidden behind the interface. Clearly, such a library is urgently required and the major challenge will consist of defining a generic and flexible enough

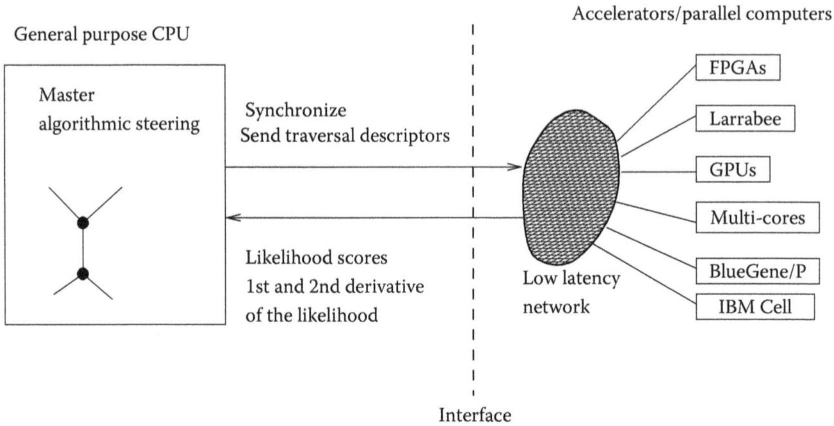

FIGURE 5.5
Combined software and hardware architecture for a PLF library.

interface. However, generality may lead to abandoning certain programming tricks and optimizations in the PLF implementations that significantly contribute to program efficiency. An initial promising initiative to devise an application programming interface (API) for the PLF together with an initial implementation is available at http://code.google.com/p/beagle-lib/. The overall API design is similar to the implementation in RAxML and also conceptually similar to the organization outlined in Figure 5.5. Ideally, the API should also be extended by maximum parsimony functions and by functions for statistical alignment under ML [7] in the future.

5.3.2.2 Scalability Issues

While the aforementioned fine-grain parallelization approach is highly efficient, and generic, that is, mostly independent of the actual search algorithm, there are some limits to scalability depending on the shape of the input alignment (see Figure 5.1). Fine-grain parallelism scales well up to 1,024 CPUs for large-scale phylogenomic analyses [51, 53] with $m \gg 1{,}000$ on well-shaped alignments, but scalability for single-gene or few-gene analyses on badly shaped datasets with $1{,}000 \leq m \leq 10{,}000$ is limited. There are significantly less alignment and probability vector columns to compute in-between synchronization or reduction events per thread or process, which means that the communication-to-computation ratio quickly becomes unfavorable.

The reason why this is an important problem on badly shaped alignments is that they typically contain thousands or tens of thousands of taxa [28]. In current collaborative analyses with biologists we are trying to analyze datasets with 38,000 and 56,000 taxa that comprise less than 10 genes. The scalability of the fine-grain approach on such datasets is limited to 8 or 16 CPUs on typical general-purpose multicore architectures such as the AMD Barcelona or the Intel Nehalem. Even when conducting tree analyses on such large datasets using the Pthreads version of RAxML on 16 cores, execution times for just a single tree search will typically range from a week to a month. This causes two problems: *First,* most HPC clusters do not allow for such extremely long execution times, and *second,* those analyses are susceptible to hardware failures and cluster unavailabilities for maintenance. While checkpointing may provide a solution, ideally we would like to improve scalability. For a single run, this can be achieved only by exploiting a distinct source of parallelism on top of the fine-grain parallelism; that is, via deployment of multigrain parallelism. The obvious candidate for a more coarse-grain parallelization of a single search is the search algorithm itself. Hence, certain steps of the search algorithm would need to be parallelized. While the respective details are outside the scope of this chapter there are two main problems inherent to such an approach: *First,* in contrast to the fine-grain parallelization of the PLF such a parallelization would be highly program specific; that is, not generally applicable, and *second,* current efficient search algorithms as implemented in GARLI or RAxML exhibit a huge degree of sequential dependencies that will be hard to resolve. This will also represent a problem for the PLF library mentioned previously.

Nonetheless, provided the current data flood that produces large well-shaped as well as badly shaped alignments, the community will have to address this challenging problem.

5.3.3 The Real World: Load Balance Issues

An issue that is often not addressed in papers on the parallelization of the PLF are the problems that arise for real-world software under the commonly used models and with the type of data biologists actually want to analyze. Typical HPC papers—some of our own papers included—often just show speedups and scalability for unpartitioned datasets of, for example, DNA or protein alignments and focus on pure proof-of-concept implementations.

The main load balance issue is caused by partitioned phylogenomic analyses that may contain partitions consisting of different data types; that is, a large phylogenomic alignment may contain partitions of morphological or binary characters, of DNA characters, secondary structure characters, and protein characters. Evidently, the number of floating point operations required to compute the per-column log likelihood at a position c for, for example, a full-tree traversal, varies as a function of the number of states s ($s = 2$ for binary data, $s = 4$ for DNA data, $s = 20$ for protein data) and has a complexity of $O(n \cdot s^2)$ (see Equation 5.1) where n is the number of taxa. The easiest way to handle this type of potential load imbalance is to distribute the data columns to threads/processes in a cyclic way, rather than

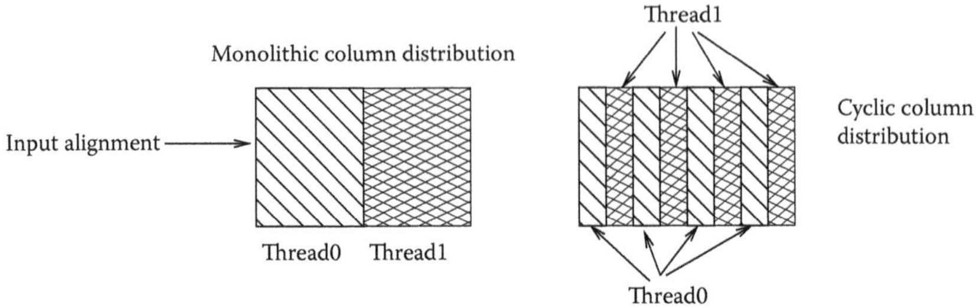

FIGURE 5.6
Monolithic versus cyclic distribution of alignment (and probability vector) columns to threads.

in a monolithic way as outlined in Figure 5.6. The major drawback of this approach in the current RAxML implementation is that every thread will have to recompute the probability transition matrix $P(t)$ locally; that is, a large number of computations are replicated for every partition. To this end, if the partitions are relatively short and if a large number of threads are used, it may happen that a thread will have to compute the transition probability matrices $P(t)$ i for all p partitions, $i = 1 \ldots p$, while only having to compute one single per-site likelihood for every partition. In this case the computation of the $P(t)$ i may actually dominate the computations. A solution for this worst-case scenario may be to also parallelize the computation of the $P(t)$ i, but this will increase the number of synchronization points in the code by a factor of about 2. Initial tests have indicated that the current strategy of computing all required $P(t)$ i locally yields better performance on all real-world datasets we have tested so far.

A related load-balance issue in partitioned analyses is that of per-partition model parameter optimization. Typically, users will chose to infer ML model parameters, such as the GTR substitution matrix or the α parameter that determines the shape of the Γ distribution separately. One will have to infer the respective parameters Q_i and α_i separately using the aforementioned iterative optimization procedures for every partition p. Moreover, one may also prefer to infer an individual set of $2n - 3$ branch lengths for every partition such that there is a total of $p \cdot (2n - 3)$ branch lengths, rather than to conduct a joint branch length estimate across all partitions.

It is important to note that per-partition branch length estimates are the prerequisite for the techniques proposed in [42]. Those techniques can be used to significantly accelerate the computation of the likelihood function and substantially reduce the memory footprint on "gappy" phylogenomic alignments. The reason for the gappyness in such alignments that typically ranges between 50% and 95% lies in the partial unavailability of sequence data for the taxa under study for specific genes; that is, sequences for a specific gene are not always available for every taxon in the dataset. While these techniques are outside the scope of this chapter, the load-balance problems

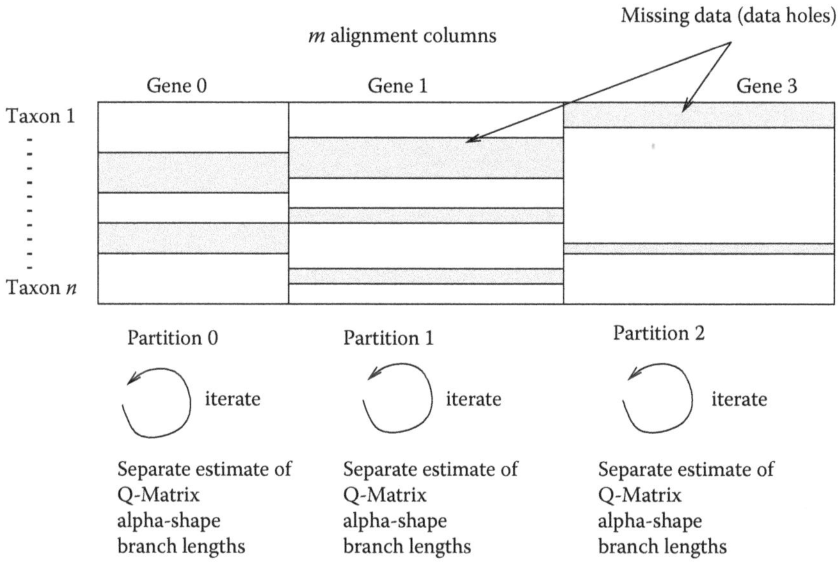

FIGURE 5.7
Schematic outline of a partitioned analysis on a phylogenomic alignment with per-partition estimates of model parameters and branch lengths.

induced by partitioned analyses represent a general problem that will be addressed in more detail at this point.

A schematic representation of a gappy phylogenomic alignment with per-partition branch length and model parameter estimates is provided in Figure 5.7.

As mentioned earlier, one important rationale for using a per-partition branch length estimate is that a significantly more efficient strategy to compute the likelihood score on gappy phylogenomic alignments can be applied [42]. The load-balance problem that arises in this context is described in more detail in [54]. If we now consider the case where we have to optimize the per-partition branch lengths of the branches for all partitions that connect a node p to a node q in the underlying tree topology, the problem emerges (see Figure 5.8). In order to implement this branch length optimization that is based on the iterative Newton–Raphson procedure we need to consider that the number of iterations required for the branch in every partition may vary. We may either chose to optimize the branch of one partition at a time or to concurrently start optimizing the branches for all partitions and keep track of the convergence condition for each single branch. While the latter method is more challenging and error prone to implement, it can significantly improve performance in the fine-grain parallelization, because we can provide more work (more columns to work on simultaneously) to each thread and dramatically reduce the number of synchronization events. Let

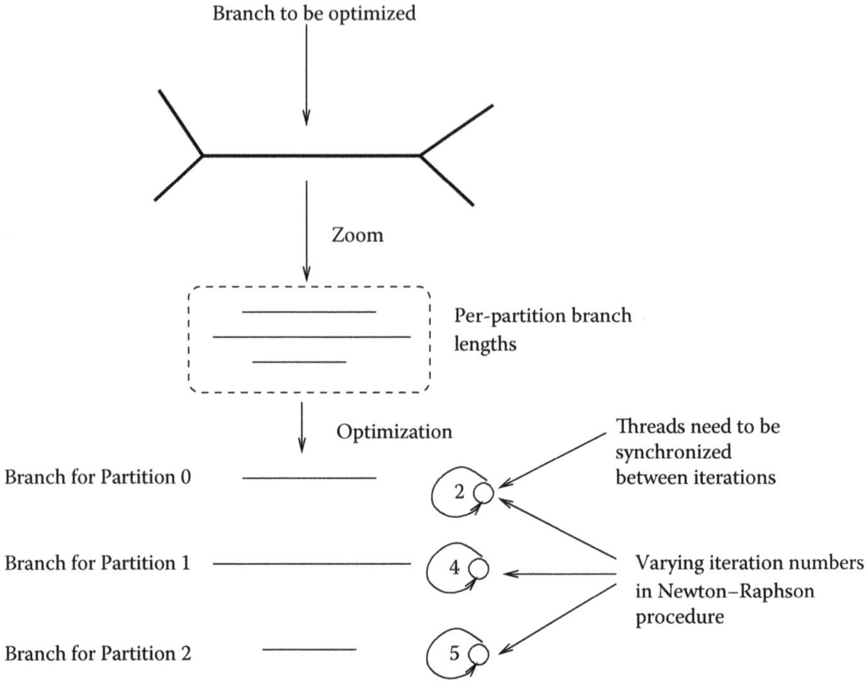

FIGURE 5.8
Outline of the application of the Newton–Raphson procedure for the optimization of one branch in a phylogenomic analysis using per-partition branch length estimates.

us consider a simple example with $p = 10$ and assume that every partition has a length of 10 characters. For the sake of simplicity we can also assume that all p branches will require five iterations of the Newton–Raphson procedure. If we have 10 threads (each thread holds one column of every partition) and optimize the branches for one partition at a time, we will need to synchronize the threads $5 \cdot 10$ times, and for all of those 50 computations every thread will just have one column to work on. If we use the approach of simultaneous optimization of all branches there will only be five synchronization events and between synchronization events, every thread will have 10 columns to work on. While this is an extreme example, in [54] we show that the simultaneous optimization of branch lengths across partitions can yield up to eight-fold improvements in execution times on current multicore architectures.

Note that while Bayesian programs do not require to explicitly optimize branch lengths and model parameters via iterative optimization methods, since this is handled by the Markov Chain Monte Carlo (MCMC) proposal mechanism, the insights obtained in [54] also apply to Bayesian analyses. In the Bayesian case, simultaneous branch length change proposals across all partitions should be applied to increase parallel efficiency.

5.4 Adaptations to Emerging Parallel Architectures

By now, a plethora of papers has been published on orchestrating the PLF on emerging parallel architectures, which I will briefly review in this section. In all cases the fine-grain parallelism in the PLF is exploited, whereas in some implementations the more coarse-grain parallelism provided by multiple ML tree searches on distinct starting trees, Bootstrap replicates [55], or Markov–Chains is used on top of the fine-grain parallelism. Since this coarse-grain type of parallelism at the level of independent tree searches is mostly straightforward to explore (in contrast to coarse-grain parallelism at the algorithmic level) I will mainly focus on exploiting fine-grain parallelism.

One key issue that the HPC community often fails to address is that of taking proof-of-concept implementations to production level; that is, it is shown on a small subset of the functionality of a widely used tool (some of the author's own work included) that scalability can be achieved by applying a certain strategy that takes into account specific characteristics of the target architecture. Unfortunately, many of these parallelizations are never taken to production level and are therefore of little or no use to the large biological user community.

There have been only few attempts [56–59] to devise explicit architectures for the PLF on FPGAs. In earlier days the main problem was the lack of support for floating-point arithmetics on FPGAs. This problem has been addressed by the introduction of digital signal processor (DSPs) on modern FPGAs that now allow for improved implementations of the PLF on FPGAs [58, 59]. In [59] we present a significantly improved implementation of the original design that implements a vector-like processor architecture. This vector-like architecture comprises 10 basic cells that act in a similar way as the worker processes/threads or synergistic processing elements (SPEs) on the Cell (see below) in the general-parallelization scheme albeit at a more fine-grain level. This improved architecture is also capable of carrying out partial tree traversals. The speedups of the dedicated PLF architecture compared to a high-end multicore machine are within the typical range (factor 5–10) for floating-point intensive computations on FPGAs. However, we have made a lot of simplifying assumptions that do not yield the current architecture practical or usable for any real-world phylogenetic inference, put aside the latency problems between CPUs and FPGAs that may nonetheless soon be resolved. The same observations apply to the work by Mak and Lam [56, 57] (see [58] for a more detailed discussion).

Thus, FPGAs may rather be viewed as prototyping devices than accelerators. We believe that a prototyping device view will help us to devise and test architectures for the PLF. Current work focuses on the development of a more versatile and fully functional PLF architecture that will be able to handle different data types (binary, DNA, Protein, etc.) and also accommodate the standard Γ model of rate heterogeneity.

Early work on general-purpose computing on GPUs (GPGPU) focused on exploiting the PLF using an, in the meantime deprecated, older version of

RAxML. The main focus of this early work on GPUs was on the exploration of the capabilities of GPUs in the pre-CUDA era using BrookGPU (http://www-graphics.stanford.edu/projects/brookgpu/). We faced problems associated with GPUs that still persist: the issue of using SP arithmetics and the data transfer bottleneck between CPU $\leftrightarrow$ GPU. Recent work on GPUs dealt with porting the likelihood kernel of BEAST [23] to a NVIDIA GPU [60]. While this paper reports impressive speedups, mainly for 61-state Codon models, the actual performance comparison does not appear to be conducted in an entirely fair way with respect to the general-purpose CPU (a 3.2 GHz Intel Core 2 Extreme) used. It is not clear how much effort was invested to optimize the C implementation on the general-purpose CPU, nor if a commercial compiler such as the Intel icc was used. It is also not entirely clear from the paper why the performance runs on the CPU were only conducted under DP, while the GPU offers DP and SP implementations. While we have found that the usage of SP leads to a performance degradation in some cases because of a 10-fold decrease in scaling events (see Section 5.2.3) this may not necessarily be the case for the implementation in BEAST. In addition, accuracy issues that may arise for input alignments with a larger number of taxa under SP are not explored because scalability is only assessed on a single phylogenomic dataset with 62 taxa. The CPU was not fully exploited, since only one core was used while the code could have been easily parallelized with OpenMP. Moreover, the code was apparently not vectorized with SSE3, a technique that can yield significant additional speedups for the likelihood kernel [45]. Thus, the question remains how much better the multicore platform would have performed if exploited to its full capabilities and if the same amount of time as for the GPU had also been invested into optimizing the C code for the CPU. While the speedups that are obtained are still very good for Codon models on GPUs, it remains an open question how much speedup a vectorized and multithreaded version of BLAS for computing Equation 5.1 (this equation represents an element-wise multiplication of the results of two dense matrix-matrix multiplications) on Codon models would yield. Some initial tests with DNA models and BLAS have shown that the overhead for calling BLAS is too large and the memory footprint of the 4×4 DNA substitution matrix is too small to achieve substantial speedups, but this may not be the case for the 61-state Codon transition matrices.

Our own assessment of GPUs, multicores, and the IBM Cell using MrBayes for DNA and protein models that is also implemented in SP shows far less spectacular results for GPUs [61]. Given the aforementioned problems with using SP on trees with many taxa, albeit this may just be a problem of the specific implementation in RAxML, a lot will depend on whether DP arithmetics will become faster on GPUs. It may well be that GPUs will lose ground compared, for instance, to the Intel Larrabee because of the currently insufficient DP performance.

Work on porting the PLF to the IBM Cell and the Sony Playstation III is described in the following papers [49, 62–64]. Here our efforts mainly focused on efficiently scheduling multigrain parallelism on the Cell, by using a combination of a fine-grain and an extremely coarse-grain approach at the level of independent tree searches. In addition, we explored various Cell-specific optimization techniques for the PLF. The results in the aforementioned papers were all based on older versions of the Cell where the performance differences between SP and DP arithmetics were still significantly larger than the current factor of two. The fine-grain parallelism in the PLF is exploited in the same way as described previously, with the sole difference that the probability vectors actually reside on the principal processing element (PPE) and only small portions (a couple of columns) of the probability vectors are shuffled back and forth between the PPE and the SPEs where the actual computations are conducted. Unfortunately, the work on the Cell mainly addressed HPC issues and we never devised a production-level implementation of RAxML for this architecture. An interesting current development is that the RoadRunner supercomputer can be programmed entirely by using the message-passing paradigm; that is, every SPE of the Cell may act as an independent MPI process. This can facilitate the development of a production-level Cell implementation by slightly modifying the fine-grain MPI-based parallelization of RAxML [51]. Nonetheless, the arithmetic operations in the likelihood functions would still need to be manually vectorized and tuned to achieve optimal performance on the Cell. An issue that may limit usage of the Playstation III for real-world phylogenetic inference is that the PPE in the PS3 does not have enough main memory (256 MB). This may not prove to be sufficient for the analysis of larger datasets that can nowadays easily exceed 1 GB of memory footprint. In addition, most biology labs would most probably not buy a significantly more expensive Cell processor and rather invest into a general-purpose multicore machine, because only a small subset of the applications typically used by evolutionary biologists has been ported to the Cell.

With respect to shared-memory nodes, shared-memory supercomputers, and general-purpose multicore systems many RAxML-specific papers deal with the usage of OpenMP [52], Pthreads [42, 54], and MPI [53] to exploit fine-grain parallelism in the PLF. We also compare performance of MPI, Pthreads, and OpenMP in [53] and find that MPI clearly performs best across all platforms (SMPs, multicores, SGI ALTIX 4700 supercomputer), but shared-memory version of MPI may be hard to install for typical users of the code. A complete transition to MPI for fine-grain parallelism would however significantly facilitate software maintenance and reduce complexity. In [61] we also provide an OpenMP implementation for MrBayes. Because of the nondeterminism of the MCMC chains, the nondeterminism of the reduction operations in OpenMP is not as critical as for RAxML. For GARLI an OpenMP parallelization is also available, but no performance study has been published so far. Finally, IQPNNI has also been parallelized using OpenMP for SMP systems [65]. IQPNNI also represents an example for a relatively straightforward parallelization of the search algorithm, which in contrast to RAxML and GARLI exhibits almost no sequential dependencies. The parallelization of the IQPNNI search algorithm is summarized in [19].

Less work has been conducted on orchestrating the PLF function on massively parallel distributed memory architectures. Most of this work has focused on the IBM BlueGene/L [51, 66, 67] but we have also assessed scalability of fine-grain parallelism with MPI on an infiniband-connected cluster of 4-way Opteron SMPs [51, 67]. Owing to the favorable configuration with relatively slow processors and a very fast dedicated network for collective operations, we have measured speedups of 890 on 1,024 nodes of the BlueGene/L. The PBPI [66] application is an analogous parallelization of a Bayesian inference algorithm on the BlueGene/L, but has remained in a proof-of-concept state. Both implementations (RAxML and PBPI) can also be executed in a multigrain mode where chains or independent tree searches can be run independently on subgroups of nodes.

5.5 Future Directions

Given the rapid development of computer architectures, it is hard to assess and predict what the best platform for executing the PLF may be. On the basis of our observations concerning loss of accuracy under SP for alignments with more than 1,000–2,000 taxa, the usage of GPUs may be critical if DP performance is not improved. The Larrabee architecture surely sounds promising, but a significant amount of recoding and reorganization of the PLF will be required to fully exploit the 512-bit wide vector instructions. The usage of GPUs for scientific computing appears to be slightly overestimated at present.

Whether a dedicated computer architecture for the PLF, a real chip, will ever become a reality is questionable, because the market for such an architecture that would be mostly dominated by Academia is most probably too small. Hence, the interest in building computer architectures for the PLF is mainly academic and tries to address the question how the ideal architecture should look like.

Overall, a lot will depend on the development of input datasets; that is, if phylogenomic datasets start growing in the number of taxa, for example, phylogenetic analyses of 2,000 taxa with 100–1,000 genes will become common, which seems to be relatively probable, we will soon reach memory limits and computational resource shortages. A recent phylogenomic analysis on a BG/L using the MPI-parallelized fine-grain version of RAxML already required 2.25 million CPU hours, while another recent collaborative phylogenomic analysis had a memory footprint of 89 GB. However, scalability is granted with the current parallelization approaches and supercomputers are available that can handle datasets that are even one order of magnitude larger than the two aforementioned studies. Significant computational savings can be achieved if tree searches and likelihood computations are conducted using the method proposed in [42]. However, the implementation of this method is algorithmically and technically challenging and for the time being RAxML is only able to optimize model parameters on a fixed tree using this method. Moreover, this method will only be efficient if phylogenomic alignments remain gappy; that is, more than 50% of the data are missing. Depending on the advances in molecular sequencing techniques the density of such large phylogenomic datasets may increase to a point where the method described in [42] will not exhibit any computational advantages any more compared to with the standard approach. However, it will still be relatively straightforward to handle well-shaped alignments computationally. In the worst case one will need to filter out some of the data, before the phylogenetic analysis to reduce future datasets to computable sizes. The development of such filtering criteria is definitely a challenge.

With respect to the PLF per se, more research is needed to understand and optimize the scaling procedures to avoid numerical underflow as well as to explore the accuracy limits of SP as a function of the number of taxa. Another issue that is directly linked with the likelihood function is that of constant increase in model complexity, that is, extensions of the likelihood model that have recently been proposed, for example, Codon models or mixture models [68]. These models require more floating point operations and more memory per alignment column and will therefore decrease the size of "computable" datasets again. Thus, there is a clear tradeoff between model accuracy (or complexity) and the size of computable input datasets.

One of the key challenges will be to devise new algorithmic concepts and new parallelization strategies for badly shaped alignments. Work on badly shaped alignments is driven by the desire to infer comprehensive trees (see, e.g., [28, 69]), such as the phylogenetic tree of plants or the tree of bacteria, with the final goal to infer *the* tree of life containing all living beings on earth. New methods to explore the rough likelihood surface and summarize collections of ML trees that do not have significantly different likelihood scores, as well as novel methods to infer support values, will be required. In addition, substantial algorithmic and HPC innovations will be required to improve scalability and execution times of phylogenetic analyses on many-taxon trees.

On a slightly different note, the development and parallelization of programs that can conduct simultaneous alignment and tree inference—current approaches only scale to 20–100 taxa—poses additional challenges.

In the final analysis, one of the keys to success of parallel computing in phyloinformatics will be to design scalable, easy-to-use, production-level codes in close collaboration with the user and HPC community.

5.6 References

1. J. Felsenstein. Evolutionary trees from DNA sequences: A maximum likelihood approach. *Journal of Molecular Evolution*, 17:368–376, 1981.
2. J. Shendure and H. Ji. Next-generation DNA sequencing. *Nature Biotechnology*, 26(10):1135–1145, 2008.
3. T.H. Ogden and M.S. Rosenberg. Multiple sequence alignment accuracy and phylogenetic inference. *Systematic Biology*, 55:314–328, 2006.
4. M. Höhl and M.A. Ragan. Is multiple sequence alignment required for accurate inference of phylogeny? *Systematic Biology*, 56(2):206–221, 2007.
5. A. Loytynoja and N. Goldman. Phylogeny-aware gap placement prevents errors in sequence alignment and evolutionary analysis. *Science*, 320(5883):1632, 2008.
6. W. Wheeler, L. Aagesen, C.P. Arango, J. Faivovich, T. Grant, C. D'Haese, D. Janies, W. L. Smith, A. Varon, and G. Giribet. *Dynamic Homology and Phylogenetic Systematics: A Unified Approach using POY*. American Museum of National History, 2006.
7. R. Fleissner, D. Metzler, and A.v. Haeseler. Simultaneous statistical multiple alignment and phylogeny reconstruction. *Systematic Biology*, 54:548–561, 2005.
8. B. Redelings and M. Suchard. Joint Bayesian estimation of alignment and phylogeny. *Systematic Biology*, 54(3), 2005.
9. W.M. Fitch and E. Margoliash. Construction of phylogenetic trees. *Science*, 155(3760):279–284, 1967.
10. L.R. Foulds and R.L. Graham. The Steiner problem in phylogeny is NP-complete. *Advances in Applied Mathematics*, 3(43–49):299, 1982.
11. S. Roch. A short proof that phylogenetic tree reconstruction by maximum likelihood is hard. *IEEE/ACM Transactions on Computational Biology and Bioinformatics*, pages 92–94, 2006.
12. A.W.F. Edwards, L.L. Cavalli-Sforza, V.H. Heywood, and J. McNeill. Phenetic and phylogenetic classification. *Systematics Association Publication*, 6:67–76, 1963.
13. D.A. Morrison. Increasing the efficiency of searches for the maximum likelihood tree in a phylogenetic analysis of up to 150 nucleotide sequences. *Systematic Biology*, 56(6):988–1010, 2007.
14. Z. Yang. PAML 4: Phylogenetic analysis by maximum likelihood. *Molecular Biology and Evolution*, 24(8):1586, 2007.
15. S. Guindon and O. Gascuel. A simple, fast, and accurate algorithm to estimate large phylogenies by maximum likelihood. *Systematic Biology*, 52(5):696–704, 2003.
16. D. L. Swofford. PAUP*: Phylogenetic analysis using parsimony (* and other methods), version 4.0b10. Sinauer Associates, 2002.
17. D. Zwickl. Genetic Algorithm Approaches for the Phylogenetic Analysis of Large Biological Sequence Datasets under the Maximum Likelihood Criterion. PhD thesis, University of Texas at Austin, April 2006.
18. A. Stamatakis. RAxML-VI-HPC: maximum likelihood-based phylogenetic analyses with thousands of taxa and mixed models. *Bioinformatics*, 22(21):2688–2690, 2006.
19. B.Q. Minh, L.S. Vinh, A.v. Haeseler, and H.A. Schmidt. pIQPNNI: Parallel reconstruction of large maximum likelihood phylogenies. *Bioinformatics*, 21(19):3794–3796, 2005.

20. G. Jobb, A.v. Haeseler, and K. Strimmer. TREEFINDER: A powerful graphical analysis environment for molecular phylogenetics. *BMC Evolutionary Biology*, 4, 2004.
21. F. Ronquist and J.P. Huelsenbeck. MrBayes 3: Bayesian phylogenetic inference under mixed models. *Bioinformatics*, 19(12):1572–1574, 2003.
22. N. Lartillot, S. Blanquart, and T. Lepage. PhyloBayes. v2. 3, 2007.
23. A.J. Drummond and A. Rambaut. BEAST: Bayesian evolutionary analysis by sampling trees. *BMC Evolutionary Biology*, 7(214):1471–2148, 2007.
24. M. Gottschling, A. Stamatakis, I. Nindl, E. Stockfleth, A. Alonso, L. Gissmann, and I.G. Bravo. Multiple evolutionary mechanisms drive papillomavirus diversification. *Molecular Biology and Evolution*, 24(5):1242–1258, 2007.
25. G.W. Grimm, S.S. Renner, A. Stamatakis, and V. Hemleben. A nuclear ribosomal DNA phylogeny of acer inferred with maximum likelihood, splits graphs, and motif analyses of 606 sequences. *Evolutionary Bioinformatics Online*, 2:279–294, 2006.
26. L. Ganzert, G. Jurgens, U. Munster, and D. Wagner. Methanogenic communities in permafrost-affected soils of the Laptev sea coast, Siberian arctic, characterized by 16s rRNA gene fingerprints. *FEMS Microbiology Ecology*, 59(2):476–488, 2007.
27. C.W. Dunn, A. Hejnol, D.Q. Matus, K. Pang, W.E. Browne, S.A. Smith, E. Seaver, G.W. Rouse, M. Obst, G.D. Edgecombe, M.V. Sorensen, S.H.D. Haddock, A. Schmidt-Rhaesa, A. Okusu, R.M. Kristensen, W.C. Wheeler, M.Q. Martindale, and G. Giribet. Broad phylogenomic sampling improves resolution of the animal tree of life. *Nature*, 452(7188):745–749, 2008.
28. S.A. Smith and M.J. Donoghue. Rates of molecular evolution are linked to life history in flowering plants. *Science*, 322(5898):86–89, 2008.
29. A. Hejnol, M. Obst, A. Stamatakis, M. Ott, G.W. Rouse, G.D. Edgecombe, P. Martinez, J. Baguna, X. Bailly, U. Jondelius, M. Wiens, W.E.G. Müller, E. Seaver, W.C. Wheeler, M.Q. Martindale, G. Giribet, and C.W. Dunn. Rooting the bilaterian tree with scalable phylogenomic and supercomputing tools. 2009. submitted.
30. N.J. Savill, D.C. Hoyle, and P.G. Higgs. RNA sequence evolution with secondary structure constraints: Comparison of substitution rate models using maximum-likelihood methods. *Genetics*, 157:399–411, 2001.
31. N. Goldman and Z. Yang. A codon-based model of nucleotide substitution for protein-coding DNA sequences. *Molecular Biology and Evolution*, 11(5):725–736, 1994.
32. Z. Yang. Maximum likelihood phylogenetic estimation from DNA sequences with variable rates over sites. *Journal of Molecular Evolution*, 39:306–314, 1994.
33. A. Stamatakis. Phylogenetic models of rate heterogeneity: a high performance computing perspective. In *Proc. of IPDPS2006*, HICOMB Workshop, Proceedings on CD, Rhodos, Greece, April 2006.
34. S. Tavaré. Some probabilistic and statistical problems in the analysis of DNA sequences. *Lectures on Mathematics in the Life Sciences*, 17:57–86, 1986.
35. B. Boussau and M. Gouy. Efficient likelihood computations with nonreversible models of evolution. *Systematic Biology*, 55(5):756–768, 2006.
36. W.H. Press, S.A. Teukolsky, W.T. Vetterling, and B.P. Flannery. Numerical recipes in C. *The Art of Scientific Computing*. Cambridge: University Press, 3(2), 1992.
37. R.P. Brent. *Algorithms for Minimization without Derivatives.* Prentice Hall, 1973.

38. Z. Yang. Maximum likelihood estimation on large phylogenies and analysis of adaptive evolution in human influenza virus A. *Journal of Molecular Evolution*, 51(5):423–432, 2000.
39. N. Goldman, J.P. Anderson, and A.G. Rodrigo. Likelihood-based tests of topologies in phylogenetics. *Systematic Biology*, 49(4):652–670, 2000.
40. K. Strimmer and A. Rambaut. Inferring confidence sets of possibly misspecified gene trees. *Proceedings of the Royal Society B: Biological Sciences*, 269(1487):137–142, 2002.
41. H. Shimodaira and M. Hasegawa. CONSEL: for assessing the confidence of phylogenetic tree selection. *Bioinformatics*, 17(12):1246–1247, 2001.
42. A. Stamatakis and M. Ott. Efficient computation of the phylogenetic likelihood function on multi-gene alignments and multi-core architectures. *Philosophical Transactions of the Royal Society Series B*, 363:3977–3984, 2008.
43. N. Metropolis, A.W. Rosenbluth, M.N. Rosenbluth, A.H. Teller, E. Teller, et al. Equation of state calculations by fast computing machines. *The Journal of Chemical Physics*, 21(6):1087, 1953.
44. J.P. Huelsenbeck, B. Larget, R. Miller, and F. Ronquist. Potential applications and pitfalls of Bayesian inference of phylogeny. *Systematic Biology*, 51(5), 2002.
45. S. A. Berger and A. Stamatakis. Accuracy and performance of single versus double precision arithmetics for maximum likelihood phylogeny reconstruction. In *Proceedings of PBC09, Parallel Biocomputing Workshop*. Springer LNCS accepted for publication.
46. J. Kurzak and J. Dongarra. Implementation of mixed precision in solving systems of linear equations on the cell processor. *Concurrency and Computation*, 19(10):1371, 2007.
47. A. Stamatakis, P. Hoover, and J. Rougemont. A rapid bootstrap algorithm for the RAxML web servers. *Systematic Biology*, 57(5):758–771, 2008.
48. A. Stamatakis, T. Ludwig, and H. Meier. RAxML-III: A fast program for maximum likelihood-based inference of large phylogenetic trees. *Bioinformatics*, 21(4):456–463, 2005.
49. A. Stamatakis, F. Blagojevic, C.D. Antonopoulos, and D.S. Nikolopoulos. Exploring new search algorithms and hardware for phylogenetics: RAxML meets the IBM Cell. *Journal of Very Large Scale Integration Signal Processing Systems*, 48(3):271–286, 2007.
50. S. Whelan and N. Goldman. A general empirical model of protein evolution derived from multiple protein families using a maximum-likelihood approach. *Molecular Biology and Evolution*, 18(5):691–699, 2001.
51. M. Ott, J. Zola, S. Aluru, and A. Stamatakis. Large-scale maximum likelihood-based phylogenetic analysis on the IBM BlueGene/L. In *Proc. of IEEE/ACM Supercomputing Conference 2007 (SC2007)*, 2007.
52. A. Stamatakis, M. Ott, and T. Ludwig. RAxML-OMP: An efficient program for phylogenetic inference on SMPs. *PaCT*, pages 288–302, 2005.
53. A. Stamatakis and M. Ott. Exploiting fine-grained parallelism in the phylogenetic likelihood function with MPI, Pthreads, and OpenMP: A performance study. In M. Chetty, A. Ngom, and S. Ahmad, editors, *Pattern Recognition in Bioinformatics*, LNCS 5265, pp. 424–435. Springer, 2008.
54. A. Stamatakis and M. Ott. Load balance in the phylogenetic likelihood kernel. In *Proceedings of ICPP 2009*, 2009. accepted for publication.

55. J. Felsenstein. Confidence limits on phylogenies: An approach using the bootstrap. *Evolution*, 39(4):783–791, 1985.
56. T.S.T. Mak and K.P. Lam. Embedded computation of maximum-likelihood phylogeny inference using platform FPGA. In *Proceedings of IEEE Computational Systems Bioinformatics Conference*, pages 512–514, 2004.
57. T.S.T. Mak and K.P. Lam. FPGA-based computation for maximum likelihood phylogenetic tree evaluation. *Field Programmable Logic and Application*, LNCS 3203, pp. 1076–1079, 2004.
58. N. Alachiotis, E. Sotiriades, A. Dollas, and A. Stamatakis. Exploring FPGAs for accelerating the phylogenetic likelihood function. In *Proceedings of HICOMB2009*, 2009. Accepted for publication.
59. N. Alachiotis, A. Stamatakis, E. Sotiriades, and A. Dollas. An architecture for the phylogenetic likelihood function. 2009. Accepted for publication.
60. M.A. Suchard and A. Rambaut. Many-core algorithms for statistical phylogenetics. *Bioinformatics*, 25(11):1370, 2009.
61. F. Pratas, P. Trancoso, A. Stamatakis, and L. Sousa. Fine-grain parallelism for the phylogenetic likelihood functions on multi-cores, Cell/BE, and GPUs. 2009. submitted.
62. F. Blagojevic, D.S. Nikolopoulos, A. Stamatakis, C.D. Antonopoulos, and M. Curtis- Maury. Runtime scheduling of dynamic parallelism on accelerator-based multi-core systems. *Parallel Computing*, 33:700–719, 2007.
63. F. Blagojevic, D.S. Nikolopoulos, A. Stamatakis, and C.D. Antonopoulos. Dynamic multigrain parallelization on the cell broadband engine. In *Proc. of PPoPP 2007*, San Jose, CA, March 2007.
64. F. Blagojevic, D.S. Nikolopoulos, A. Stamatakis, and C.D. Antonopoulos. RAxML-Cell: Parallel phylogenetic tree inference on the cell broadband engine. In *Proc. of International Parallel and Distributed Processing Symposium (IPDPS2007)*, 2007.
65. B.Q. Minh, L.S. Vinh, H.A. Schmidt, and A.v. Haeseler. Large maximum likelihood trees. In *Proceedings of the NIC Symposium 2006*, pages 357–365, 2006.
66. X. Feng, K.W. Cameron, C.P. Sosa, and B. Smith. Building the tree of life on terascale systems. In *Proc. of International Parallel and Distributed Processing Symposium (IPDPS2007)*, 2007.
67. M. Ott, J. Zola, S. Aluru, A.D. Johnson, D. Janies, and A. Stamatakis. Large-scale phylogenetic analysis on current HPC architectures. *Scientific Programming*, 16(2–3):255–270, 2008.
68. N. Lartillot and H. Philippe. A Bayesian mixture model for across-site heterogeneities in the amino-acid replacement process. *Molecular Biology and Evolution*, 21(6):1095–1109, 2004.
69. P.A. Goloboff, S.A. Catalano, J.M. Mirande, C.A. Szumik, J.S. Arias, M. Källersjö, and J.S. Farris. Phylogenetic analysis of 73060 taxa corroborates major eukaryotic groups. *Cladistics*, 25:1–20, 2009.

6

Parallel Bioinformatics Algorithms for CUDA-Enabled GPUs

Yongchao Liu, Bertil Schmidt, and Douglas Maskell

6.1 Introduction

Bioinformatics has evolved into a compute-intensive and data-intensive research area driven by advances in both computer hardware and software algorithms. Therefore, many important biological problems are facing challenges both in runtime and memory consumption because of the exponential growth of biological databases. Problem examples include sequence alignments and motif discovery.

Nowadays, incorporating multiple processor cores into a single silicon die has become a commonplace to improve computational performance by means of parallelism. As more and more cores are being incorporated into a single chip, the era of many-core processors is around the corner, which indicates that the future mainstream processors are parallel systems with their parallelism continuing to scale with Moore's law. The emergence of many-core architectures, such as general-purpose graphics processor unit (GPGPU), especially compute unified device architecture (CUDA)-enabled GPUs [1, 2], provides the opportunity to significantly reduce the runtime of many bioinformatics algorithms on commonly available and inexpensive hardware with more powerful high-performance computing power, which are generally not provided by conventional general-purpose processors. However, while demonstrating great compute power, many-core GPUs impose many design constraints and challenges to achieve peak performance. These factors make many-core GPUs less flexible. In general, they would not be able to outperform conventional general-purpose processors for certain application domains.

In this chapter, we describe several effective techniques to fully exploit the compute capability of many-core CUDA-enabled GPUs. These techniques serve at two different scales: the system scale and the device scale. At the system scale, a hybrid computing framework is suggested to overlap the computation of the central processing unit (CPU) and GPU. At the device scale, two approaches, based on intertask and intratask parallelization, respectively, are described to leverage the computational power of CUDA for different application conditions. In particular, we use three techniques to reduce the requirements for global memory bandwidth: coalesced subject sequence arrangement pattern, coalesced global memory access pattern, and cell block division method. On the basis of these techniques, we have parallelized three algorithms on CUDA-enabled GPUs for sequence alignments and motif discovery: CUDASW++, MSA-CUDA, and CUDA-MEME.

6.2 Techniques for Many-Core GPUs

6.2.1 Hybrid Computing Framework

CUDA-enabled GPUs are generally used as additional boards to a general-purpose workstation or PC. To maximize computational performance, it is preferable to overlap the computation of the CPU and GPU to fully exploit the compute capability. This framework is most suitable for the cases in which the compute-intensive task assigned to GPU requires multiple passes to complete, and the computing results in each pass can be directly used as input for the following tasks running on the CPU. Hence, when each pass is finished on the GPU, the CPU can directly conduct the following tasks assigned to it with no necessity to wait for the whole completion of the task on the GPU. Therefore, by overlapping the computation of the CPU and GPU, the runtime is shortened. Figure 6.1 shows the basic structure of a hybrid computing framework of overlapping GPU–CPU computation. The framework mainly consists of four components: a main thread invoking the CUDA

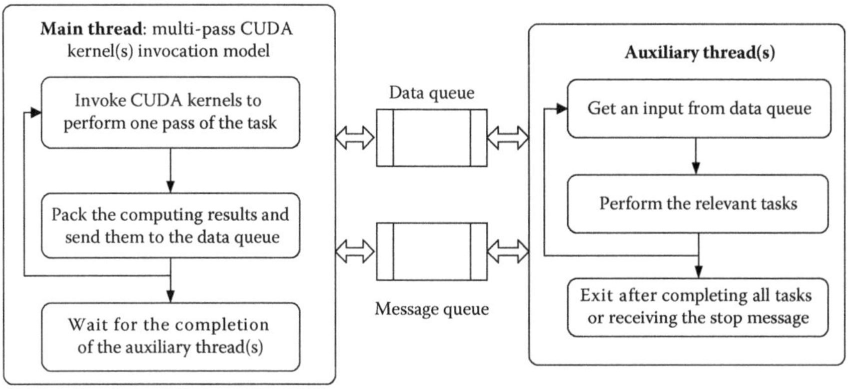

FIGURE 6.1
Basic structure of the hybrid computing framework.

kernel(s), one or more auxiliary threads performing the following relevant tasks according to the input data, a data queue storing the input data for the auxiliary thread(s), and a message queue facilitating the communication between the threads.

6.2.2 Intertask and Intratask Parallelization

We have investigated two basic parallelization approaches to map a number of bioinformatics algorithms to CUDA programming model: intertask parallelization and intratask parallelization.

- *Intertask parallelization:* Each task is assigned to exactly one thread, and *dimBlock* tasks are performed in parallel by different threads in a thread block.
- *Intratask parallelization:* Each task is assigned to one thread block, and *dimBlock* threads in the thread block cooperate to perform the task in parallel, exploiting the inherent parallel characteristic of a task.

Depending on different applications, the definition of *task* refers to different meanings. For instance, for sequence database search, a task refers to the computation of the optimal local alignment of a query sequence and a subject sequence, whereas for multiple sequence alignment, a task refers to the pairwise distance computation of a sequence pair. In general, our results have shown that intertask parallelization achieves better performance but occupies more device memory than intratask parallelization. However, because intratask parallelization occupies significantly less device memory, it is possible to deal with larger problem sizes. Sometimes, these two parallelization approaches are combined in order to meet difference problem sizes.

6.2.3 Coalesced Subject Sequence Arrangement

This technique is particularly suitable for biological database search. Subject sequences are prestored in the device memory of GPUs, depending on the utilized parallelization (intertask or intratask) approach. Two corresponding arrangement patterns for subject sequences in the database are therefore designed to achieve the coalesced access to global memory.

As a preprocessing step, subject sequences are presorted in the increasing order according to their lengths. Intertask parallelization arranges the sorted subject sequences in an array like a multilayer bookcase (see Figure 6.2 (a)). All symbols of a sequence are restricted to be stored in the same column from the top to bottom. All sequences are arranged sequentially in the increasing order of length from left to right and top to bottom in the array. Intratask parallelization sequentially stores the sorted subject sequences in an array row by row from the top-left corner to the bottom-right corner (see Figure 6.2 (b)). All symbols of a sequence are restricted to be stored in the same row from left to right. Using these arrangement patterns for both parallelization methods, access to the subject sequences can be coalesced for all threads in a half-warp.

6.2.4 Coalesced Global Memory Access

To gain maximum bandwidth for global memory access, all threads in a half-warp need to access the global memory in a coalesced pattern. A prerequisite for coalescing is that the words accessed by all threads in a half-warp must lie in the same segment, where the segment size is subject to the device compute capability. The memory spaces, referred to by the same variable names (not referring to same addresses), for all threads in a half-warp have to be allocated in the form of an array to keep them contiguous in address. In this memory array, consecutive memory slots must be allocated for consecutive threads in a thread block to achieve coalescing. Figure 6.2 also presents two global memory allocation patterns of a basic type vector

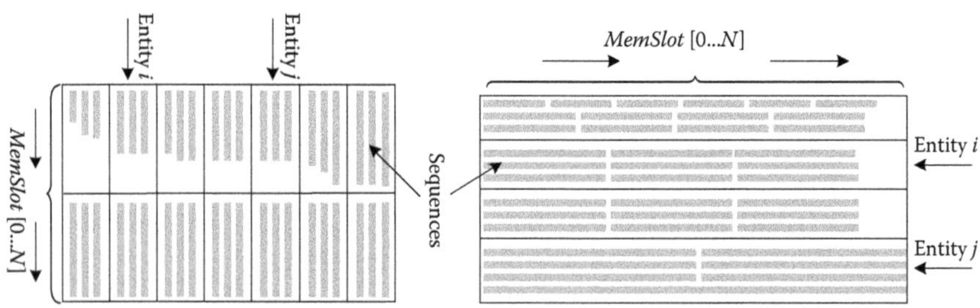

(a) For intertask parallelization (b) For intratask parallelization

FIGURE 6.2
Coalesced arrangement pattern for database search and coalesced global memory allocation patterns for processing entities.

variable of size N for M processing entities (i.e., threads or thread blocks), depending on the parallelization approach.

Intertask parallelization exploits the pattern shown in Figure 6.2 (a), where a memory slot is indexed from top to bottom. When accessing the *MemSlot* array using the same index for all threads in a half-warp, these simultaneous memory accesses are coalesced into one or two memory transactions depending on the compute capability of devices. Intratask parallelization exploits the pattern shown in Figure 6.2 (b), where a memory slot is indexed from left to right. It is able to obtain the coalesced accesses by using the common global memory access pattern; that is, successive threads access the successive addresses in a memory slot.

6.2.5 Cell Block Division Method

In this chapter, we use the cell block division method to further reduce the number of access to global memory when performing the Smith–Waterman (*SW*) algorithm. The concepts derived from this method can also be used in many other cases to reduce the bandwidth requirements, but the specific implementations highly depend on the specific algorithms. The following description of this method is based on an SW-based sequence database search.

The alignment matrix is divided into cell blocks of size $n \times n$ (or $n \times 1$). We define *qlen* and *slen* to, respectively, denote the lengths of a query sequence and a subject sequence. For simplicity, assume that *qlen* and *slen* are multiples of n (if not, the sequence is padded with an appropriate number of dummy symbols). Without cell block division, the computation of one DP-matrix cell (including the computation of the corresponding values in the H, E, and F matrices) requires two global memory accesses (one load operation and one store operation for the intermediate results).

However, when using the cell block division method, the computation of n cells in one column (or row) in a cell block requires only one load operation and one store operation on the global memory instead of n load operations and n store operations.

Because one global memory access typically takes hundreds of clock cycles, the cell block division method leads to a significant reduction of the total runtime owing to a reduction in the global memory accesses. However, the size of cell block is limited by the number of registers (or the amount of shared memory) available per thread.

6.3 SW Database Search

The CUDASW++ software suite [3] is designed for protein sequence database search using the SW algorithm running on many-core GPUs. In CUDASW++, each subject sequence is aligned to the query sequence using the score-only SW algorithm with affine gap penalties (see Chapter 1).

CUDASW++ uses two stages: the first stage exploits intertask parallelization and the second exploits intratask parallelization. A subject sequence length *threshold* is introduced to separate these two stages. All subject sequences with length less than or equal to *threshold* are aligned to the query sequence in the first stage. All alignments of subject sequences of length greater than *threshold* are carried out in the second stage (*threshold* = 3072 is used in CUDASW++). Furthermore, the techniques described in Section 6.2 are used to optimize performance: coalesced subject sequence arrangement, coalesced global memory access, and cell block division method.

Constant memory is exploited to store the gap penalties, scoring matrix, and the query sequence. Before searching for a query sequence against the database, the query sequence is loaded into constant memory. The 64-KB memory capability of the constant memory makes it possible to accommodate much longer query sequences. CUDAWSW++ supports query sequences of length up to 59 K. As mentioned earlier, as long as all threads in a half-warp read the same address in constant memory, the access is as fast as reading from registers. Placing the query sequence in constant memory provides a significant performance improvement as all threads in a warp on the common execution path read the same query sequence address. The scoring matrix is loaded into shared memory, as the performance of constant memory degrades linearly if multiple addresses are requested by threads. This is because threads may frequently access different addresses in the scoring matrix. The integer functions $max(x, y)$ and $min(x, y)$ in the CUDA runtime library are used to map them to a single instruction on the device.

The performance of CUDASW++ is benchmarked by searching for six sequences of lengths from 464 to 5,478 against Swiss-Prot release 56.6. The tests of the single-GPU version are carried out on the GTX 280 graphics card installed on a PC with an AMD Opteron 248 2.2 GHz processor and 1 GB RAM, and the multi-GPU version on the GTX 295 graphics card installed in the same PC. Maximal performance is achieved for a thread block size of 256 threads and a grid size equal to the number of streaming multiprocessors for both the single-GPU and multi-GPU versions. The scoring matrix BLOSUM45 is used with a gap penalty of 10–2k. For the single-GPU version, it achieves a relatively constant performance for all query sequences, with a highest performance of 9.7 GCUPS (giga-cell updates per second). For the multi-GPU version, the performance increases as the lengths of query sequences become longer, because of the overhead incurred mainly by the database loading from host memory to GPU and the host threads scheduling. It achieves a highest performance of 16.1 GCUPS.

We next compare the performance of CUDASW++ with other publicly available implementations for protein database search: SWPS3 [4], SW-CUDA [5], and NCBI-BLAST [6] (version 2.2.19). All the following

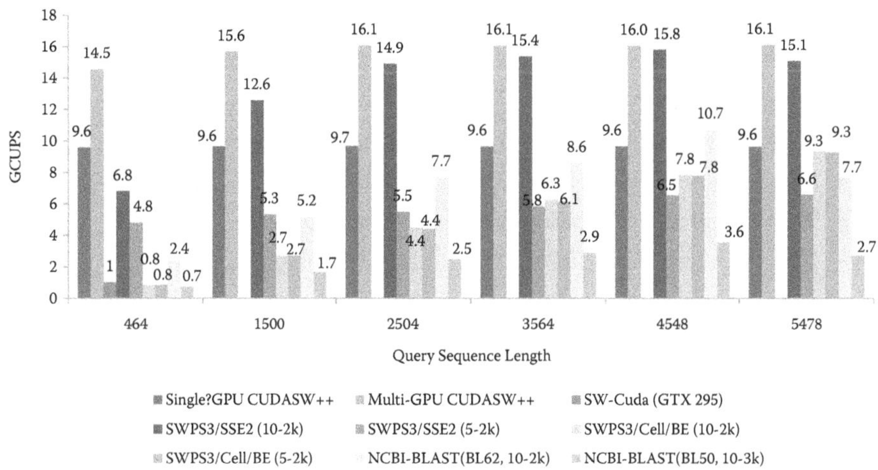

FIGURE 6.3
Performance comparison between CUDASW++ and other publicly available implementations.

tests are performed against Swiss-Prot release 56.6. SWPS3 for x86/SSE2 is tested on a Linux workstation with two Intel Xeon 3.0 GHz dual-core processors by running four threads, and SWPS3 for Cell/BE is tested on a stand-alone PlayStation 3 (PS3). SWPS3 is a vectorized SW implementation with striped query profile layout [7]. The scoring matrices BLOSUM50 and BLOSUM62 are used for the tests. Figure 6.3 presents the performance comparison between CUDASW++ and the other three publicly available implementations.

CUDASW++ (available from http://cudasw.sourceforge.net/) is targeted for CUDA-enabled GPUs with compute capability 1.2 and higher and supports query sequences of length up to 59K, far longer than the maximum sequence length 35,213 in Swiss-Prot release 56.6.

6.4 Multiple Sequence Alignment

In this subsection, we describe how to parallelize the three stages of the ClustalW [8, 9] pipeline for multiple sequence alignment using CUDA. The three stages are described in more detail in Chapter 1. To develop a parallel version, it is imperative to understand the associated time complexities. Given an input dataset $S = \{S_1, \ldots, S_n\}$ of n sequences with average length l_{ave}, the time complexities of the three stages are

- Distance matrix computation (Stage 1): $O(n^2 l_{ave}^2)$
- Guided Tree (Stage 2): $O(n^3)$

- Progressive alignment (Stage 3): $O(nl_{ave}^2 + n^2 l_{ave})$

On the basis of these complexities, we can make the following observations about the runtime behavior of ClustalW:

- Stage 1 generally occupies a large proportion of the total runtime.
- Stage 1 generally has a longer runtime than Stage 3.
- If $n > l_{ave}$, Stage 2 has a longer runtime than Stage 3.
- If $n > l_{ave}^2$, Stage 2 has a longer runtime than Stage 1.

As mentioned in Chapter 1, the distance matrix computation requires the actual optimal alignment path for each pair of input sequences, which can be found in linear space by computing a trace-back with a divide-and-conquer approach. However, sequential implementation of the linear-space trace-back algorithm uses recursion. Unfortunately, CUDA currently does not support recursion. Therefore, we have developed a new stack-based iterative implementation. MSA-CUDA [10] uses this implementation for both pairwise alignments in Stage 1 and profile–profile/sequence alignments in Stage 3.

We have also investigated intertask and intratask parallelization for pairwise distance computation. Coalesced global memory access patterns are exploited for both parallelization approaches. The cell block division method is only exploited in the forward score-only pass using the SW algorithm for the intertask parallelization.

The comparison of time complexities of all stages indicates that the runtime of Stage 2 can dominate the overall runtime of ClustalW for large sequence datasets. Hence, it is important to speed up Stage 2 for large sequence datasets to get a good overall speedup. The basic algorithm of the neighbor-joining (NJ) method is described in Chapter 1.

As mentioned in Chapter 1, NJ computes a phylogenetic tree by iteratively picking and joining two nodes, whose joining minimizes the sum of all branch lengths of the resulting new tree. A pair of nodes i and j is selected if their joining minimizes $S_{i,j} = (n - 2) \times D_{i,j} - (R_i + R_j)$, where n is the number of valid nodes (i.e., the remaining nodes), D is the distance matrix of valid nodes, R_i is the sum of all values in the ith row of D, and R_j is the sum of all values in the jth row of D. It is easily seen that for each pair of nodes i and j, the calculation of $S_{i,j}$ is independent from the other pairs; that is, there is no data dependency between the computations of any pair of nodes. Because the distance matrix is a symmetric square matrix, the data of the cells in the upper triangle (or lower triangle) suffices for NJ tree reconstruction. Therefore, a basic and straightforward method is to map the upper triangle of the distance matrix to a 2D grid of thread blocks and then group the cells in the upper triangle into many equally shaped cell blocks including several equally shaped small cell matrixes, so that all the cells can be tackled in a coherent way. One thread block in the grid is designed to logically correspond to one cell block in the distance matrix and every thread in a thread

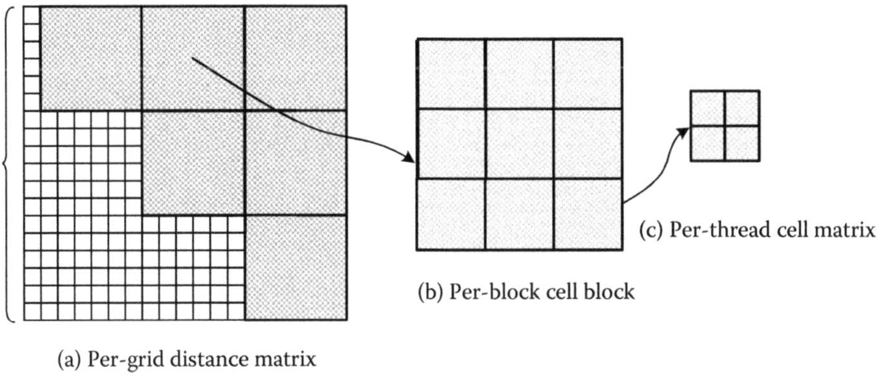

FIGURE 6.4
Basic and straightforward grid mapping method: (a) the distance matrix is mapped to a grid of thread blocks; (b) one thread block is assigned to process one cell block; (c) one thread in a thread block is assigned to process one cell matrix in the corresponding cell block.

block is assigned to process one small cell matrix in the corresponding cell block. This parallelization scheme is illustrated in Figure 6.4.

However, if the entire distance matrix is loaded into GPU device memory without any modification, half of the GPU memory space is wasted because of the symmetry of the distance matrix. To improve the GPU device memory utilization, a memory compaction approach can be applied. In this approach, the distance matrix is considered as a 2D coordinate space. Through coordinate mapping in the distance matrix, up to a half of the memory size is saved compared with the basic method. For n sequences, in the basic method, the size of the memory occupied by the distance matrix is $(n + 1) \times (n + 1) \times 4$ bytes (assuming single precision floating point is used), whereas in the compact memory method, the size is reduced to $(n + 1) \times (MidPoint + 1) \times 4$ bytes, where *MidPoint* is equal to $(n + 1)/2$ (see Figure 6.5 for an example).

As can be seen from Figure 6.5, the compact method maps the bottom half of the upper triangle to the lower triangle of the top half. Considering the matrix as a 2D coordinate space on the Cartesian plane, where the origin is located on the left-top corner, the horizontal (x) coordinates increase from the left to the right and the vertical (y) coordinates increase from the top to bottom. The coordinate mapping rules are as follows:

For each cell (x,y) in the upper triangle of the distance matrix,

- If $y \leq MidPoint$, the coordinate is not modified.
- If $y > MidPoint$, the coordinate is mapped to $(n + 1 - x, n + 1 - y)$.

Hence, a coordinate transformation is required for the cells whose y-coordinates are greater than *MidPoint* when accessing data in the distance matrix. After the coordinate transformation, the cells that are in the same

	0	1	2	3	4	5	6
0	0	0	0	0	0	0	0
1	*	0	d_1	d_2	d_3	d_4	d_5
2	*	*	0	d_6	d_7	d_8	d_9
3	*	*	*	0	d_{10}	d_{11}	d_{12}
4	*	*	*	*	0	d_{13}	d_{14}
5	*	*	*	*	*	0	d_{15}
6	*	*	*	*	*	*	0

(a)

	0	1	2	3	4	5	6
0	0	0	0	0	0	0	0
1	d_{15}	0	d_1	d_2	d_3	d_4	d_5
2	d_{14}	d_{13}	0	d_6	d_7	d_8	d_9
3	*	*	*	0	d_{10}	d_{11}	d_{12}

(b)

FIGURE 6.5
Examples of the distance matrices in both algorithms: (a) the original distance matrix used in the basic algorithm; (b) the new compact memory distance matrix in the improved algorithm.

row in the original distance matrix are still in the same row in the mapped distance matrix.

At the initialization stage, the distance matrix (computed in Stage 1) is copied from host to GPU device memory. After a node pair has been selected, the values of relevant cells in the distance matrix have to be updated for the successive iteration. If the whole matrix is entirely reloaded, the time overhead would be fairly high because of the relatively narrow memory bandwidth between the host and the GPU. To significantly reduce the amount of data transferred, only the changed valid cells are updated. In this way, only the data values of one column and one row in the distance matrix need to be transferred from host to GPU every iteration, which makes the data transfer overhead negligible.

Shared memory is used to store the temporary results of each thread block. Each thread in one thread block compares and selects the node pair (*imin, jmin*), whose joining into a new node gives the smallest branch length among the node pairs allocated to it, and then stores the selected node pair and its value $S_{imin,jmin}$ into the storage space in the shared memory. Texture memory is exploited to store the distance matrix and the row and column sums of all valid nodes.

After reconstructing the NJ tree, the NJ tree is rerooted to calculate the weights of sequences and to traverse the rooted tree to identify the alignment steps for Stage 3. The unrooted NJ tree is rerooted using a "mid-point" method [11]. The root is placed at the position where the means of branch lengths on either side of the root are identical.

Using the conventional sequential C code, all tree nodes can be stored in a vector and the relationship between nodes is maintained through pointers. To parallelize the rerooting using CUDA, the tree nodes must be transferred from host memory to GPU device memory while still maintaining the tree structure. However, pointers will be invalidated while transferring due to the changes of memory address spaces. In this case, we substitute vector indices

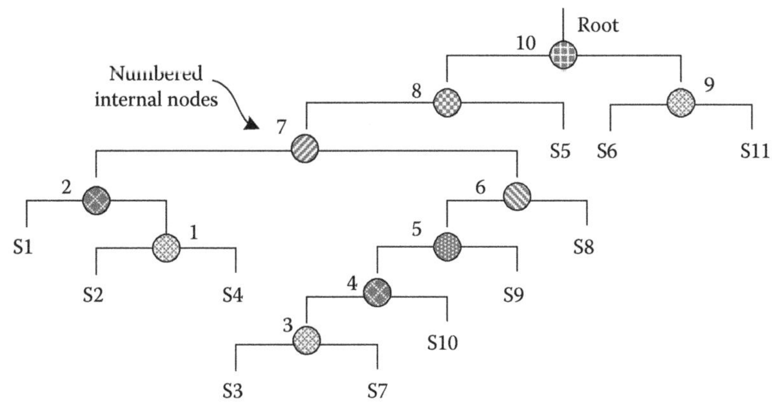

FIGURE 6.6
Example of a rooted guided tree produced using the NJ method.

for pointers to maintain the relationship between nodes, where each node object stores the indexes of itself, its parent, and its left and right children.

For the CUDA implementation of rerooting the NJ tree, one thread block corresponds to one node that is selected as the reference, and is assigned to compute the difference value of the branch length means of leaf nodes on the left and on the right of this node. Every thread in a thread block is assigned to perform the computation on a separate subset of leaf nodes. For each leaf node in a subset, the corresponding thread identifies on which side of the selected node this leaf node lies, and then computes the distance between this leaf node and the selected node. Shared memory is exploited by each thread to store the results for the corresponding subset of leaf nodes. Texture memory is used to store the tree nodes.

Stage 3 aligns larger and larger groups of sequences using pairwise alignment following the branching order of the rooted guided tree from the leaves up to the root. Every leaf node of the guided tree corresponds to a sequence and each internal node corresponds to an alignment produced from the aligned sequences in the left subtree and in the right subtree. The alignment corresponding to an internal node can be launched if and only if the alignments corresponding to the roots of its left and right subtrees have been performed. Obviously, the alignments at the same level of the guided tree can be performed in parallel but even alignments that are not at the same level could also be parallelized. For example in Figure 6.6, all alignments corresponding to internal nodes with the same patterns can be performed in parallel.

Initially, the rooted guided tree is depth-first traversed in post order to number all the internal nodes and build the dependency relationship with their left and right subtrees. All internal tree nodes are stored in a vector in traversal-order. For all tree nodes, three auxiliary vectors are used to record

	0	1									10
			Indices of numbered internal nodes →								
Left child	NIL	0	0	0	3	4	5	2	7	0	8
Right child	NIL	0	1	0	0	0	0	6	0	0	9
Aligned flags	1	0	0	0	0	0	0	0	0	0	0

FIGURE 6.7
Three initial auxiliary vectors storing the dependency relationship with their left and right subtrees and the aligned flags.

the indices of their left children, their right children, and a flag indicating whether the corresponding alignment has been performed. For a leaf node, the indices of its left and right children are set to 0. For an internal node, if one child is a leaf, then the index of this child is also set to 0. The dummy subtree numbered as 0 is always defined aligned because it corresponds to an input sequence for an alignment. Figure 6.7 presents the three initial auxiliary vectors for the rooted guided tree shown in Figure 6.6.

In our CUDA implementation, the progressive alignment is conducted iteratively in a multipass way. For each pass, firstly, all undone alignments that are able to be performed in this pass are identified by checking the flag words of their left and right children stored in the flag-vector. If both of its left and right children have been aligned, this alignment is added to the ready alignment list managing all the alignments to be performed in this pass; otherwise, this alignment has to wait until both of its children have been aligned. After the completion of the ready alignment list, the pairs of profiles corresponding to those alignments are constructed. Second, the pairwise alignments of all pairs of profiles are performed on the GPU in parallel. Third, gaps are added to the sequences corresponding to each pair of profiles by tracing back its optimal alignment. Finally, all the alignments performed in this pass will set their flag words in the flag vector to indicate that they are aligned.

As illustrated in Figure 6.6, the guided tree is seldom well balanced and the numbers of alignments that can be performed in one pass decreases as the alignments move up to the root of the tree. Therefore, MSA-CUDA uses the following parallelization strategy. When the number of alignments to be performed in one pass is relatively large, the intertask parallelism method is utilized, and when it is relatively small, the intratask parallelism method is superior. Thus, a combination of intertask and intratask parallelism is used to compute all the alignments to be performed in one pass. A *threshold* determines the branches of the program flow. If the total number of alignments or the remaining number of alignments after one or more passes is still more than a *threshold*, the intertask parallelization

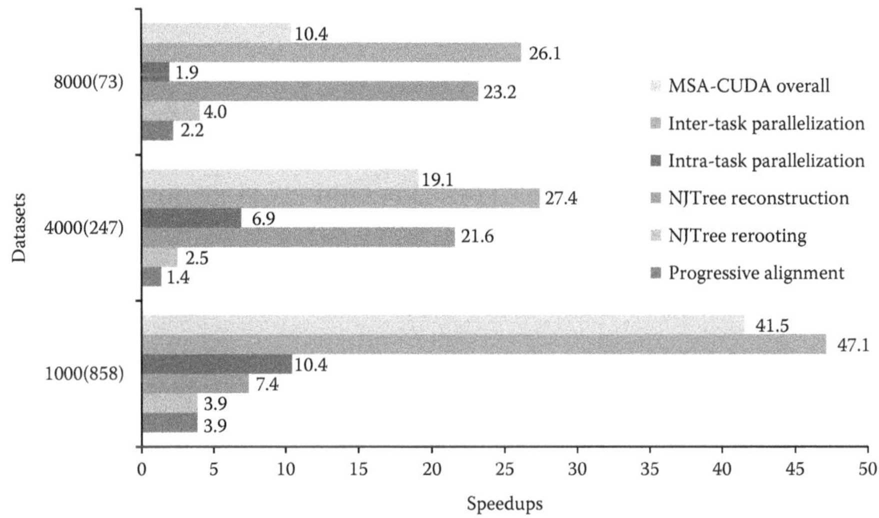

FIGURE 6.8
Speedups of MSA-CUDA compared with sequential ClustalW.

method is used, and when the total number of alignments or the remaining number of alignments is less than the *threshold*, the intratask parallelization method is used to compute those remaining alignments.

The tests of MSA-CUDA are carried out on the GTX 280 graphics card installed in the PC with an AMD processor. The sequential ClustalW (version 2.0.9) program is profiled on a desktop PC with a Pentium 4 3.0 GHz processor running the Linux OS. Three protein sequence datasets are used to evaluate the performance of MSA-CUDA. The datasets consist of sequences selected from the *Human immunodeficiency virus* dataset downloaded from NCBI [12], as given below:

- A: 1,000 sequences of average length 858;
- B: 4,000 sequences of average length 247;
- C: 8,000 sequences of average length 73.

Figure 6.8 shows the speedups of MSA-CUDA compared with sequential ClustalW. The graph clearly shows that the intertask parallelization outperforms the intratask parallelization for all datasets. Thus, if there are sufficient tasks and available large device memory capacity on the GPU, MSA-CUDA chooses intertask parallelization for Stage 1. Dataset A achieves higher speedups than datasets B and C in Stage 1 because of the larger amount of computation performed.

Speedups for the NJ tree reconstruction substage generally increase with the number of input sequences, but grow more slowly for the larger datasets. Consequently, dataset C achieves the highest speedup in this stage.

Speedups for Stage 3 are relatively low and vary largely, because

1. Building of the profiles of each alignment is performed sequentially on the CPU, which reduces the speedups achieved in the parallelized parts.
2. The speedup heavily depends on the topology of the guided tree, which influences the number of alignments that can be processed in parallel.
3. The lengths of the profiles of an alignment also have impact on performance. Generally, larger datasets and longer sequences mean better performances.

6.5 Motif Discovery

MEME (Multiple expectation maximization [EM] for Motif Elicitation) is an established and popular tool for motif discovery in DNA and protein sequences [13, 14]. MEME relies on an EM approach to find, which is time consuming for large datasets. Therefore, we have developed CUDA-MEME, a parallelization of motif discovery with MEME using CUDA.

Input is a set of related DNA or protein sequences $S = \{S_0, S_1, \ldots, S_{n-1}\}$ and a motif width W. The motif finding problem is to find a string of length W (a so-called *motif*) that occurs a certain number of times in the input dataset. Figure 6.9(a) shows an example with the motif ATCCG occurring exactly once in four input DNA sequences. Depending on the distribution of occurrences in the input sequences there are three different motif search methods:

- Exactly one occurrence per sequence (*OOPS*)
- Zero-or-one occurrence per sequence (*ZOOPS*)
- Any number of occurrences

In this chapter we only focus on OOPS and ZOOPS.

Of course, the occurrences of the motifs in the sequences are in general not exact, but approximate; that is, a certain number of mismatches are allowed (see Figure 6.9(b)). Therefore, MEME uses a statistical motif model. A motif is represented as a letter frequency matrix Ψ; that is, for a motif width W and an alphabet $\Sigma = \{x_0, x_1, \ldots, x_{A-1}\}$ with A letters, Ψ is of size $A \times (W + 1)$. The matrix value $\Psi_{i,j}$ is defined as

- Probability of x_i appearing at position $j - 1$ in the motif for all $0 \le i \le A - 1$ and $1 \le j \le W$

(a)	(b)
ACAC**ATCCG**GTT	ACAC**TTCCG**GTT
ATCCGAATTCTC	**AAGCG**AATTCTC
GGGTTTG**ATCCG**	GGGTTTG**ATCTC**
A**ATCCG**CGCGCT	A**CTCAG**CGCGCT

FIGURE 6.9
(a) The motif ATCCG occurring exactly once in every input sequence; (b) the motif ATCCG occurring once in every input sequence with up to two mismatches.

	0	1	2	3	4	5
A	0.16	0.50	0.25	0.00	0.25	0.00
C	0.24	0.25	0.00	0.75	0.50	0.25
G	0.24	0.00	0.00	0.25	0.00	0.75
T	0.36	0.25	0.75	0.00	0.25	0.00

FIGURE 6.10
A letter frequency matrix for the motif ATCCG occurring approximately in Figure 6.9(b). Note that column zero models the background.

- Probability of x_i appearing outside the motif for all $0 \le i \le A - 1$ and $j = 0$

Figure 6.10 shows an example of a letter frequency matrix for the example shown in Figure 6.9(b).

The EM algorithm [15] in MEME is carried out from an initial *starting point* model $\Psi^{(0)}$. It then runs for a fixed number of iterations or until convergence to find a model $\Psi^{(q)}$ with maximal posterior probability. During the search for a given motif width, MEME performs a *multistart search*, where it returns a number of initial models. The multistart search of a given motif width W consists of two stages:

- *Starting point search stage*: Iterates over all initial models derived from the actual W-length substrings occurring in the input sequence dataset. Firstly, the log-likelihood ratio of each possible initial model is computed. Second, a P-value is calculated, which represents the probability of a random string, generated from the background letter frequencies, having the same score or higher. Initial models with the highest-statistical significance are selected.
- *EM*: An EM algorithm is executed for a fixed number of iterations or until convergence from each of the highest-scoring initial models and then the best motif model is chosen.

Profiling of MEME shows the starting points search as the runtime bottleneck. It typically takes more than 98% of the overall runtime. Therefore, we focus on parallelizing the starting point search stage, which is explained in more detail in the following text.

Given the input sequences $S = \{S_0, S_1, \ldots, S_{n-1}\}$ from Σ, and the motif width W. L_i denotes the length of sequence S_i, $S_{i,j}$ denotes the substring of length W starting at position j in sequence S_i, and $S_i(j)$ denotes the jth letter in S_i, for all $0 \le i \le n-1$ and $0 \le j \le L_i - W$. The starting point search algorithm performs an independent computation from each W-length substring $S_{i,j}$ to determine a set of initial models. It consists of three steps:

- *Step 1:* Compute the probability score $P(S_{i,j}, S_{k,l})$ against each substring $S_{k,l}$, which is the probability that a motif starts at position l in S_k as well as at position j in S_i.
- *Step 2:* Identify a substring $S_{k,maxk}$ with the global maximum score for each sequence S_k as a possible starting point.
- *Step 3:* Sort the highest-scoring substrings in the decreasing order of score, and align them to identify the initial models for the given motif width by computing their statistical significance.

The probability score $P(S_{i,j}, S_{k,l})$ is defined by Equation 6.1, where *sbt* denotes the letter frequency matrix of size $A \times A$.

$$P(S_{i,j}, S_{k,l}) = \sum_{w=0}^{W-1} sbt[S_i(j+w)][S_k(l+w)] \tag{6.1}$$

Computing the probability score between each pair of substrings $S_{i,j}$ and $S_{k,l}$ directly using Equation 6.1 results in redundant calculations. To reduce this redundancy, Equation 6.2 can be used instead. Using Equation 6.2, the scores $\{P(S_{i,j}, S_{k,l})\}$ of S_i, for $1 \le j \le L_i - W$ and $1 \le l \le L_k - W$, in the jth iteration can be computed using the probability scores $\{P(S_{i,j-1}, S_{k,l-1})\}$ computed in the $(j-1)^{th}$ iteration. Only $P(S_{i,j}, S_{k,0})$ needs to be computed individually using Equation 6.1. The number of operations for each P-computation is therefore reduced from $O(W)$ to $O(1)$.

$$\begin{aligned} P(S_{i,j}, S_{k,l}) = {} & P(S_{i,j-1}, S_{k,l-1}) + sbt[S_i(j+W-1)][S_k(l+W-1)] \\ & - sbt[S_i(j-1)][S_k(l-1)] \end{aligned} \tag{6.2}$$

On the basis of the hybrid computing framework described in Section 6.2, we have parallelized the starting point search using CUDA. It consists of four components: main thread, auxiliary thread, task queue, and message queue. The main thread invokes the CUDA kernel(s), and the auxiliary thread conducts the alignment of top substrings and identifies the initial models. The task queue stores the sorting and alignment tasks to be processed, and the message queue facilitates the communication between the two threads.

The starting point search parallelization takes advantage of the fact that for a given W-length substring $S_{i,j}$ ($0 \le i \le n-1$ and $0 \le j \le L_i - W$), the computation of scores $\{P(S_{i,j}, S_{k,l})\}$ is independent of each other for any sequence S_k ($0 \le k \le n-1$ and $0 \le l \le L_k - W$). We have again used the general concept of intertask and intratask parallelization, which has been introduced in Section 6.2.2, to compute the scores for a given width W as part of the hybrid computing framework:

- *Intertask parallelization:* Each thread block is assigned to compute the scores of all W-length substrings in one sequence S_i against another sequence S_k, and all threads in a thread block cooperate to complete the computation. In this case, the task assignment can be arranged into a square matrix of size $n \times n$, where S_i and S_k are indexed from left to right horizontally and from top to bottom vertically, respectively. Cell (i,j) represents the task of computing the scores of all W-length substrings in S_i against S_k. The total number of thread blocks is n^2 and the tasks are assigned to all thread blocks sequentially along the matrix from left to right and then from top to bottom.
- *Intratask parallelization:* Working in the same way as the sequential algorithm. For one substring $S_{i,j}$, the scores against all the sequences are computed by p thread blocks, where for each sequence S_k, the set of all the W-length substrings $\{S_{k.l}\}$ is roughly equally divided and distributed to p thread blocks.

In most cases intertask parallelization achieves higher performance than intratask parallelization, except for datasets with a few long sequences. As observed in Equation 6.2, the score computation for the substring $S_{i,j}$ depends on the scores for the substring $S_{i,j-1}$. The scores of the substring $S_{i,j}$ (corresponding to the jth iteration of sequence S_i) against all substrings $\{S_{k,l}\}$ in S_k are stored in a score vector $[P(S_{i,j}, S_{k,l})]$ in global memory. Both methods exploit two score vectors $[P(S_{i,j-1}, S_{k,l})]$ and $[P(S_{i,j}, S_{k,l})]$ to store the scores for the $(j-1)^{th}$ and j^{th} iterations of S_i, using a simple cyclic vector swapping method. In this method, the score vector $[P(S_{i,j-1}, S_{k,l})]$ serves as input and $[P(S_{i,j}, S_{k,l})]$ as output for the jth iteration of S_i. For the $(j+1)^{th}$ iteration, the score vectors $[P(S_{i,j-1}, S_{k,l})]$ and $[P(S_{i,j}, S_{k,l})]$ are swapped.

As mentioned above, during the starting point search from a given substring $S_{i,j}$ only one globally highest-scoring substring $S_{k,maxk}$ for each sequence S_k is selected. In this case, both parallelization approaches are designed to determine the highest-scoring substrings while computing the scores. Owing to the different design details, these two methods exploit different determination methods, respectively. For each invocation of the CUDA kernel(s), intratask parallelization performs the computation of probability scores $\{P(S_{i,j}, S_{k,l})\}$ for one substring $S_{i,j}$ against all the sequences and stores all the scores for the next invocation. During the score computation, each thread block selects the highest-scoring substring $S_{k,maxk}$ from the set of substrings $\{S_{k,l}\}$ assigned to it for each sequence S_k and then outputs them when exiting. After the CUDA kernel(s) return(s), the highest-scoring substrings $\{S_{k,maxk}\}$for all the sequences $\{S_k\}$ are determined by simply comparing the outputted highest-scoring substrings of each thread block. After completing each pass, the substring $S_{i,j}$ and its corresponding highest-scoring substrings are packed as a task and added to the task queue. Because intertask parallelization works in a multipass way and is based on a task assignment matrix, it does not guarantee that all the highest-scoring substrings $\{S_{k,maxk}\}$ for a substring $S_{i,j}$ against all the sequences$\{S_k\}$ are computed in one pass. In this case, an unprocessed starting point (USP) buffer is exploited to store all the unprocessed highest-scoring substrings for a set of substrings $\{S_{i,j}\}$. After completing one pass, sequence-level parallelization combines the returned highest-scoring substrings in this pass into the USP buffer and then performs an analysis procedure to check whether all the highest-scoring substrings for each substring $S_{i,j}$ of S_i against all the sequences have been computed. Once having determined all the highest-scoring substrings $\{S_{k,maxk}\}$ for all the sequences $\{S_k\}$ with respect to S_i, it removes those highest-scoring substrings with respect to S_i from the USP buffer, packs them as a task, and then adds this task to the task queue. It iteratively performs the above analysis procedure until no task is available to be added to the task queue, and then returns to invoke the CUDA kernel(s) to perform the remaining computation.

On the right side of the hybrid computing framework, the auxiliary thread always waits until either the task queue or the message queue is not empty. When the task queue is not empty, the auxiliary thread retrieves a task from the task queue, sorts those highest-scoring substrings in order of decreasing score and then aligns different number of top substrings to identify the initial models. Before retrieving a task from the task queue, the auxiliary thread checks the message queue to see whether there is a message from the main thread to itself. If yes, the auxiliary thread performs the corresponding operations, and otherwise continues accessing the task queue.

CUDA-MEME is benchmarked on the GTX 280 graphics card installed in the desktop PC with an AMD processor and 2 GB RAM running the Linux OS. The sequential MEME (version 3.5.4) is also profiled on the same computer. For all the tests, the minimum and maximum motif widths are set to 6 and 50, respectively, and other parameters use the default values. Input datasets containing a varying number of DNA sequences are used to evaluate the performance of CUDA-MEME:

- The mini-drosoph dataset (with 4 sequences of an average length of 124 824)
- Three datasets of human promoter regions consisting of 100, 200, and 400 sequences of lengths 5,000 bps each (called HS_100, HS_200 and HS_400, respectively).

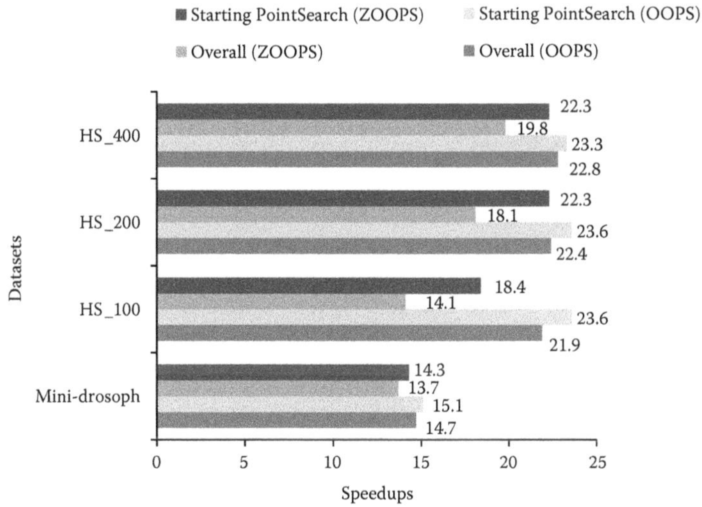

FIGURE 6.11
CUDA-MEME speedups for OOPS and ZOOPS models.

On the basis of the hybrid computing framework, the main thread and the auxiliary thread run concurrently on the host. The tests exploit intertask parallelization for the three human promoter regions datasets and intratask parallelization for the mini-drosoph dataset. The choices of grid size and thread block size are considering that the CUDA kernel occupies only a small quantity of device resources and that the maximum number of active threads per SM is 1,024. For intratask parallelization, a grid consisting of thread blocks whose number is equal to or less than the number computed by multiplying the number of SMs by (1,024/*dimBlock*) is bound to the kernel and launched, where the number of threads in a thread block *dimBlock* is set to 64. For substring-level parallelization and the parallel alignment, each thread block is comprised of 256 threads. Figure 6.11 shows the speedups of CUDA-MEME using the OOPS and ZOOPS models for all the datasets. CUDA-MEME is available for download at http://sites.google.com/site/yongchaosoftware/Home/cuda-meme.

6.6 Conclusion

In this chapter, we have described several techniques for algorithm design on CUDA-enabled GPUs. These techniques serve at two levels: the system level and the device level. At the system level, a hybrid computing framework is suggested to fully exploit the computational power of the system by overlapping the computation of GPU and CPU. At the device scale, we have suggested intertask and intratask parallelization approaches from the macroscopic view to leverage the power of the CUDA-enabled GPUs for different application conditions, and three techniques from the microscopic view to ameliorate the performance by reducing the bandwidth requirements of global memory. On the basis of these techniques, three parallel algorithms, running on many-core CUDA-enabled GPUs, for sequence alignments and motif discovery have been presented: CUDASW++, MSA-CUDA, and CUDA-MEME.

Our results on GPU show that it is possible to improve the performance of bioinformatics algorithms by making full use of the compute characteristics of the underlying commodity hardware. The very rapid growth of both biological databases and available transcription data demands even more powerful high-performance sequence alignments and motif discovery solutions in the near future. Hence, our results are especially encouraging because GPU performance grows faster than Moore's law as it applies to CPUs.

6.7 References

1. Lindholm E., Nickolls J., Oberman S., Montrym J., NVIDIA Tesla: A unified graphics and computing architecture. *IEEE Micro*, 28, 39–55, 2008.
2. Nickolls J., Buck I., Garland M., Skadron K., Scalable parallel programming with CUDA. *ACM Queue*, 6, 40–53, 2008.
3. Liu Y., Schmidt B., Maskell D.L., CUDASW++: optimizing Smith–Waterman sequence database searches for CUDA-enabled graphics processing units. *BMC Research Notes*, 2(73), 2009.
4. Szalkowski A., Ledergerber C., Krahenbuhl P., Dessimoz C., SWPS3—Fast multi-threaded vectorized Smith–Waterman for IBM Cell/B.E. and x86/SSE2. *BMC Research Notes*, 1, 107, 2008.
5. Manavski S.A., Valle G., CUDA compatible GPU cards as efficient hardware accelerators for Smith–Waterman sequence alignment. *BMC Bioinformatics*, 9(Suppl 2), S10, 2008.
6. Altschul S.F., Madden T.L., Schaffer A.A., Zhang J., Zhang Z., Miller W., Lipman D.J., Gapped BLAST and PSI-BLAST: A new generation of protein database search programs. *Nucleic Acids Research*, 25(17), 3389–3402, 1997.
7. Farrar M.S., Optimizing Smith–Waterman for the cell broadband engine, http://farrar.michael.googlepages.com/smith-watermanfortheibmcellbe. Accessed February 15, 2010.
8. Thompson J.D., Higgins D.G., Gibson T.J., CLUSTALW: Improving the sensitivity of progressive multiple sequence alignment through sequence weighting, position-specific gap penalties and weight matrix choice. *Nucleic Acids Research*, 22, 4673–4680, 1994.
9. Larkin M.A., Blackshields G., Brown N.P., Chenna R., McGettigan P.A., McWilliam H., Valentin F., Wallace I.M., Wilm A., Lopez R., Thompson J.D., Gibson J.D., Higgins D.G., Clustal W and Clustal X version 2.0. *Bioinformatics Applications Note*, 23(21), 2947–2948, 2007.

10. Liu Y., Schmidt B., Maskell D.L., MSA-CUDA: Multiple sequence alignment on graphics processing units with CUDA, *20th IEEE International Conference on Application-Specific Systems, Architectures and Processors (ASAP)*, 121–128, 2009.
11. Thompson J.D, Higgins D., Gibson T.J., Improved sensitivity of profile searches through the use of sequence weights and gap excision. *Computer Applications in the Biosciences*, 10, 19–29, 1994.
12. NCBI Homepage, http://www.ncbi.nlm.nih.gov. Accessed February 15, 2010.
13. Bailey T.L., Elkan C., Unsupervised learning of multiple motifs in biopolymers using expectation maximization. *Machine Learning*, 21, 51–80, 1995.
14. Bailey T.L., Williams N., Misleh C., Li W.W., MEME: Discovering and analyzing DNA and protein motifs. *Nucleic Acid Research*, 34, W369–W373, 2006.
15. Dempster A.P., Laird N.M., Rubin D.B., Maximum likelihood from incomplete data via the EM algorithm. *Journal of the Royal Statistical Society, Series B (Methodological)*, 39(1), 1–38, 1977.

7

CUDA Error Correction Method for High-Throughput Short-Read Sequencing Data

Haixiang Shi, Weiguo Liu, and Bertil Schmidt

7.1 Introduction

One of the great challenges of the Human Genome Project is the DNA sequencing problem, which is of the great importance due to the fact that it will be of great help in exploring the human biology. With the knowledge of DNA, one can prevent viruses attacking human health, disorders, and diseases. However, previous standard methods of sequencing have been labor intensive, inefficient, and expensive; for example, the total sequencing output was only about 200 million base pairs (bps) in the year 1998.

To improve the throughput of sequencing machines significantly, several so-called second-generation DNA sequencing technologies have recently been introduced [1–3]. Examples of such sequencers are products from 454 Life Sciences/Roche, Solexa/Illumina, and Applied Biosiciences/SOLiD. The massive throughput of these sequencers can be illustrated using the Illumina Genome Analyzer IIx (IGA-IIx) as an example. The IGA-IIx can currently generate an output of up to 25 billion bps within a single run. This output is expected to increase to around 100 billion bps by 2010.

Although their throughput is drastically higher, there is a significant difference between the reads produced by second-generation sequencers and traditional (Sanger) sequencers: the length of produced reads is significantly shorter. For example, the length of reads produced by the IGA-IIx is between 35 and 100 bps while read length of a traditional Sanger sequencers is typically between 500 and 1,000 bps. Therefore, the output produced by second-generation sequencers is also referred to as *high-throughput short-read* (HTSR) data.

Established methods and tools for DNA fragment assembly (e.g., Arachne [4]) have been designed and optimized for Sanger shotgun sequencing; that is, they assume read lengths of around 500–1,000 bps and coverage of 6- to 10-fold. Consequently, they are generally applicable to process HTSR because of

- Scalability (i.e., the ability to process the much larger amount of reads) and
- Much shorter read length

Therefore, several de novo assemblers for HTSR data have been recently introduced. They can be divided into two categories: overlap graph-based approaches and de Bruijn graph-based approaches. Edena [5] and Taipan [6] use an overlap graph, while Euler-SR [7, 8], Velvet [9], ALLPATHS [10], and ABYSS [11] are examples of de Bruijn graph approaches.

HTSR graph-based assembly approaches generally use an exact overlap of length *k* to generate a link in the graph and are therefore highly sensitive to sequencing errors. Hence, correcting as many base-pair errors as possible in the input read data before graph construction can significantly improve both assembly quality and runtime. To demonstrate the usefulness of error correction as a preprocessing step, we have generated three datasets of 0.6 million random reads each with read length of 70 from the genome sequence of *Saccharomyces cerevisiae* chromosome V using a per-base error rate of 1%, 2%, and 3%, respectively. We have then executed the SHREC error correction algorithm [12] on each dataset. Afterward we compared the assembly results produced by Edena for the original datasets and the error-corrected datasets in terms of N50-values. The results shown in Figure 7.1 clearly indicate that error correction can greatly improve assembly results.

Unfortunately, error correction, as a preprocessing step, is highly time consuming. The profiling results in Table 7.1 show that the error correction step in Euler-SR can take up to 72% of the overall runtime in the whole assembly process. It can also be seen that the percentage goes up with increasing error rates. This means the time spent in the error correction surges up as more errors need to be corrected.

In this chapter, we demonstrate how the compute unified device architecture (CUDA) programming model can be used to accelerate error correction for HTSR data on CUDA-compatible graphic processor units (GPUs). Our

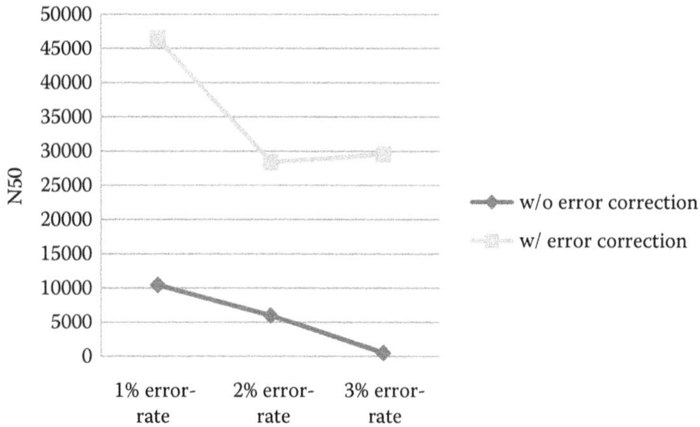

FIGURE 7.1
Comparison of N50-values produced by Edena for the three datasets with and without prior error correction.

TABLE 7.1
Runtime of Error Correction for Euler-SR with Simulated Read Dataset*

Error Rate (%)	(s)	Percentage (%)
1	1,527	52
2	2,506	63
3	3,324	72

Note: *Generated from *Saccharomyces cerevisiae* chromosome V with an error rate range between 1% and 3%.

parallel error correction algorithm is based on the so-called spectral alignment problem (SAP). To take advantage of the CUDA memory hierarchy we employ a Bloom filter data structure to implement hashing of *k*-mers. We test the performance of our implementation in terms of sensitivity, specificity, and accuracy using several read datasets. Furthermore, speedups are presented in comparison to the sequential SAP implementation of Euler-SR [7, 8].

7.2 Spectral Alignment Approach to Error Correction

Sequencing errors can produce erroneous excessive computing, for example, in the Euler-SR algorithm; the number of erroneous edges is several times larger than the number of real edges. Error correction is therefore an important preprocessing step for many de novo assemblers. The approach to error

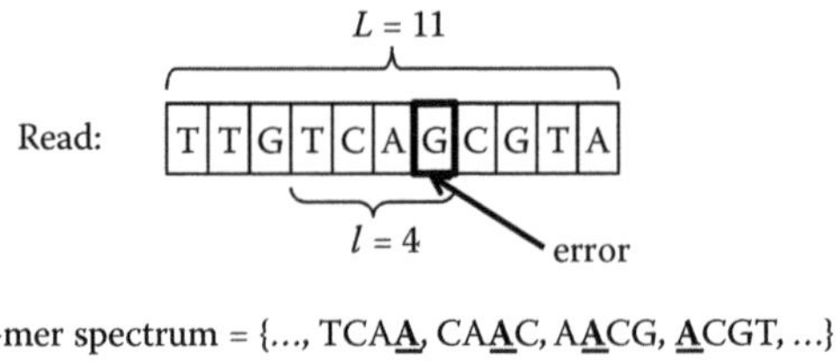

FIGURE 7.2
Changing the single error at position 6 in the given read from G to A results in l corresponding matches in the spectrum.

correction used in this chapter is based on the *SAP* [13, 14], which can be described as follows:

Given are a set of k reads $R = \{r_1, \ldots, r_k\}$ over the alphabet {A, C, G, T} where each read is of length L and a tuple-length l with $l < L$. SAP considers every l-mer of each read (i.e., every substring of length l) and compares them to the spectrum $T(G)$. The spectrum $T(G)$ consists of all l-mers of the reference genome G, where the reads originate from. If a read is error free, all its l-mers will have a corresponding exact match in $T(G)$. If a read has a single error (mutation) at position j, the corresponding $min\{l, j, L-j\}$ overlapping l-mers have (in most cases) a lower number of corresponding exact matches $T(G)$. However, by mutating position j back to the correct base pair, all l-mers have a corresponding match. This is the basic idea of SAP, which is illustrated in Figure 7.2.

We use the following terminology for the definition of the SAP. An l-mer of a read is called *solid* if it has an exact match in a given l-mer spectrum T and *weak* otherwise. A read R is called a T-string if all its l-mers have an exact match in the spectrum T. SAP can now be defined as follows:

Definition (SAP): Given a read r_i and an l-mer spectrum T, find a T-string r_i^* in the set of all T-strings that minimizes the distance function $d(r_i, r_i^*)$.

Depending on the error model of the utilized sequencing technology the distance function d() can be either edit distance (suitable for 454 Life Sciences/Roche) or hamming distance (suitable for Solexa/Illumina). We focus on Illumina technology, and therefore, the latter approach is chosen in this work.

In a de novo assembly project the reference genome G is generally not known beforehand. Therefore, the spectrum $T(G)$ of all correct (or trusted) l-mers needs to be approximated from the available read data. This is usually done by introducing the additional parameter m (multiplicity). The ideal spectrum $T(G)$ is then replaced by the approximated spectrum $T(R,m)$, where $T(R,m)$ consists of all l-mers that occur at least m times in R. It should be mentioned that the use of an approximate spectrum is not always ideal, because

1. Some l-tuples that are in $T(G)$ might not necessarily be in $T(R,m)$ because of low coverage.
2. Some l-tuples that are in $T(R,m)$ might not necessarily be in $T(G)$ because of the same error occurring several times.

These cases can be minimized by an optimal choice of the parameters m and l. This choice depends on the average number of reads covering an l-mer of the sequenced genome.

7.3 Parallel Error Correction with CUDA

7.3.1 Bloom Filter Data Structure and Spectrum Computation

A very frequent operation in SAP-based error correction is the *spectrum membership test;* that is, testing whether $s \in T$ for an l-mer s and a given spectrum T. This test has to be done for a large number of l-mers and a fixed spectrum and can be efficiently performed by hashing. An efficient way to store a frequently accessed hash table in CUDA is to use the read-only texture memory. We have found that the fastest and most space-efficient way to implement this membership test with CUDA is to use probabilistic hashing based on the space-efficient Bloom filter data structure. Another interesting application of the Bloom filter data structure in bioinformatics is the word matching stage in basic local alignment search tool deoxyribonucleic acid (BLASTN) on an field-programmable gate array (FPGA) (see Chapter 8 for more details).

A Bloom filter represents a set of given keys in a bit-vector [15]. Insertion and querying of keys are supported using several independent hash functions. Bloom filters gain their space efficiency by allowing a false-positive answer to membership queries. Space savings often outweigh this drawback in applications where a small false-positive rate can be tolerated, particularly when space resources are at a premium. Both criteria are met for the CUDA error correction algorithm (and also for BLASTN word matching in Chapter 8). In the following section we briefly review definition, programming, querying, and false-positive probability (FPP) of Bloom filters.

A Bloom filter is defined by a bit-vector of length b, denoted as $BF[1..b]$. A family of k hash functions $h_i: K \rightarrow A$, $1 \le i \le k$, is associated to the Bloom filter, where K is the key space and $A = \{1,\ldots,b\}$ is the address space. K is the set of all l-mers over the alphaber {A, C, G, T} in this paper.

For a given set I of n keys, $I = \{x_1,\ldots,x_n\}$, $I \subseteq K$, the Bloom filter is programmed as follows. The bit-vector is initialized with zeros; that is, $BF[i] := 0$ for all $1 \le i \le b$. For each key $x_j \in I$, the k hash values $h_i(x_j)$, $1 \le i \le k$, are computed. Subsequently, the bit-vector bits addressed by these k values are set to one; that is, $BF[h_i(x_j)] := 1$ for all $1 \le i \le k$. Note that, if one of these values addressed a bit that is already set to one, that bit is not changed.

For a given key $x \in K$, the Bloom filter is queried for membership in I in a similar way. The k hash values $h_i(x)$, $1 \le i \le k$, are computed. If at least one of the k bits $BF[h_i(x)]$, $1 \le i \le k$, is zero, then $x \notin I$. Otherwise, x is said to be a member of I with a certain probability. If all k bits are found to be one but $x \notin I$, x is said to be a false positive (see Figure 7.3).

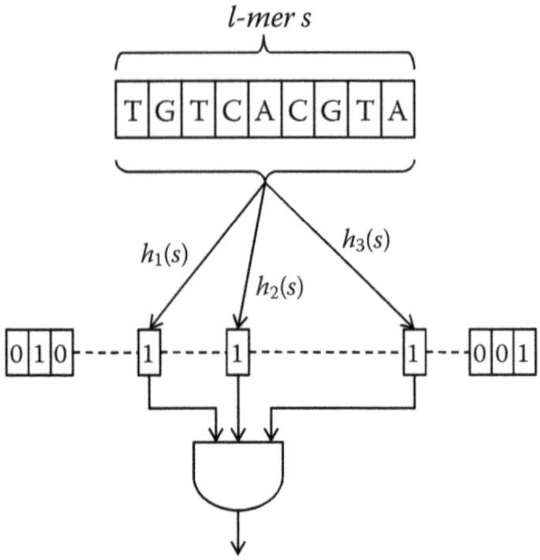

FIGURE 7.3
Bloom filter data structure for querying the spectrum T for membership of the l-mer s.

The presence of false positives arises from the fact that the k bits in the bit-vector can be set to one by any of the n keys. Note that a Bloom filter can produce false-positive but never false-negative answers to queries. The *FPP* of a Bloom filter is given by Equation 7.1.

$$FPP = \left(1 - \left(1 - \frac{1}{m}\right)^{kn}\right)^k \approx \left(1 - e^{-\frac{kn}{m}}\right)^k \tag{7.1}$$

Obviously, *FPP* decreases as the bit-vector size m increases, and increases as the number of inserted keys n increases. It can be shown that for a given m and n, the optimal number of hash functions k_{opt} is given by $(m/n) \cdot \ln(2)$. The corresponding FPP is then approximately $0.6158^{m/n}$.

Hence, in the optimally configured Bloom filter, the false-positive rate decreases exponentially with the size of the bit-vector. Furthermore, to maintain a fixed FPP, the bit-vector size needs to scale linearly with the inserted key set. In our Bloom filter implementation using one-dimensional linear texture memory we have chosen $k = 8$ and $m = 64 \cdot n$, which leads to FPP = 3.63e–08.

Before correcting errors with the SAP approach, the spectrum $T(R,m)$ consisting of the set of all l–mers that have a multiplicity of at least m needs to be computed in a preprocessing step. The spectrum is represented by the Bloom filter $B(T(R,m))$, which is subsequently transferred to the CUDA texture memory to be used for parallel error correction. The computation of $B(T(R,m))$ is done sequentially on the host central processing unit (CPU). Our implementation uses m Bloom filters, $B_1, \ldots, B_m$, to represent the multiplicity m. For each l-mer, the algorithm queries the Bloom filters successively in descending order. Once a positive query is found for some k, B_{k+1} is programmed for the given l-mer and the algorithm continues with the next l-mer.

7.3.2 Parallel Error Correction Using CUDA

The parallel CUDA algorithm for error correction with SAP uses the following idea. An error (i.e., mutation) at position j in read r_i creates $min\{l, j, L-j\}$ erroneous l-mers. Therefore, the correction of r_i at position j can be associated to the transformation of $min\{l, j, L-j\}$ weak l-mers into solid ones. The parallel SAP error correction searches for such corrections using a voting procedure as follows.

SAP voting procedure: Given an l-mer spectrum T and a read r_i, let r_i^j denote the jth l-mer of r_i; that is, $r_i[j\ldots j+l]$, the substring of length l starting at position j of r_i. Firstly, all weak l-mers of r_i are identified; that is, all $0 \le j \le L-(j+1)$ with $r_i^j \notin T$. Every nucleotide position of each weak l-mer r_i^j is then mutated to check whether the mutated l-mer is solid; that is, the l-mers $t[k,c] = r_i[j\ldots j+k-1]\cdot c\cdot r_i[j+k+1\ldots j+l]$ ("$\cdot$" denotes string concatenation) for all $0 \le k \le l-1$ and $c \in \{A, C, G, T\}\backslash\{r_i[j+k]\}$ are created and tested for spectrum membership. If $t[k,c] \in T$, the corresponding counter in the voting matrix, $V(r_i)[j+k][c]$, is incremented by one. $V(r_i)[][]$ is of size $L \times 4$, where $V(r_i)[pos][char]$ represents the read r_i with the nucleotide at position pos mutated to $char$, denoted as $r_i[pos][char]$. The counter value $V(r_i)[pos][char]$ represents the number of l-mers that are weak in r_i but are solid in $r_i[pos][char]$. A large value in the voting matrix is therefore an indicator for an error at the corresponding read position. For each read r_i that is not a T-string, the maximum position $[p_{\max}][c_{\max}]$ in $V(r_i)[][]$ is determined. The read $r_i[p_{\max}][c_{\max}]$ is then created. If $r_i[p_{\max}][c_{\max}]$ is a T-string, then r_i is considered to be corrected by $r_i[p_{\max}][c_{\max}]$. Otherwise, r_i is trimmed or discarded. Figure 7.4 outlines an example for the read and spectrum used in Figure 7.2.

The outlined voting procedure only considers a single mutation error; that is, $\Delta = 1$. To consider several errors (i.e., $\Delta \ge 2$) within the same read, the same approach can be used where up to Δ mutations are considered with each l-mer.

CUDA parallelization: The parallelization approach of the voting procedure with CUDA exploits the fact that $V(r_i)[][]$ can be computed independently for each read $r_i \in R$. Hence, we use a CUDA kernel to represent the sequential processing necessary for the voting of an individual read r_i. The kernel is then invoked using a thread for each read $r_i \in R$. The time complexity of the kernel is determined by Δ ($\Delta \ge 1$), the number of corrections within a weak l-mer.

Our CUDA kernel for correcting *exactly* Δ mutations consists of two phases. The first phase is the Δ-mutations voting algorithm. It identifies all l-tuples of the given read that are not in the spectrum (i.e., the Bloom filter). All possible Δ-point mutations of these l-tuples are then queried for membership in the spectrum. If successful, corresponding counters in the voting matrix are incremented.

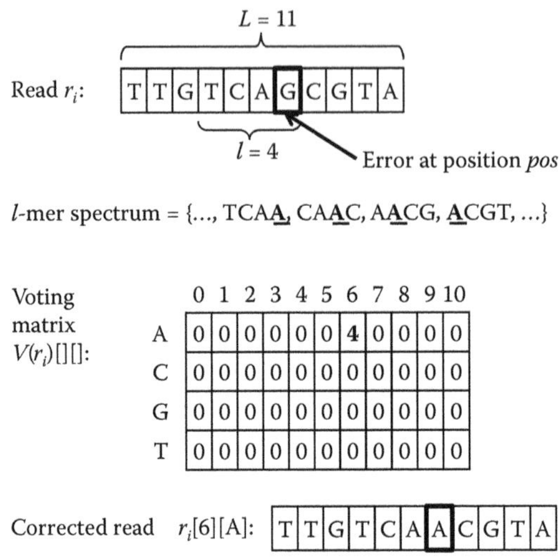

FIGURE 7.4
The single error at position *pos* = 6 in the read r_i results in a high value in the corresponding position in the associated voting matrix $V(r_i)[6][A]$.

After the voting matrix is computed by each thread for a given read, errors can be fixed based on high values in the voting matrix. In certain cases; for example, when $\Delta + 1$ errors are close to each other, the Δ-mutation voting algorithm cannot correct the errors. However, it is still possible to identify the read as erroneous and to trim it at certain positions or discard it. This is done using a fixing/trimming/discarding procedure, which is the second phase of our CUDA kernel.

The time complexity of the kernel is dominated by the first phase. The operation that determines the runtime of the Δ-mutation voting algorithm is the Bloom filter membership test. The overall amount of membership queries by a single thread is $(L - l) \cdot p \cdot O(l^{\Delta})$, where p is the number of l-tuples of the read that do not belong to the spectrum.

Our CUDA algorithm for correcting *up to* Δ mutations uses a filtration approach to reduce the amount of reads that are corrected with a large Δ value. In the first step, Δ-mutation voting and Δ-mutation fixing/trimming/discarding is performed on the GPU only for $\Delta = 1$. In the next step, the CUDA kernel for $\Delta = 2$ is executed only for the set of reads that have been trimmed or discarded during the $\Delta = 1$ computation. This approach can then be continued for larger values of Δ.

The per-thread memory requirement for storing the voting matrix $V(r_i)[][]$ can be reduced to $l \times 4$ bytes using cyclic indexing. The per-thread memory amount required for storing voting matrix and read is therefore reduced to $4 \times l + L$ bytes. Therefore, shared memory could be used to store this data. However, this would limit the number of threads per block to 128 (already

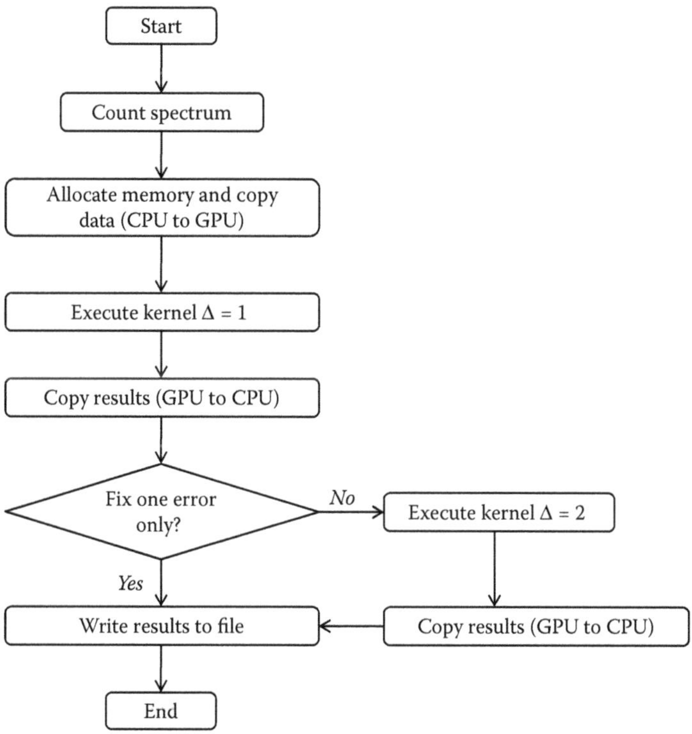

FIGURE 7.5
CUDA error correction algorithm flow chart.

for relatively small values of l and L, such as $L = 35$ and $l = 20$). Furthermore, it would negatively affect the maximum number of thread blocks per multiprocessor (MTBPM). In general, there is a trade-off between partitioned-based spatial-merge (PBSM) usage and MTBPM: increased PBSM usage decreases MTBPM. Lower MTBPM results in a lower warp occupancy and efficiency. The CUDA occupancy calculator tool recommends a usage of less than 4 KB PBSM for our implementation. Therefore, we have decided not to store the voting matrix in PBSM but in local memory. Furthermore, if the number of reads exceeds the total number of threads used in the kernel our implementation allows processing several reads per thread.

7.3.3 Execution Example

Figure 7.5 illustrates all steps of our CUDA error correction (CUDA-EC) implementation for $\Delta = 1$ and $\Delta = 2$ using a flow chart.

CUDA-EC first performs the precomputation on the CPU host. The precomputed spectrum and the reads are then transferred to device and the CUDA kernel is executed. The individual steps are as follows:

1. *Precomputation of spectrum list:* Using the input parameters l and m, a spectrum list is compiled with all l-mers occurring at least m times in the read dataset.
2. *Hashing the spectrum list into Bloom filter:* Before transferring the spectrum to the GPU, all l-mers in the spectrum list are hashed into Bloom filter. We then bind the Bloom filter to the texture memory on the GPU. Each item in the spectrum list will be added to the Bloom filter using the hash functions given below.

```
//Step 1: get hash value
hash =hash_function_list[j](tempTuple,TUPLE_SIZE) %
(table_size * char_size);
//Step 2: add to the bloom filter
hash_table[hash / char_size] |= bit_mask[hash % char_size];
```

   The size of the Bloom filter is dependent on the number of hashed items. As discussed in Section 7.3.1, we use $k = 8$ and $m = 64 \cdot n$, with FPP = 3.63*e*–08. Assume the total number of l-mers in the spectrum is N, then the Bloom filter requires 8N bytes in total. As the Bloom filter is stored as a read-only 1D char array, we can fit the Bloom filter to GPU texture memory for fast fetch.

3. *Bind the Bloom filter to 1D texture on the GPU*:

```
// allocate data on device
unsigned char *dev_hash;
cudaMalloc( (void**)&dev_hash, table_size );
// copy memory to device
cudaMemcpy( dev_hash, hash_table, table_size,
cudaMemcpyHostToDevice );
// bind texture
cudaBindTexture(0, tex, dev_hash );
```

4. *Allocate memory on the GPU and transfer the reads from CPU to GPU*: Reads are allocated in 1D character array and copied to device memory.
5. *Parallel error correction on GPU*: By applying each-thread-one-read policy, the CUDA threads are mapped to the read data using thread indices. With each thread processing one read, there will be no non-coalesced global memory accesses. Each thread may also be used to fix several reads depending on the number of reads. The results are stored in the output reads array with flags to differentiate between fixed and discarded reads.

6. *Transfer from GPU to CPU and write results to files:* The reads are transferred from GPU to CPU and are written to two separate output files, one for fixed and one for discarded reads.

The command line to run the error correction application program is as follows:

```
/cuda-ec -f {inputfilename} -t {tuplesize} -o {fixed-filename}
-d {discarded-filename} -r {read_length}[-maxTrim {maximum_
trim}] [-minVotes {minimum votes}] [-minMult {multiplicity}]
[-search {num_error_to_fix]]
```

The required parameters are defined as follows:

- -f {inputfilename}: path and name of the read input file in FASTA format
- -t {tuplesize}: length of a tuple (or *l*-mer)
- -o {fixed-filename}: name of output fixed file
- -d {discarded-filename}: name of output discarded file
- -r {read_length}: length of the input reads

The optional parameters are defined as follows:

- -maxTrim: maximum number of trimmed character allowed at the beginning and at the end of a read (default 20)
- -minVotes: minimal number of votes required for error correction (default 2)
- -minMult: multiplicity (default 6)
- -search: number of error to be fixed in each read (default $\Delta = 1$)

7.4 Performance Evaluation

We have evaluated the performance of our CUDA-EC approach for datasets with varying coverage, error rate, and read length using simulated Illumina-style datasets as well as two real Illumina datasets. The simulated datasets have been produced by generating random reads with a given error rate from a reference genome sequence. To test scalability, we have selected reference genomes of various lengths (ranging from 0.58 Mbp to 4.71 Mbp). Three datasets have been created for each reference genome sequence using per-base error rates of 1%, 2%, and 3%, respectively. The features of the simulated input datasets are summarized as follows in the format IDs: Reference genome (GenBank ID), Genome length, Coverage, read length, number of reads. (Note that the ID A*i* indicates a per-base error rate of *i*%.)

- SA1, SA2, SA3: S.cer 5 (NC_001137), 0.58 M, 70×, 35, 1.1 M
- SB1, SB2, SB3: S. cer7 (NC_001139), 1.1 M, 70×, 35, 2.2 M
- SC1, SC2, SC3: H.inf (NC_007146), 1.9 M, 70×, 35, 3.8 M
- SD1, SD2, SD3: E.col (NC_000913), 4.7 M, 70×, 35, 9.4 M

The real datasets consist of 3.5 M unambiguous reads (i.e., they do not contain any nondetermined nucleotide) of length 35 each and have been downloaded from http://www.genomic.ch/edena.php and of 8.2 M unambiguous reads of length 36 downloaded from http://sharcgs.molgen.mpg.de/download.shtml. The former has been obtained experimentally by Hernandez et al. [5] using the Illumina Genome Analyzer for sequencing the *Staphylococcus aureus* strain MW2 (*H. Aci*). The latter has been tested by Dohm et al. [16] for sequencing *Helicobacter acinonychis*. We have estimated the error rate of the two real dataset as 1.0% and 1.6%, respectively, by aligning each read to the reference genome using RMAP [17]. The real datasets are summarized in the format IDs: Reference genome (GenBank ID), Genome length, Coverage, read length, number of reads, estimated per-base error rate.

- RA: S. aureus (NC_003923.1), 2.8 M, 43×, 35, 3.5 M, 1%
- RB: Helicobacter (NC_008229), 1.6 M, 190×, 8.2 M, 1.6%

To evaluate the time efficiency of CUDA-EC, we have measured the runtime of these datasets on an NVIDIA GeForce GTX 280 with CUDA version 2.0. The card is connected to an AMD Opteron dual-core 2.2 GHz CPU with 2 GB RAM running Linux Fedora 8 by the PCIe 2.0 bus. The performance of CUDA-EC is compared with the single-threaded C++ code running on the same CPU from the SAP error correction implementation of in Euler-SR (available at http://euler-assembler.ucsd.edu). The code is a serial implementation of the Δ-mutation error correction algorithm. However, different from our parallel method, it stores the spectrum in a sorted vector and then calls the standard template library (STL) function "`std::lower_bound()` for membership queries. Runtime comparisons between the sequential and the CUDA implementation for all datasets have been performed using the default parameters $l = 20$ and $m = 6$. The CUDA timings include precomputation time on the CPU, CUDA kernel time, and CPU–GPU data transfer time. CUDA kernels are executed using 256 thread-blocks and 256 threads per block. The Euler-SR code is compiled using GNU GCC 4.1.2 with the full optimization (-O3) enabled. Figures 7.6 and 7.7 show the speedup for the simulated datasets for $\Delta = 1$ and $\Delta = 2$, respectively.

Speedups are shown for the parallel part only (i.e., the voting procedure running on the GPU) and for the complete application (i.e., parallel voting plus sequential precomputation of the Bloom filter on the CPU). Figure 7.8 shows the corresponding speedups for the two real datasets.

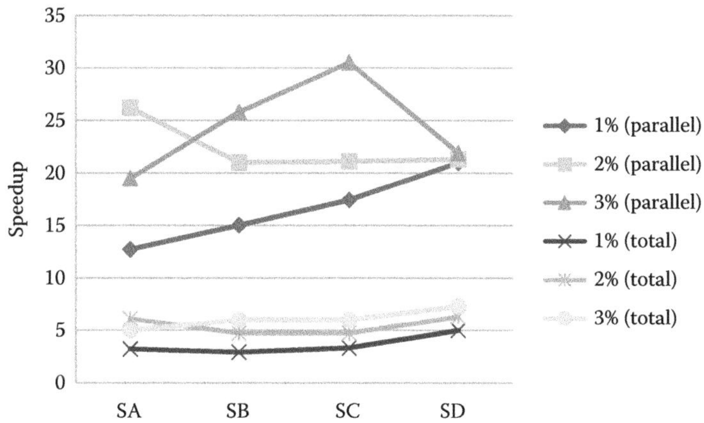

FIGURE 7.6
Speedups for the simulated datasets using $\Delta = 1$ ((parallel) is the speedup for the parallelized part only; while (total) is the speedup for the complete application).

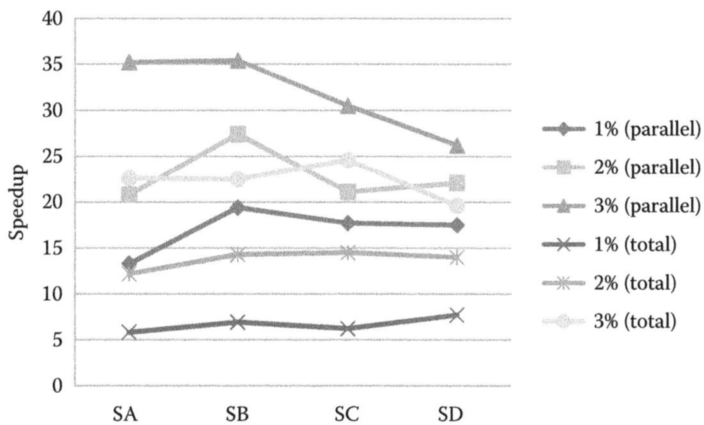

FIGURE 7.7
Speedups for the simulated datasets using $\Delta = 2$.

The speedups indicate the following trends

1. The speedup increases for higher error rates.
2. The speedup increases for $\Delta = 2$.

Trend 1 can be explained as follows. The voting algorithm tests each l-mer r_i^j for spectrum membership and therefore contains a corresponding data-dependent conditional branch (if $r_i^j \notin T$ then) that is executed for each l-tuple of the given read. This leads to inefficiencies in the CUDA implementation due to the single-instruction multiple-thread (SIMT) execution model. Threads for which this statement is true execute another $O(l)$ membership queries. Threads for which this statement is false need to wait for these

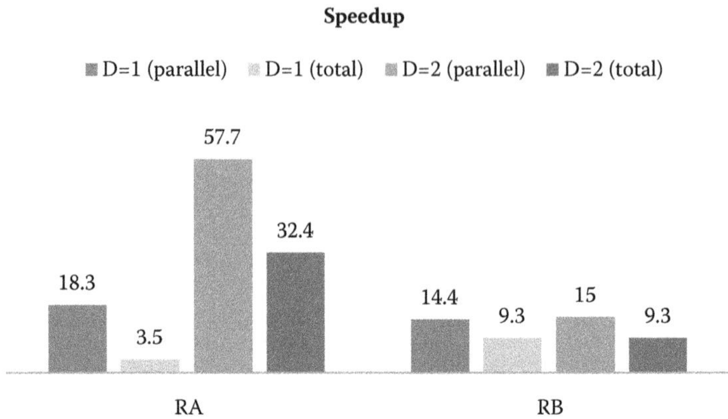

FIGURE 7.8
Speedups for the real datasets.

threads within the same warp to finish these tests. The number of error-free reads is generally decreasing for higher error rates. Thus, the number of idle threads per warp is decreasing for higher error rates, which in turn improves the efficiency of the CUDA implementation.

Trend 2 is due to of the filtration approach used in the parallel error correction algorithm. The double-mutation voting algorithm is only applied to the subset of reads that contain at least two errors (i.e., all reads that have not been be fixed by the single-mutation voting algorithm). Therefore, the data-dependent conditional branch is true in most threads within a warp, resulting in a higher-parallel efficiency compared to $\Delta = 1$.

We have further analyzed the accuracy of our CUDA implementation in terms of

- *Identification*; that is, identifying reads as erroneous or error free
- *Correction*; that is, correcting reads that have been identified as erroneous

The identification of erroneous reads can be defined as a *binary classification test*. The corresponding definitions of true positive (*TP*), false positive (*FP*), true negative (*TN*), and false negative (*FN*) are as follows.

- *TP*: erroneous read that is fixed, trimmed, or discarded by CUDA-EC
- *FP*: error-free read that is fixed, trimmed, or discarded by CUDA-EC
- *TN*: error-free read that is kept unchanged by CUDA-EC
- *FN*: erroneous read that is kept unchanged by CUDA-EC

Sensitivity and specificity measures are then defined as: sensitivity = $TP/(TP + FN)$; specificity = $TN/(TN + FP)$.

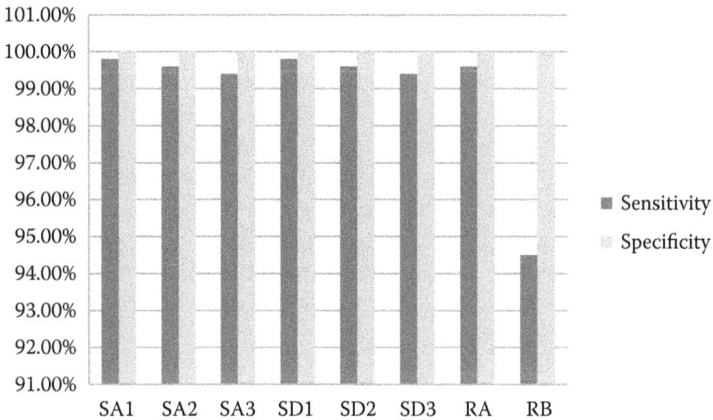

FIGURE 7.9
Performance of the read identification classification test measured in sensitivity and specificity for selected datasets after executing the CUDA voting algorithm.

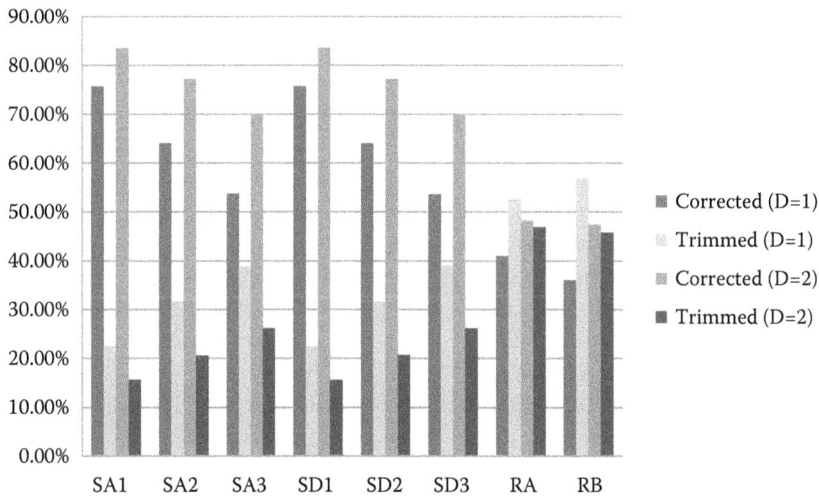

FIGURE 7.10
Percentage of reads in *TP* that have been corrected, trimmed, or discarded for selected datasets.

Figure 7.9 shows the specificity and sensitivity measures for selected datasets. It can be seen that the algorithm identifies erroneous reads with very high accuracy. We have further analyzed the reads that have been classified as *TP*. The amount of corrected/trimmed reads relative to the number of discarded reads is shown Figure 7.10.

Figure 7.10 shows that in 1-error and 2-error corrections, the percentage of corrected/trimmed reads decreases compared to the discarded reads for higher error rates. This can be explained by the larger number of erroneous reads with more than one error for higher error rates. These reads cannot be corrected with the single-mutation voting algorithm, and will therefore be either trimmed or discarded. For 2-error corrections, it can be seen that the percentage of corrected reads increases and the trimmed and discarded reads decrease for higher error rates compared with 1-error correction.

A further observation is that the percentage of corrected/trimmed reads is lower for the real dataset. The likely reason for this is that errors in real reads are not as evenly distributed as in our simulated reads. It has been reported [18] that error rates in Illumina reads range from 0.3% at the beginning of a reads to 3.8% at the end of reads.

7.5 Conclusion and Future Work

The emergence of new HTSR sequencing technologies establishes the need for new tools and algorithms that can process massive amounts of short reads in reasonable time. In this chapter we have addressed this challenge by writing scalable CUDA error correction software for modern many-core architectures, which is an important but time-consuming preprocessing step for many de novo assembly tools. To derive an efficient CUDA implementation we have used a space-efficient Bloom filter for hashing to take advantage of the CUDA memory structure. Our performance evaluation on a commodity GPU shows speedups around one order of magnitude for various datasets at high-correction accuracy. Our CUDA-EC implementation is available at http://cuda-ec.sourceforge.net.

A weakness of the described error correction method is that the l-mer spectrum of the reference genome $T(G)$ is only approximated by $T(R,m)$; that is, the set of all l-mers with multiplicity $\geq m$ in the input read dataset. Our future work includes the incorporation of base-call quality scores to the spectrum construction to improve this approximation. The speedup of the current CUDA implementation is reduced by the sequential precomputation of the Bloom filter on the CPU. Therefore, another part of our future work is to investigate more efficient methods for the sequential preprocessing stage. Furthermore, it would be interesting to compare the SAP-based error correction approach to other approaches such as the recently introduced SHREC method [12], which uses a suffix tree of all input reads to identify and correct sequencing errors.

7.6 References

1. Mardis, E.R. 2008. The impact of next generation sequencing on genetics. *Trends in Genetics* 24(3), 133–141.
2. Pop, M., and Salzberg, S.L. 2008. Bioinformatics challenges of new sequencing technology. *Trends in Genetics* 24(3), 142–149.
3. Strausberg, R.L., Levy, S., and Rogers, Y.H. 2008. Emerging DNA sequencing technologies for human genomic medicine. *Drug Discovery Today* 13(13/14), 569–577.
4. Batzoglou, S., et al. 2002. ARACHNE: A whole-genome shotgun assembler. *Genome Research* 12(1), 177–189.
5. Hernandez, D., et al. 2008. De novo bacterial genome sequencing: Millions of very short reads assembled on a desktop computer. *Genome Research* 18(5), 802–809.
6. Schmidt, B., Sinha, R., Beresford-Smith, B., and Puglisi, S. 2009. A fast hybrid short read fragment assembly algorithm. *Bioinformatics* 25(17), 2279–2280.
7. Chaisson, M.J., and Pevzner, P.A. 2008. A short read fragment assembly of bacterial genomes. *Genome Research* 18(2), 324–330.
8. Chaisson, M.J., Brinza, D., and Pevzner, P.A. 2009. De novo fragment assembly with short mate-paired reads: Does the read length matter? *Genome Research* 19(2), 336–346.
9. Zerbino, D.R., and Birney, E. 2008. Velvet: Algorithms for de novo short read assembly using de Bruijn graphs. *Genome Research* 18(5), 821–829.
10. Butler, J., et al. 2008. ALLPATHS: De novo assembly of whole-genome shotgun microreads. *Genome Research* 18(5), 810–820.
11. Simpson, J.T., Wong, K., Jackman, S.D., Schein, J.E., Jones, S.J., and Birol, I. 2009. ABySS: A parallel assembler for short read sequence data. *Genome Research* 19(6), 1117–1123.
12. Schröder, J., Schröder, H., Puglisi, S., Sinha, R., and Schmidt, B. 2009. SHREC: A short-read error correction method. *Bioinformatics* 25(17), 2157–2163.
13. Chaisson, M.J., Tang, H., and Pevzner, P.A. 2004. Fragment assembly with short reads. *Bioinformatics* 20(13), 2067–2074.
14. Pevzner, P.A., Tang, H., and Waterman M.S. 2001. An Eulerian path approach to DNA fragment assembly. *Proceedings of the National Academy of Science* 98(17), 9748–9753.
15. Bloom, B. 1970. Space/time trade-offs in hash coding with allowable errors. *Communications of the ACM* 13(7), 422–426.
16. Dohm, J.C., Lottaz, C., Borodina, T., and Himmelbauer, H. 2007. SHARCGS, a fast and highly accurate short-read assembly algorithm for de novo genomic assembly. *Genome Research* 17(11), 1697–1706.
17. Smith, A.D., Xuan, Z., and Zhang, M.Q. 2008. Using quality scores and longer reads improves accuracy of Solexa read mapping. *BMC Bioinformatics* 9(128), 2008.
18. Dohm, J.C., Lottaz, C., Borodina, T., and Himmelbauer, H. 2008. Substantial biases in ultra-short read data sets from high-throughput DNA sequencing. *Nucleic Acid Research* 36(16), e105.

8

FPGA Acceleration of Seeded Similarity Searching

Arpith C. Jacob, Joseph M. Lancaster, Jeremy D. Buhler, and Roger D. Chamberlain

Biological sequence comparison studies the relationship between DNA or protein sequences. A pairwise alignment algorithm matches fragments in a sequence of unknown function, termed the *query*, to similar fragments in a *reference* sequence from a large database. Biologically relevant matches may represent genes, structural domains, regulatory elements, or other sequence features that provide clues to the biochemical function and structure of the query. Biosequence databases, such as GenBank from the U.S. National Center for Biological Information (NCBI), provide an annotated list of reference sequences that form the basis for comparative analysis. High-throughput comparison is widely used to annotate functional elements in newly sequenced genomes, to assemble sequence reads against a reference genome, to compare related genomes, or to analyze sequence reads from microbial communities.

The best match between two sequences, as determined by some score metric, can be computed using the Smith–Waterman algorithm [1], which uses dynamic programming to efficiently examine an exponential alignment space. Even so, its computational cost, which is proportional to the product of the lengths of the two sequences, is prohibitively expensive for comparing two genomes or a query to a large database. To address this problem the biological community has developed heuristic methods that produce satisfactory alignments, though not necessarily the best as found by Smith–Waterman, at a fraction of the computational expense. One such heuristic, *seeded alignment,* implemented in the Basic Local Alignment Search Tool (BLAST) family of algorithms [2, 3] is a sequence filtering technique that limits Smith–Waterman comparison to subsets of the database that are highly likely to be similar to the query. BLAST first identifies pairs of short, fixed-length substrings in the query and reference sequences known as *seeds.* Progressively expensive computations *filter* these seeds and their corresponding reference sequences so that full Smith–Waterman comparison is done on only a small fraction of the database. Unlike the classical Smith–Waterman algorithm, BLAST is capable of identifying many alignments between two sequences, which may represent distinct sequence features. Owing to its speed advantage, BLAST has been widely adopted by the biological community as a replacement for the Smith–Waterman algorithm.

While BLAST is roughly two orders of magnitude faster than Smith–Waterman, it too is subject to ever-increasing computational demands driven by low-cost sequencing and the emerging data-intensive field of metagenomics. New technologies can sequence larger genomes at a cost order of magnitude lower than the technology used during the Human Genome Project a decade ago. Publicly available databases have seen an exponential growth over the last decade as a large number of organisms have been sequenced and new genes have been identified. Figure 8.1 shows the growth of the GenBank Non-Redundant DNA database [4], increasing at the rate of 1 billion bases per month. The National Human Genome Research Institute envisions an era of affordable individual human genome sequencing for less than $1,000 by 2014, potentially enabling the sequencing and analysis of any person's genome.

Metagenomics is an emerging field that studies the genetic diversity of a complex mixture of microbial species in an ecosystem through large-scale sequencing. The generated DNA reads are analyzed by comparing them against reference databases using sequence analysis tools such as BLAST. Already, several times as many genes have been generated from metagenomic samples in a few years as from complete genomes in a decade of sequencing. Comparative metagenomics studies the influence of environmental factors on microbial communities by comparing samples from different environments. The bottleneck in all these analyses is the BLAST computation. As sequencing costs drop further, the metagenomics approach can be applied to study any microbial community on earth.

Given this situation, most large-scale projects use a cluster of commodity workstations for high-throughput sequence comparison [5–7]. In a cluster,

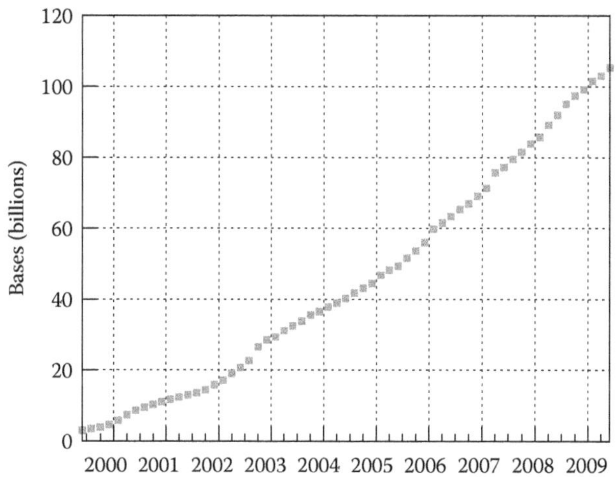

FIGURE 8.1
Growth of the GenBank Non-Redundant DNA database over the last decade.

the workload is distributed among nodes that perform largely independent computations, together producing a linear speedup. As an example, the BLAST web server made available by NCBI for processing query sequences from the community used a Linux cluster of around 200 CPUs in 2004. On a typical weekday in 2004 the cluster processed 1.4×10^5 query sequences, and NCBI projected doubling the number of nodes to keep up with demand [8]. While clusters are widely used, they have important limitations. They require a considerable initial investment, are expensive to maintain, occupy large amounts of floor space, have significant power requirements, and need effective cooling solutions.

Special-purpose architectures on graphic processor units (GPUs) and field-programmable gate arrays (FPGAs) address some of these limitations. These technologies are well suited for data-intensive algorithms that operate on large volumes of data, and have ample fine-grained parallelism. FPGAs have programmable logic, interconnects, and specialized arithmetic units on-chip, that can be used to build high-computational density architectures specialized to an application. In addition, FPGAs have customizable, massively parallel on chip memory elements that can be used as an efficient "cache." Many data-intensive applications have seen orders of magnitude better performance on FPGAs than on general-purpose microprocessors; for these applications FPGAs consume less energy and cost less than an equivalent workstation cluster.

A large body of existing work accelerates the Smith–Waterman computation on FPGAs [9, 10]. Unfortunately, these accelerators are not suitable for large-scale sequence comparison because they are not performance competitive with heuristic tools. For example, Smith–Waterman accelerated 100-fold is about as fast as a software implementation of BLAST on a modern workstation. In fact, Smith–Waterman is just one of many stages in a modern heuristic pipeline; seed generation, the stage that identifies word matches between the query and a reference sequence is the bottleneck in the BLAST computation. Seeded alignment is more challenging to accelerate because, unlike Smith–Waterman, the various stages are not fully data parallel; therefore, we must rely on pipeline parallelism and novel hardware-friendly memory data structures to achieve an appreciable speedup.

In this chapter we will describe an FPGA accelerator for sequence comparison that exploits the characteristics of the *streaming* model. A stream program operates on a large data stream of items using a chain of self-contained computation stages. Each stage reads a data item; performs a predictable, finite set of operations on it; and sends the item to the next stage in sequence, or alternatively discards it. Because there is no reuse of data items in a stage, and the chain of stages usually does not have feedback or other interactions, stream programs can achieve high throughput on GPUs and FPGAs. In addition, stream programs exhibit abundant data, task, and pipeline parallelism that can be efficiently exploited on FPGAs. Data parallelism (commonly found in loops) refers to the execution of the same set of instructions on distinct data items with either no dependencies or those that are statically defined. Task parallelism refers to a multiplicity of independent tasks that do not exchange data and can be mapped to independent processing units. Finally, pipeline parallelism is available in a linear chain of stages that execute simultaneously. The BLAST streaming architecture we will describe is organized as a pipeline of linear stages that operates on a stream of database sequences. Each stage acts as a filter, passing only reference sequences that match the query on to the next stage in the pipeline.

An often ignored, yet critical task, is the validation of a sequence comparison accelerator, that is, comparing the quality of its output against BLAST. Biologists have come to accept the results of BLAST as a de facto standard (even over Smith–Waterman) and have little reason to trust an accelerator that deviates from the original algorithm. Preserving the behavior of each stage while making it more hardware friendly is a recurring theme in this chapter. We validate the results of the implementation on large-scale, realistic biological datasets to build confidence in our accelerator.

The rest of this chapter is organized as follows: Section 8.1 introduces two versions of the BLAST algorithm: one for pairwise (DNA) comparison (BLASTN) and another for comparing protein sequences (BLASTP). We describe specialized hardware architectures for each version in Section 8.2 that capitalize on their unique properties. We validate the speedup of the accelerator and quality of its results in Section 8.3. We conclude in Section 8.4 with general principles that can be applied to design accelerators for seeded pipelines of other sequence analysis tools.

```
    Query: CVRAERAMQEEFYLELKEGLLEPLAVTERL----AIIS
           | |    +| |    ++|  || +||   |    +|||
Reference: CSR--ELIQHELDQVVEE--LEKIAVVNLLRHRRSIIS
```

FIGURE 8.2
An alignment between a query and a reference protein. Pairs of biologically similar and identical residues are marked by pluses and vertical lines, respectively. Two word matches, or seeds, of length 3 are highlighted in the alignment.

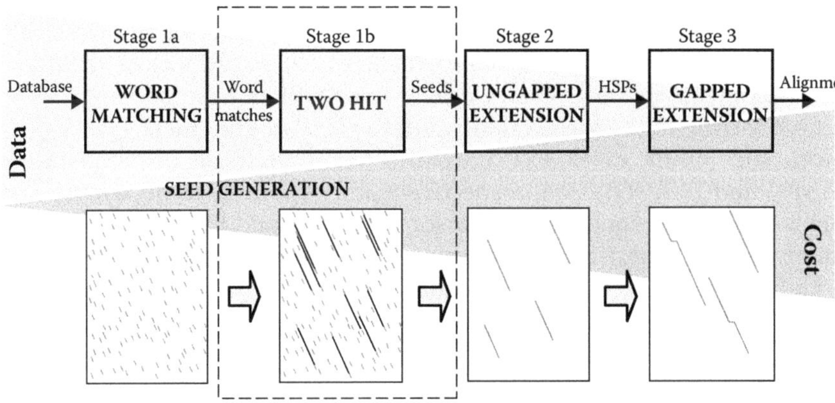

FIGURE 8.3
The BLAST heuristic is a three stage pipeline that filters reference sequences using progressively more expensive computations. Seeded alignment is illustrated using four dot-plots, with the query and reference sequence positions on the X- and Y-axes, respectively. Features are marked along antidiagonals. Stage 1 identifies word matches and two-hits, stage 2 high-scoring ungapped alignments, and stage 3 gapped alignments.

8.1 The BLAST Algorithm

An alignment is a side-by-side comparison of pairs of sequence characters, referred to as nucleotides in a DNA sequence; amino acids in a protein sequence; or, collectively, residues. Figure 8.2 shows an example of an alignment between two protein sequences. A good alignment maximizes the number of pairs of identical or biologically similar residues while keeping dissimilar pairs and unaligned residues called *gaps* to a minimum. The score of an alignment is computed by adding the individual score of paired residues, found in a table 8, and the penalty of introducing gaps.

BLAST compares a query sequence Q to every reference sequence D in a database, identifying one or more statistically significant alignments. The BLAST computation is divided into three stages: seed generation, ungapped extension, and gapped extension (see Figure 8.3). The key observation exploited by the BLAST heuristic is that strong alignments between a query and a reference sequence are likely to contain pairs of consecutive residues in each sequence that are highly similar. The seed-generation stage identifies these *word matches*, or *seeds*, and focuses the search around them. Figure 8.2 highlights two seeds, though only the one on the right is an identical word match. Seed generation is split into two substages with a *two-hit* unit that is used only for protein comparison. Ungapped extension extends a seed by aligning residues surrounding the word match without introducing gaps, producing a *high-scoring segment pair* (HSP). HSPs whose alignment score exceeds a threshold are passed on to the next stage. Finally, gapped extension applies the full Smith–Waterman algorithm to extract the highest scoring alignment centered on a seed. The BLAST stages employ progressively more expensive computations, but each stage also discards close to 90% of its input. In what follows, when we refer to an HSP or a gapped alignment, we are referring to the seed contained in the HSP or alignment.

From the family of BLAST programs we will focus on accelerating BLASTN and BLASTP, used, respectively, for nucleotide-to-nucleotide and protein-to-protein sequence comparison. BLASTN and BLASTP are similar in many aspects, but differ in the way seeds are identified in the first stage. We will now describe the BLAST algorithm in detail, with special attention given to the differences between the two.

8.1.1 Seed Generation

Seed generation produces tuples (q, d) that represent a word match at position q in the query and d in the reference sequence. Word matches may undergo further processing, as in the case of BLASTP, after which they are passed as seed matches to the ungapped extension stage.

In the case of BLASTN, a seed is simply an exact word match of length w between the query and the reference sequence. The typical word length is 11, a good compromise between speed and quality of results. Requiring an exact match is appropriate for DNA sequences, for which mismatched residue pairs hold relatively little information about sequence similarity. Many amino acids in proteins, however, share similar chemical properties and so have a high likelihood of being substituted by each other. BLASTP therefore uses inexact matching that generates seeds in a two-step process: *word matching* (Stage 1a) and *two-hit* (Stage 1b). Word matching in BLASTP finds pairs of words of fixed length, usually 3, in the two sequences that are similar according to a biologically meaningful score table δ. Word pairs that satisfy the condition $\sum_{i=1}^{\omega} \delta(Q[q+i], D[d+i]) \geq T_{1a}$ are classified as matches. Here T_{1a} is a neighborhood threshold score selected by the user. The default threshold value is 11.

When a short word length is used, word matching generates a large number of matches purely by chance that have no relation to biologically meaningful alignments. A two-hit stage is therefore employed to filter this stream by exploiting the observation that a high-scoring ungapped alignment is likely to contain multiple word matches in close proximity. A seed match in

FIGURE 8.4
Part of the neighborhood of the first word MNT of a protein computed using inexact matching. The neighborhood of a DNA word is simply the word itself.

BLASTP occurs only when two word matches (q_1, d_1) and (q_2, d_2) are found such that (a) they are on the same ungapped alignment represented by a so-called *diagonal* $d_1 - q_1 = d_2 - q_2$, and (b) they lie within a user-defined window of Y residues, that is, $w \le d_1 - d_2 < Y$. The word match (q_2, d_2) is designated as the seed in BLASTP. Though word matching in BLASTN also generates false positives, the two-hit filter is less effective, and so is not used.

For efficient computation, the word-matching stage uses a lookup table containing a precomputed *neighborhood* of the query sequence. The neighborhood is a list of all possible database words that match query words, along with their positions q in the query. In the case of exact matching there are $|Q| - w + 1$ database words that match the query. In BLASTP, all possible database words are first compared with every query word, and pairs that score at least T_{1a} become part of the neighborhood. The neighborhood of a query word is shown in Figure 8.4.

8.1.2 Ungapped Extension

Ungapped extension investigates query and reference sequence pairs that contain seed matches. Residues in the two sequences on either side of a seed match are aligned, permitting matches and substitutions. As gaps are disallowed, this stage is less expensive than full Smith–Waterman. Aligning a residue pair contributes a score, which may be negative, and is determined by the table δ, to a running total. The highest-scoring forward and backward extension from the seed match constitutes its HSP. Note that this HSP threads through the seed match and contains it. We can therefore have multiple, possibly overlapping HSPs in the same query-reference pair associated with distinct seeds. If the score of an HSP is above a user-defined threshold, it is passed along to the next stage (more precisely, the location of its seed is passed on).

To guarantee finding the highest-scoring subsequence pair, residue pairs must be scored till the end of the sequences, which is computationally expensive. BLAST therefore terminates extension early using an *X-drop* mechanism that finds a high-scoring substring pair, though not necessarily the highest. The X-drop approach notes the score of the best HSP after each residue pair extension. Extension is terminated early if the score of the HSP containing the current pair of residues falls X below that of the highest-scoring HSP. The X-drop procedure is able to reduce useless computations on the majority of seed matches that lie on poorly scoring ungapped alignments.

8.1.3 Gapped Extension

This stage uses the Smith–Waterman recurrence to compute the highest-scoring gapped alignment that passes through a seed. BLAST's implementation includes minor variations from the classical Smith–Waterman algorithm. BLAST performs gapped forward and backward extension from the first pair of residues in the input HSP's seed. The recurrence is modified to ensure that the alignment always threads through this pair. Here too, an X-drop procedure is employed for early termination. If a gapped alignment scores above a threshold value, it is reported to the user. We refer the interested reader to Chapter 4 for details on the Smith–Waterman recurrence.

8.1.4 Execution Profile of the BLAST Algorithm

Before building a specialized architecture for a stream program, it is important to study its execution profile and data-consumption characteristics. We used the GNU profiler to generate runtime statistics of BLAST on typical nucleotide and protein datasets. Table 8.1 shows these results. In BLASTN more than 80% of the application runtime is spent in seed generation, so its corresponding hardware architecture is critical for an improved application performance. Seed generation dominates despite requiring an order of magnitude less time to process an input than in later stages. This can be attributed to the large disparity in the input data volume—seed generation processes the entire database, while the extension stages operate on a small filtered subset. In fact, the filter is so effective that gapped extension, computationally the most involved stage in the pipeline, is active less than 1% of the runtime. Protein search, in contrast, spends significant time in all three stages.

Application disk input/output (I/O) and postprocessing are less expensive than the stages in the BLAST pipeline. We must be careful, however, and not completely ignore them. If the BLAST pipeline is accelerated enough, I/O or postprocessing, may become the bottleneck. The accelerator platform must support high-throughput data transfer from disk to FPGA, and the software functions may need to be executed on a small multicore system to keep up.

Next, we study the filtering characteristics of the stages. We denote the *match rate* in word matching, ungapped, and gapped extension respectively as the number of seeds, HSPs, and alignments crossing a stage's threshold score per database residue. Every stage, with the exception of two, discards more than 95% of its input. Gapped extension in BLASTN accepts 14% of its input, but as the final stage in the pipeline, it has no bearing on the computational expense.

TABLE 8.1
Execution Profile of BLAST

		Word Matching	Two-Hit	Ungapped Extension	Gapped Extension
BLASTN	*% time*	83.89%	—	15.88%	0.22%
	Match Rate	0.0205	—	0.0000619	0.141
BLASTP	*% time*	30.96%	19.29%	15.85%	33.60%
	Match Rate	3.873×	0.043	0.003	0.031

Source: Jacob, A. et al., *ACM Transactions on Reconfigurable Technology and Systems*, 1(2):1–44, 2008. With permission. © 2008 ACM, Inc.
Note: The match rate of word matching and two-hit is specified as seeds per database residue. For ungapped and gapped extension the match rate units are HSPs and alignments per database residue, respectively.

Word matching in BLASTP is a net expander of data, generating on average over three matches to the query per residue of the database. In contrast, most database words do not match the query in BLASTN. We exploit these observations to design both an efficient filtering data structure for BLASTN and a high-throughput table lookup architecture for BLASTP. The two-hit stage, though a simple computation, must be capable of accepting multiple word matches per unit time if it is to keep up with its input rate.

8.2 A Streaming Hardware Architecture for BLAST

We now describe the architecture of Mercury BLAST, our FPGA implementation of the BLAST algorithm on the high-speed stream processing *Mercury* platform [11]. A Mercury BLAST search compares a set of queries against a database of reference sequences. Initially, user-programmable parameters are sent to the FPGA, after which the first query is loaded on-chip. Reference sequences of the database separated by a special character are streamed in a single pass from the disk through the hardware. Each stage of the BLAST pipeline runs in parallel, with sufficient buffering to smooth bursty matches likely to occur in biological sequence comparison. The results of the final FPGA stage are delivered to a host CPU for postprocessing. This procedure is repeated for every query sequence in the set.

The BLASTN deployment has seed generation and ungapped extension running in hardware; gapped extension remains in software and can easily keep up with its workload. In contrast, all stages of BLASTP are deployed in hardware. We use an ungapped extension design that is similar in both cases; seed generation, however, has major differences. Next we give a high-level overview of the various hardware architectures. Implementation details and the process of selecting appropriate parameter values can be found in the referenced papers [12–15].

8.2.1 Seed Generation

As mentioned in Section 8.1.1, every word in the database is looked up in the neighborhood table to retrieve matching query positions. Due to its size, this table must be stored in memory external to the FPGA device. External memory has relatively low bandwidth into the FPGA, so the lookup operation becomes a bottleneck and an impediment to any meaningful acceleration of the BLAST application. Fortunately, we can exploit the unique characteristics of nucleotide and protein comparisons to alleviate this problem. In the former, we use a Bloom filter, an on-chip, highly parallel, set membership testing data structure to efficiently discard a majority of database words that do not match the query. Thereafter, a low memory bandwidth lookup stage is sufficient for words that pass the prefilter. Because the longer word lengths used in BLASTN would require a prohibitively large direct lookup table, we instead use a space-efficient hash table in our memory lookup stage.

In protein search, every database word is highly likely to match the query, making set membership testing redundant—external memory references are inevitable. To support high-throughput word matching in BLASTP we must replicate the lookup table computation using as many external memory devices as possible; we can attain a reasonable speedup using two. The short word length used in BLASTP allows the use of a direct lookup table to store the query neighborhood.

8.2.1.1 Nucleotide Seed Generation Architecture

A *Bloom filter* [16] is a highly parallel, space-efficient data structure invented by Burton Bloom that tests for an exact match to any element in a set. Originally used for spell checking, recently it has found use in network packet processing, content summarization, and database caching. The filter is ideal for a hardware implementation because it uses simple arithmetic operations and can be deeply pipelined for a high-throughput architecture. We employ Bloom filters as a specialized "cache" to efficiently determine if a database word matches some word in the query (although it cannot give the location of the match).

A Bloom filter consists of k hash functions that in our implementation probe their corresponding memory bit-vectors. Each hash function is a many-to-one linear transformation from the larger nucleotide word space to a smaller memory address space. A Bloom filter has two operating modes: *program* and *test*. In the initial state the bit-vectors are cleared. When programming, a query word is hashed to generate k addresses that indicate the locations in the corresponding bit-vectors to be set. This is repeated for every word in the query. A database word matches the query in a test operation if

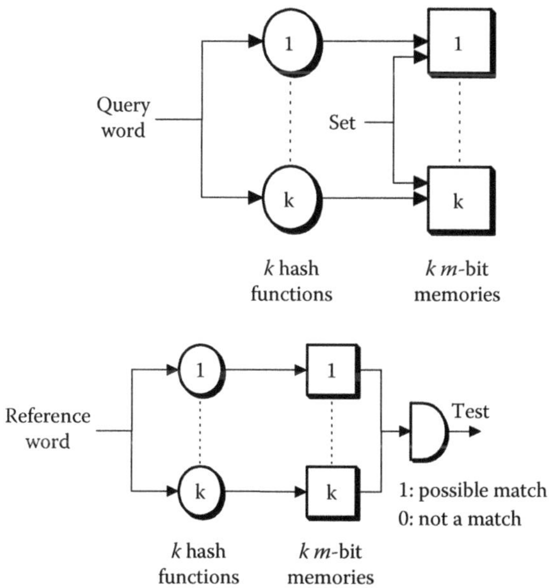

FIGURE 8.5
The Bloom filter acts as a specialized cache to discard reference words that do not match query words. (Adapted from Jacob, A. and Gokhale, M., *HPRCTA'07: Proceedings of the 1st International Workshop on High-Performance Reconfigurable Computing Technology and Applications*, New York, 2007, pp. 31–37. With permission. © 2007 ACM, Inc.)

the bit locations computed by the hash functions are all set. The operation of a Bloom filter is shown in Figure 8.5.

A Bloom filter may produce false positives but never produces false negatives. The false positive rate is given by $f = (1 - e^{-Nk/m})^k$, where N is the number of query words programmed, k is the number of hash functions, and m is the size of each bit-vector. A high false-positive rate will result in a poor prefilter, overwhelming the subsequent lookup stage. To restrict the false-positive rate to a low value, we empirically selected $m = 32,768$ and $k = 6$ for a query size of 17 kbases (34 kbases when the query is reverse complemented). Each bit-vector is stored in two dual-ported memories on our FPGA and is shared between two independent test operations. A single Bloom filter uses only 96 Kbits of storage, an order of magnitude less than that required for the entire neighborhood table. We can therefore replicate the filter setup to support sixteen parallel test operations on-chip. The key to our successful acceleration of BLASTN lies in reducing the subsequent lookup rate to about one in sixteen database words, enabling us to process over 10^9 database residues per second.

The *hash lookup* stage retrieves query positions that match a database word. As a large fraction of possible 11 residue database words do not match the query, we can reduce the storage requirement for the query neighborhood by using a hash table instead of a direct lookup table. We have two important considerations for the hash lookup architecture. First, the number of external memory probes for each database word must be reduced so that, on average, one database word can be processed every clock cycle. Second, the hashing functions must be computable using limited on-chip resources. We use a *near-perfect* hashing scheme, that is, one with almost no collisions, that is a variant of Tarjan and Yao's displacement hashing [17].

The hash table is organized into *primary* and *collision* tables. Each query word s is mapped into the primary table at the address.

$$h_1(s) = A(s) \oplus \tau[B(s)]$$

In the case of a collision, it is mapped to

$$h_2(s) = C(s)$$

in the collision table. The symbol $\oplus$ represents the XOR operation. We choose hardware friendly H_3 hash functions [18], which are essentially linear transformations over the field of integers modulo 2, for A, B, and C. These functions are unique for every query sequence: A and B are selected so that the pair $[A(s), B(s)]$ is distinct for every query word s. Our near-perfect hashing scheme attempts to resolve collisions using a small displacement table τ of integers, which is also unique for every query sequence. When two query words s_1 and s_2 have $A(s_1) = A(s_2)$ but $B(s_1) \neq B(s_2)$, we try to choose distinct values for $\tau[B(s_1)]$ and $\tau[B(s_2)]$ so that $h_1(s_1) \neq h_1(s_2)$. This technique is able to resolve most collisions using 8 Kbits of on-chip storage; nevertheless, on occasion we must still probe the collision table. Finally, a *duplicate* table is used to store excess matching query words. In an 1 MB SRAM we are able to store the neighborhood of query sequences ranging in size up to 17 kbases (34 kbases reverse complemented).

8.2.1.2 Protein Seed Generation

An important decision we made after studying the filtering characteristics of BLASTP was to increase the word length and neighborhood threshold from 3 and 11, respectively, to 4 and 13. As a consequence, the number of expected matches to a database word is halved to two, greatly reducing the workload for downstream stages. To validate this parameter change we tested BLASTP with the new parameters, comparing the GenBank Non-Redundant protein database (2,321,957 sequences; 787,608,532 residues) against the entire *Escherichia coli k12* proteome (4,242 sequences; 1,351,322 residues). With a word length of 4, more than 99.60% of the gapped alignments reported with the shorter word length were still located.

Seed generation for protein comparison uses a straightforward *direct lookup* table architecture to retrieve matching query positions. For our chosen word size $w = 4$, the table is too large to fit in on-chip memory, so we use external

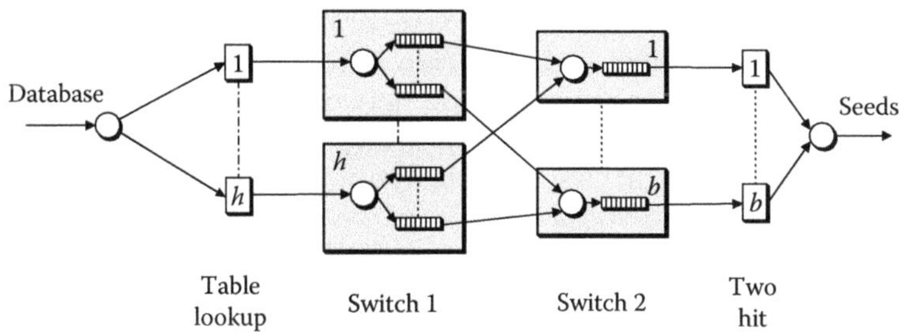

FIGURE 8.6
Seed generation logic, showing routing of word matches. (From Jacob, A. et al., *ACM Transactions on Reconfigurable Technology and Systems*, 1(2):1–44, 2008. With permission. © 2008 ACM Inc.)

SRAM. The memory is organized into a primary lookup table with 20^w entries, one for every possible database word (the alphabet is 20 amino acids). Matching query positions are stored in compressed form in each 32-bit entry of the table. Excess positions are appended in a duplicate table. Our implementation supports a maximum protein sequence length of 2,048.

To implement the two-hit filter, we use an array, which stores the database position of the most recently encountered word match, on its diagonal. On arrival of a new word match (q_1, d_1), its corresponding diagonal $d-q$ in the array is referenced to see if it is a two-hit. As we are streaming the database over a query of maximum length 2,048, we need a sliding window of only 4,096 diagonals for the active computation; these are stored in eight on-chip memories on our FPGA. Our two-hit architecture is a three-stage pipeline (with data forwarding to eliminate hazards) able to process one word match per clock.

Recall that seed generation is expected to emit two word matches per database position, but the two-hit design can process only one of them at a time. To avoid a bottleneck in the two-hit stage we must replicate it. A naive method is to replicate the two-hit block, with each block having a copy of the entire diagonal array. Keeping all arrays coherent, however, requires a multicycle sequential update operation, which seriously degrades throughput. Rather than replicate, we partition the diagonals across b two-hit units so that a word match (q, d) is processed by two-hit unit j if $d - q \equiv j - 1 \pmod{b}$. As a two-hit computation depends only on its diagonal, each two-hit unit now operates independently, and maintaining coherency is not an issue. Furthermore, word matches in biological sequences tend to be clustered within a band of diagonals; the modulo scheme naturally distributes these matches more evenly among the two-hit units.

Finally, we increase the throughput of seed generation by using a number h of parallel lookup stages that access independent external memories. The complete architecture is illustrated in Figure 8.6. An important component is the switching architecture to route matches from one of the h lookup stages to a number b of two-hit units. Each lookup stage uses a switching network

(Switch 1) to route its word matches to one of the b two-hit input queues. A two-hit stage then receives its input from one of h queues, one from every lookup stage, using a second switching network (Switch 2). The switching network allows us to implement a high-throughput, load-balanced seed generation architecture by varying b and h. In our implementation we use $h = 2$ lookup stages and $b = 4$ two-hit units.

An unintended consequence of using multiple lookup stages concerns the order of word matches entering the two-hit and the subsequent stages. As the time to process a database word varies with the number of matches to the query, the independently operating lookup stages may lose synchronization and generate matches that are out of order with respect to their database positions. The downstream stages expect in-order word matches to guarantee performance, and in the case of two-hit, to maintain correctness. To alleviate this problem we limit the maximum number of matching query positions per database word to at most 15 (this has negligible effect on sensitivity). Additionally, we augment the two-hit heuristic to handle out-of-orderness as follows: if a word match is at most Y database residues before the most recently seen match on the diagonal, discard it, as it is likely a part of a cluster of matches; else, it is likely part of a distinct HSP, so accept it. Using this additional check, we are able to reduce the negative effect on sensitivity.

8.2.2 Ungapped Extension

As described in Section 8.1.2, ungapped extension as implemented in software uses the X-drop heuristic, where extension proceeds indefinitely until the running score falls X below a previous high. The length of an extension is data dependent—in the extreme case it proceeds till the end of the query or reference sequence. For the ungapped extension loop, we desire a fully unrolled, pipelined architecture that is able to accept a seed match every clock. To fully unroll the loop we must know a priori the length of every extension. We therefore decided to only inspect a fixed window of L residues in the two sequences that is centered on a seed. Furthermore, unlike in software, extension proceeds in one pass from the start to the end of the window.

Selecting L and the threshold T_2 presents an interesting tradeoff. On the one hand, increasing L produces a better filter as the stage is able to distinguish statistically significant HSPs from random noise. We can increase T_2 in tandem to reduce the workload of the downstream stage without a loss in sensitivity. Unfortunately, increasing the size of the window also increases the resource requirements. To find suitable values for these parameters, we compared sequences from the *E. coli* proteome against the GenBank Non-Redundant protein database. We found that the X-drop heuristic terminates 95% of unsuccessful seed extensions, which are mostly noise, within a 60 residue window. We settled on the parameter values $L = 64$ and $T_2 = 16$, which minimize resource requirements while maximizing the filtering rate of this stage. In a small

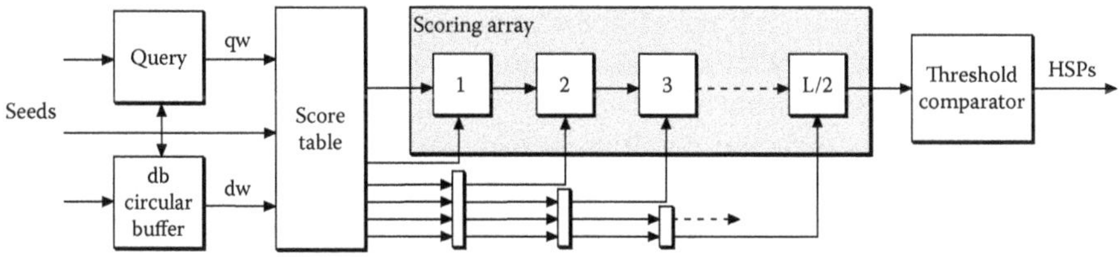

FIGURE 8.7
Overview of the architecture of stage 2. (From Jacob, A. et al., *ACM Transactions on Reconfigurable Technology and Systems*, 1(2):1–44, 2008. With permission. © 2008 ACM, Inc.)

fraction of cases the window is too small to determine the score of the best HSP of a seed, that is, the best HSP extends beyond the window boundaries; we simply pass these inputs along to the next stage to avoid a loss in sensitivity.

An overview of the architecture for this stage is shown in Figure 8.7. We store the entire query on-chip, but a circular buffer is used for the database stream. Care must be taken to ensure that the circular buffer always maintains the window of database residues required by every input seed, including those that are still in the pipeline of seed generation. A locally connected array with $\frac{L}{2}$ processing elements implements the dynamic programming recurrence shown in Algorithm 1.

Algorithm 1 Ungapped extension loop

$\Gamma_0 \leftarrow \gamma_0 \leftarrow 0$
for $i \leftarrow 1, L$ **do** ▷ *Iterate across window [QW, DW]*
  if $i < SEED_{start}$ **then** ▷ *Extend HSP by one residue pair*
    ▷ *If HSP's score is negative force to zero and start new HSP at i + 1*
    $\gamma_i \leftarrow \max\{\gamma_{i-1} + \delta(QW[i], DW[i]), 0\}$
  else
    $\gamma_i \leftarrow \gamma_{i-1} + \delta(QW[i], DW[i])$ ▷ *HSP cannot restart once the seed is reached*
  end if

  if $i > SEED_{end}$ **then**
    $\Gamma_i \leftarrow \max\{\Gamma_{i-1}, \gamma_i\}$
  else
    $\Gamma_i \leftarrow 0$ ▷ *Record score of best HSP that crosses the seed* ($i > SEED_{end}$)
  end if
end for

Each processor implements two steps of the recurrence and uses saturation arithmetic to reduce resource requirements. The score of every residue pair in the window, $\delta(QW[i], DW[i])$, is precomputed and pipelined to its corresponding processor $\frac{i}{2}$. The processor $\frac{i}{2}$ computes the score of the best HSP γ_i

terminating at position i. If the HSP terminating at i has a negative score, the computation restarts the HSP at $i + 1$. The processor also tracks Γ_i, the score of the best HSP terminating at or before i. The pipeline registers shown in the figure are used to keep the δ values in lock-step with the scores computed in the processors. We use two new constraints, shown in Lines 7 and 13 of Algorithm 1, to force an HSP to contain its seed. This ensures that a "weak" ungapped alignment will not be masked by a "stronger" one when both lie on the same diagonal (but are associated with distinct seeds). The array outputs Γ_L, the score of the best HSP in the entire window, which is compared to the threshold score.

8.2.3 Gapped Extension

Owing to the excellent filtering properties of the first two stages in the BLAST pipeline, most query-reference sequence pairs have already been eliminated, so much so that we can use a software implementation of this stage for BLASTN without introducing a bottleneck. While we must still accelerate this stage in BLASTP, we have at our disposal hundreds of clock cycles per input. Consequently we do not fully unroll the computational loop of this stage, unlike ungapped extension.

At its core, gapped extension in software implements the familiar Smith–Waterman computation but with a speed optimization that restricts activity to an irregular pattern of cells around the seed; an example is shown in Figure 8.8a. To design a hardware-friendly architecture, we choose to restrict the Smith–Waterman computation to a band of fixed-size *antidiagonals* centered on the seed, that is, sets of cells (i, j) that have the same value of $i + j$. Figure 8.8b illustrates an example of the rectangular band of antidiagonals with length λ and width ω. The regular activity pattern reduces the complexity of control circuitry while still providing significant time savings over full Smith–Waterman. Similar to ungapped extension, we have a tradeoff between area, throughput, and the filtering capacity when fixing the band geometry. A larger band improves the discriminating power of this stage; the size of the parallel array, however, grows with ω, and lengthening λ decreases the throughput. Using empirical measurements we balance this tradeoff using parameter values $\omega = 65$ and $\lambda = 1{,}601$.

The design of the banded Smith–Waterman array is similar to standard implementations and will not be expounded here; instead we will highlight the unique properties of our design. In a standard implementation, the size of the array is equal to the length of the query and computation proceeds vertically, consuming a single database residue every clock cycle (see Figure 8.8). In contrast, we use an array of ω processing elements that simultaneously computes all cells on the same antidiagonal using a stair-step computation pattern that proceeds diagonally along the length of the band. This pattern requires shifting in of $\omega + \lambda/2$ query and database residues, one of each every two clock cycles—the former in even and the latter

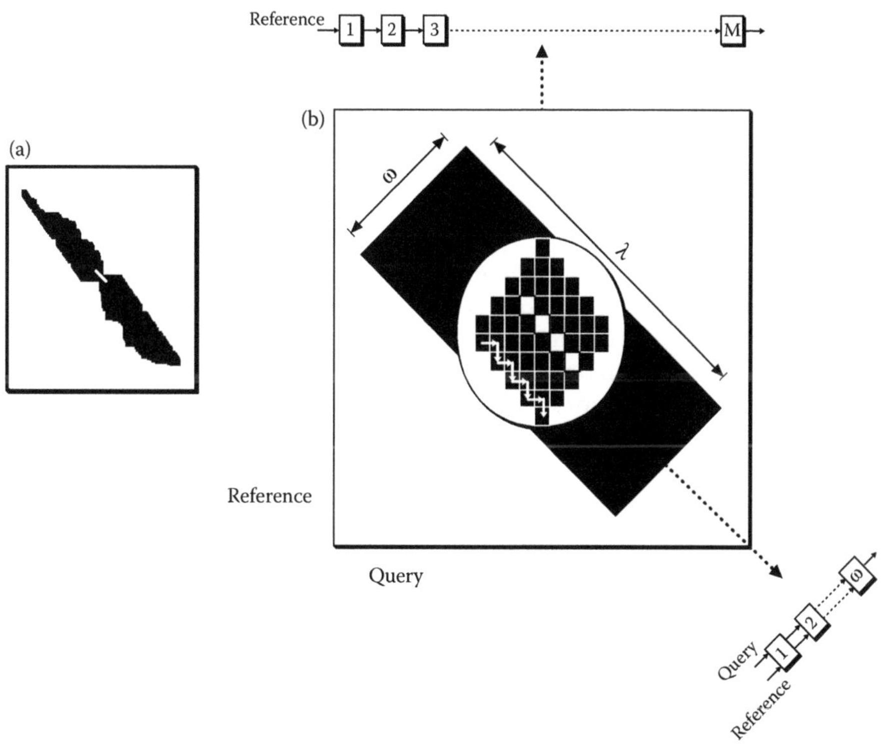

FIGURE 8.8
Example of gapped extension in (a) NCBI and (b) hardware-accelerated BLASTP, with cells computed by each method shaded. The query and reference sequence positions are along the x- and y-axes respectively. Computation is centered on a seed match shown in white. The canonical Smith–Waterman array shown at the top uses M processors, where M is the length of the query, and streams in a residue of the database every clock. The banded array shown below uses ω processors, where ω is the width of the band, and has a stair-step computation pattern that streams both query and reference sequence residues through it. (From Jacob, A. et al., *ACM Transactions on Reconfigurable Technology and Systems*, 1(2):1–44, 2008. With permission. © 2008 ACM, Inc.)

in odd cycles. The best gapped alignment computed must contain the input seed; that is, the gapped alignment must start before the antidiagonal at the center of the seed and terminate after it. We enforce the former constraint by not resetting the score of negatively scoring alignments to zero, as does the standard Smith–Waterman recurrence, after the center of the seed is crossed. To enforce the latter constraint we record only the score of gapped alignments that cross the seed. As a speed optimization, extension is terminated early if we observe only negative scores in all cells of two consecutive antidiagonals. The worst-case latency for gapped extension of a seed is 5 + ω + λ clocks, but early termination contributes to a latency savings averaging 56%.

8.3 Results

We have coded the BLAST hardware accelerator in VHDL and implemented it on the Mercury system. This is a prototyping platform that provides high-throughput data transfer from disk to reconfigurable logic at more than 800 MB/sec. Our system uses two dual-core 2.4 GHz AMD Opteron processors with 16 GB of memory running 64-bit Linux (CentOS 4). Two Xilinx Virtex-II 6000–6 FPGAs, each with three synchronous 1 MB SRAM modules, are connected via the PCI-X bus to the host.

We integrated the BLAST FPGA accelerator with NCBI BLAST version 2.2.9. The software initializes the hardware stages with the search parameters, loads a query and its associated memory tables, and streams the database through the FPGA in a single pass. Output from the hardware is collected and processed by the unmodified, ungapped, and gapped extension stages in software. These pre- and postprocessing activities run concurrently in different threads of execution. The tight integration of our hardware accelerator with the original NCBI codebase preserves the user interface, including command-line options and I/O formats, allowing Mercury BLAST to be used as a drop-in replacement in existing bioinformatics analysis pipelines.

Hardware BLASTN can support a DNA query of length up to 17 kbases (34 kbases including both the sequence and its reverse complement) while hardware BLASTP supports proteins up to 2,048 residues long; very large sequences must be split into smaller overlapping chunks. The software performs a query packing optimization to efficiently process small queries. Here, smaller sequences are packed using a bin-packing approximation algorithm into a single composite query over which the search is executed. This reduces the number of passes of the database stream through the hardware, significantly decreasing the overall search time.

We compared hardware BLAST to the software, using large comparisons that are typical of genome annotation. In addition to the speedup, we report the quality of the accelerator output, given by the sensitivity of the hardware, measured as the fraction of software baseline's alignments detected by hardware BLAST. For this test, alignments from the same query-reference sequence pair that overlap more than 50% of their bounding rectangles are considered to be the same. All sequences in our experiments were filtered for low-complexity regions, and BLAST was run with default parameters, except for a lower E-value threshold of 10^{-5}, which is reasonable for large-scale comparison. Runtimes reported exclude time spent formatting the output for printing.

8.3.1 BLASTP

Owing to its high resource requirements, hardware BLASTP uses both of the Virtex-II FPGAs on the prototyping system: seed generation and ungapped extension runs on the first, and gapped extension runs on the second. The three stages were synthesized to run at 110 MHz, 85 MHz, and 90 MHz, respectively. They occupy 63% of the slices and 77% of the on-chip block RAM memories on the first FPGA and 33% of the slices and 48% of the block RAMs on the second. Most of the block RAM memories in the design are used to hold the score table for the extension stages. All three stages, however, fit on a single newer generation FPGA device.

The baseline system we used to run NCBI BLASTP is an eight-node compute cluster, with each node having two 2.4 GHz AMD Opteron processors and 4 GB of memory. The runtime of the baseline system is the total of the individual execution times on each node, which gives the single core performance. We used a recent version of NCBI BLASTP (2.2.17) for speed comparisons, which is more than twice as fast as the version we have integrated with hardware BLAST. For sensitivity measurements we compared against the version integrated with our accelerator.

To evaluate the performance of our accelerator, we ran the following two experiments, typically performed on proteins predicted from a newly sequenced genome.

1. *E. coli* K12 proteome (1.35 Mresidues) versus GenBank Non-Redundant (NR) protein database (1.39 Gresidues);
2. *B. thetaiotaomicron* proteome (1.85 Mresidues) versus GenBank NR.

Table 8.2 shows hardware-accelerated BLASTP executing more than an order of magnitude faster than the baseline system. Moreover, in both experiments our hardware had a sensitivity more than 99.40%.

TABLE 8.2

Execution Time of Hardware-Accelerated BLASTP Compared to the Baseline System

Experiment	Baseline Time	Hardware Time	Speedup
E. coli vs. NR	28.7 h	1.9 h	15.11×
B. theta vs. NR	40.5 h	2.7 h	15.29×

Source: Jacob, A. et al., *ACM Transactions on Reconfigurable Technology and Systems*, 1(2):1–44, 2008. With permission. © 2008 ACM, Inc.

8.3.2 BLASTN

A single Virtex-II FPGA is sufficient to implement the BLASTN design. The seed generation and ungapped extension stages operate at clock frequencies of 140 MHz and 60 MHz, respectively, while occupying 40% of the FPGA slices and 93% of the on-chip block RAM memories. Scalability of the BLASTN implementation is limited by the memories available to implement the Bloom filter stage.

For comparison we ran the baseline experiments using the newer, enhanced version of NCBI BLAST on a single core of a 3.0 GHz Pentium D CPU with 1.5 GB of RAM. Note that this CPU is more powerful than the AMD Opteron used in the previous section. Furthermore, we are comparing the more powerful CPU to the older Virtex-II FPGA, which is two generations behind the currently available state of the art.

We performed two experiments that used the following datasets:

1. 3,975 randomly sampled human cDNA sequences (9 Mbases after removing known repeats and Ns) from release 21 of the NCBI RefSeq cDNA library, against all other vertebrate cDNAs (586 Mbases after removing known repeats and Ns).
2. Human chromosome 22 (hg18, 21 MBases after removing known repeats and Ns) against the entire mouse genome (mm8, 1.5 Gbases after removing known repeats and Ns).

Table 8.3 shows the speedup of the hardware accelerator, which ranges from 5× to more than an order of magnitude over the software baseline. The hardware is able to find 98.6% and 99.0% of the alignments, respectively, for the two experiments.

The order-of-magnitude speedup of the hardware design and its validation on large-scale DNA and protein comparisons gives us confidence in its use as a replacement for a small workstation cluster running NCBI BLAST. We expect the design to scale on newer generations of FPGA devices with an increased number and storage capacity of off-chip memories used for the lookup table. This will allow the packing of more sequences into a composite

TABLE 8.3

Execution Time of Hardware Accelerated BLASTN Compared to the Baseline System

Experiment	Baseline Time	Hardware Time	Speedup
Human cDNA vs. RefSeq	101 min	20 min	5.05x
Human Chr22 vs. Mouse	218 min	19 min	11.47x

query, reducing the number of passes of the database stream through the hardware, thus decreasing overall execution time.

8.4 Conclusions

Many alternate FPGA designs that accelerate the BLAST computation have been published. Unfortunately, most accelerate one or only a few of the stages in the pipeline, which is insufficient to appreciably speedup the entire application. An alternate design for seed generation is to store the lookup table on-chip. The low capacity of on-chip memories, however, limit this design to short queries or word lengths that generate small neighborhoods. Using a shorter word length, however, increases the workload for downstream stages. A second option is to compute a word match to the query online without using a precomputed neighborhood. This approach exposes more parallelism and scales better with newer generation FPGA devices. It is useful if large external memories are unavailable on the accelerator platform. Nevertheless, both these designs will require a large fraction of reconfigurable logic on modern FPGAs, limiting the acceleration of ungapped and gapped extension on the same device.

We believe the streaming paradigm is well suited for sequence analysis applications because they use self-contained integer arithmetic operations with simple control structures, operate on a localized portion of the input, show little data reuse, and have no feedback between stages. In the case of BLAST, we were able to exploit pipeline parallelism by streaming in reference sequences over a linear chain of stages running concurrently on an FPGA and a microprocessor core. It is important to study the profile of the stream program and consider the implications of design choices of a stage on the entire pipeline's throughput. For example, using a word length of four in BLASTP's seed generation module greatly reduces the workload for two-hit and ungapped extension; fully unrolling the loop in ungapped extension allows this stage to keep up with input seeds; and a multicycle latency gapped extension stage that conserves resources is sufficient to keep up with its input rate. Hardware-accelerated BLASTP has its bottleneck in the seed generation stage, as does BLASTN, though the latter is currently limited by input system bandwidth.

An important goal we have stressed throughout our design process is the maintenance of the quality of results as compared to the de facto standard implementation. This is especially necessary when accelerating well-accepted heuristics; replacing them by more hardware-friendly ones is difficult to justify to the user community. Any deviation from these heuristics, for example, using more favorable parameter values, requires measurement of the output quality on substantial size datasets. In keeping with this goal, we have for the most part stuck to a faithful duplication of the original BLAST algorithm.

The principles we have illustrated and the architecture presented in this chapter extend to other seeded sequence comparison tools. Biological sequences can be organized into families that have a similar function or structure. An alignment of sequences in the family reveals patterns of conservation or divergence and contains more useful information for a search than any one of its members in isolation. A sequence family is represented by an ordered list of columns of its alignment, termed a *position-specific scoring matrix* (PSSM). Each column in a PSSM describes the distribution of residues at one position in the alignment. For example, a PSSM for a family of proteins may show that 50% have residue R in their first column, 30% have residue Q, and so forth. Search tools such as PSI-BLAST [3] and IMPALA [19] can compare families represented by PSSMs to sequences and help identify additional members of a family. PhyloNet [20] compares PSSMs to each other to discover similarity between families.

The design of hardware BLASTP extends to comparison using a PSSM. The main difference is in the scoring function δ, which will now operate on a residue and a PSSM column or on pairs of PSSM columns. The neighborhood in the seed generation stage is redefined as all ω length sequence words that score at least T when compared to some ω contiguous columns in the PSSM. When both the query and database consist of PSSMs, as in PhyloNet, vector quantization is used to map residue distributions in each column to a small number of characters. This will result in a search similar to pairwise sequence comparison, but with a larger alphabet than that of proteins. In both these cases the neighborhood is computed offline so the seed generation design will remain unchanged. Ungapped and gapped extension will have to be modified to use the new scoring procedure.

A *hidden-Markov model* (HMM) is a generalization of a PSSM that more accurately describes an alignment of members in a family. The power in this representation lies in its probabilistic representation and formal treatment. Analogous to Smith–Waterman for pairwise comparison, the Viterbi algorithm uses dynamic programming to compare an HMM and a sequence, generating a probability that the sequence is a member of the family. A recent heuristic, HMMERHEAD [21], employs a multistage filter to accelerate search with Viterbi. Neighborhood generation for the first stage is similar to that described for PSSMs. The second stage directly applies ungapped extension to seeds and is likely to require replication to keep up with its high workload. We can use the modulo input distribution scheme similar to that of the two-hit stage in BLASTP. A third stage in HMMERHEAD then uses two-hit on seeds that pass ungapped extension. The definition of a two hit is modified to include pairs of seeds that are within a window of Y residues in the database and within Z diagonals of each other. Owing to the upstream ungapped extension stage, it may be possible to use a two-hit design that does not require replication.

Acceleration of BLAST is challenging because the heuristic already has a time complexity that is linear in the size of the database and the software implementation is memory bound. We have used the streaming paradigm with a specialized hardware design to expose and exploit parallelism and a large number of distributed on-chip memories as a "cache" to improve memory performance. Our implementation accelerates the entire BLAST pipeline, which is important, as application runtime is distributed among all the stages. We have integrated our FPGA-microprocessor accelerator with the existing NCBI BLAST codebase. End-to-end performance measurements show that DNA and protein BLAST comparisons achieve a 5–15× speedup over the software on a modern workstation. Furthermore, close to 99% of the software alignments were also detected by the hardware. We are in the process of increasing the length of the composite query four-fold using higher capacity external memories and current-generation FPGAs, thus requiring one-fourth as many passes of the database and producing an additional 4× speedup.

8.5 References

1. T. F. Smith and M. S. Waterman. Identification of common molecular subsequences. *Journal of Molecular Biology*, 147(1):195–197, March 1981.
2. S. F. Altschul, W. Gish, W. Miller, E. W. Myers, and D. J. Lipman. Basic local alignment search tool. *Journal of Molecular Biology*, 215:403–410, 1990.
3. S. F. Altschul, T. L. Madden, A. A. Schäfer, J. Zhang, Z. Zhang, W. Miller, and D. J. Lipman. Gapped BLAST and PSI-BLAST: A new generation of protein database search programs. *Nucleic Acids Research*, 25:3389–3402, 1997.
4. National Center for Biological Information. Growth of GenBank, 2008. http://www.ncbi.nlm.nih.gov/Genbank/genbankstats.html Accessed February 20, 2010.
5. A. E. Darling, L. Carey, and W. C. Feng. The design, implementation, and evaluation of mpiBLAST. In *4th Int'l Conf. on Linux Clusters*, 2003.
6. H. Lin, X. Ma, P. Chandramohan, A. Geist, and N. Samatova. Efficient data access for parallel BLAST. In *Proc. 19th Int'l Parallel and Distributed Processing Symposium*, 2005.
7. H. Rangwala, E. Lantz, R. Musselman, K. Pinnow, B. Smith, and B. Wallenfelt. Massively parallel BLAST for the Blue Gene/L. In *High Availability and Performance Computing Workshop*, 2005.
8. S. McGinnis and L. Thomas Madden. BLAST: At the core of a powerful and diverse set of sequence analysis tools. *Nucleic Acids Research*, 32:20–25, 2004.
9. D. T. Hoang. Searching genetic databases on Splash 2. In *IEEE Workshop on FPGAs for Custom Computing Machines*, pages 185–191, 1993.
10. Y. Yamaguchi, T. Maruyama, and A. Konagaya. High speed homology search with FPGAs. In *Pacific Symposium on Biocomputing*, 7:271–282, 2002.

11. R. D. Chamberlain, R. K. Cytron, M. A. Franklin, and R. S. Indeck. The *Mercury* system: Exploiting truly fast hardware for data search. In *Proc. of Int'l Workshop on Storage Network Architecture and Parallel I/Os*, pages 65–72, September 2003.
12. J. D. Buhler, J. M. Lancaster, A. C. Jacob, and R. D. Chamberlain. *Mercury* BLASTN: Faster DNA sequence comparison using a streaming hardware architecture. In *Reconfigurable Systems Summer Institute*, 2007.
13. A. Jacob, J. Lancaster, J. Buhler, B. Harris, and R. D. Chamberlain. *Mercury* BLASTP: Accelerating protein sequence alignment. *ACM Transactions on Reconfigurable Technology and Systems*, 1(2):1–44, 2008.
14. P. Krishnamurthy, J. Buhler, R. Chamberlain, M. Franklin, K. Gyang, A. Jacob, and J. Lancaster. Biosequence similarity search on the *Mercury* system. *Journal of VLSI Signal Processing Systems*, 49(1):101–121, 2007.
15. J. Lancaster, J. Buhler, and R. D. Chamberlain. Acceleration of ungapped extension in *Mercury* BLAST. *Microprocessors and Microsystems*, 33(4):281–289, 2009.
16. B. Bloom. Space/time trade-offs in hash coding with allowable errors. *Communications of the ACM*, 13(7):422–426, May 1970.
17. R. E. Tarjan and A. C. C. Yao. Storing a sparse table. *Communications of the ACM*, 22(11):606–611, 1979.
18. M. V. Ramakrishna, E. Fu, and E. Bahcekapili. Efficient hardware hashing functions for high performance computers. *IEEE Transactions on Computers*, 46:1378–1381, 1997.
19. A. A. Schaffer, Y. I. Wolf, C. P. Ponging, E. V. Koonin, L. Aravind, and S. F. Altschul. IMPALA: Matching a protein sequence against a collection of PSI-BLAST-constructed position-specific score matrices. *Bioinformatics*, 15:1000–1011, 1999.
20. T. Wang and G. D. Stormo. Identifying the conserved network of cis-regulatory sites of a eukaryotic genome. *Proceedings of the National Academy of Science USA*, 102:17400–17405, 2005.
21. E. Portugaly, S. Johnson, M. Ninio, and S. R. Eddy. Improved HMMER-HEAD for better sensitivity. *RECOMB 07 Poster*, 2008.

9

Seed-Based Parallel Protein Sequence Comparison Combining Multithreading, GPU, and FPGA Technologies

Dominique Lavenier and Van-Hoa Nguyen

9.1 Introduction

Despite the increasing diversity of genomic data now available through many different biotechnologies, DNA or protein sequences remain one of the main materials for bioinformatics studies. The basic treatment performed on these data is often a comparison process for detecting any kind of similarities. Traditionally, given a single request, the scan of large databases aims to report all related sequences. On the other hand, more specific applications, such as genome annotation, for instance, have a large set of sequences (proteome) to compare with a complete genome. In both cases, the heart of the algorithms, from a computational point of view, is the same: detection of similarities between strings of characters.

For almost two decades, the sizes of the genomic banks have steadily increased, nearly doubling every 18 months. From 1999 to 2009, for example, the size of UniProtKB/TrEMBL database [1] has been multiplied by 43 (release July 11, 1999, 199,794 entries). Today it contains 8,594,382 sequence entries comprising 2,774,832,018 amino acids (release 40.4, June 16, 2009). Practically, during the last 10 years, TrEMB has grown by a factor 1.9 every 18 months. From the DNA size, the situation is similar. In 1999, GenBank [2] (release 111, April 1999) contained 2.56 billions of nucleotides. Today, Genbank (release 171, April 2009) contains 103 billions of nucleotides. Its size has been multiplied by 40.

Furthermore, recent progresses in biotechnologies, and specifically fast improvements of sequencing machines, have revolutionized the genomic research field [3]. The equivalent (in raw data) of the human genome can now be generated in a single day. Billions of nucleotides spread in millions of very short fragments (25–70 nucleotides) are thus available, allowing a large spectrum of new large-scale applications to be set up: genome resequencing, metagenomic analysis, molecular bar coding, and so on. Bioinformatics treatments related to these new types of data often deals, in their earlier steps, with intensive sequence comparison.

Hence, together, exponential growth of the databases and next-generation sequencing (NGS) technology make the processing of this avalanche of data a more and more challenging task. This is also currently strengthened by the relative stagnation of the microprocessor clock frequencies that cannot help any longer, as it was the case for more than 20 years, to compensate the exponential growth of the genomic data. Today, to keep things on track, the use of parallelism is essential.

As a main task in bioinformatics, the parallelization of genomic sequence comparison algorithms has been widely investigated. When large volumes of data need to be processed, a straightforward way is to split data into smaller packets and to dispatch the computation on independent processing units. This parallelization scheme fits well with platforms of cluster and is commonly implemented in most bioinformatics research centers. Main advantages of this approach are (1) an efficient parallelization: each node works independently and does not require intensive communication with other nodes; (2) good scalability: computation may be deployed on large clusters or targets grid environments.

Nonetheless, other alternatives exist for parallelizing this basic bioinformatics treatment. They rely on internal potential parallelism of the algorithms.

As opposed to cluster or grid implementations, where each processing unit works independently on different data, comparison of two sequences—or a group of sequences—is shared between different processing units tightly interconnected. This fine-grained implementation targets specific hardware platforms such as reconfigurable accelerators (field-programmable gate array [FPGA]) or graphical processing units (GPU). Their great advantages, compared to cluster machines, are their lower cost and their high performance. Standard computers enhanced, for example, with two recent GPU boards or one medium-level FPGA board can then be 10–20 times faster.

These two schemes of parallelization (cluster/grid vs. GPU/FPGA), however, are not antagonists and can be combined to provide optimal use of computer resources. A few nodes of a general purpose cluster can be advantageously equipped with such accelerators. When intensive comparisons are required, the system automatically assigns these nodes to these specific processes, freeing the rest of the machines for other tasks.

The class of genomic sequence comparison algorithms implemented on these accelerators is mainly related to dynamic programming methods. Two main reasons can be emphasized: (1) algorithms are very time consuming, yielding a real need for speeding them up; (2) computations are very regular and fit well with highly parallel hardware structures. Description of such parallelization techniques can be found in [4].

Another class of algorithms, designed with a powerful heuristic based on the use of seeds to limit the search space, allows the computation to be drastically reduced, compared to dynamic programming algorithms. Two famous software, widely adopted by the scientific community, are FASTA [5] and basic local alignment search tool (BLAST) [6–7]. Unfortunately, very few attempts to parallelize them onto hardware accelerators have been done [8–13]. Again, two reasons can be proposed: (1) these algorithms are very fast, and the pressure to speed them up is lower; (2) computations are not regular and are much better suited to sequential processors than to parallel machines. With the NGS data surge, the first reason will rapidly become obsolete and fast solutions are now needed to parallelize these softwares at any levels, from transistor to grid! The second reason may represent a serious bottleneck for parallelization: algorithms have been designed for sequential machines and cannot be directly mapped to parallel hardware. They need to be redesigned, at a fine-grained level, to benefit from current technologies such as GPU or FPGA.

This chapter presents a parallel seed-based algorithm for comparing protein banks, and its instantiation into two technologies: GPU boards and reconfigurable accelerators. The algorithm has been thought to express the maximum of parallelism and to be easily speeded up by specific hardware platforms. As opposed to BLAST or FASTA, it does not aim to scan databases. It takes as input two banks and performs an all-by-all sequence comparison. Speedup from 10 (GPU) to 30 (FPGA) is measured compared to the latest optimized NCBI BLAST version.

The chapter is organized as follows: Section 9.2 presents the principle of the parallel seed algorithm, called parallel local alignment search tool (*PLAST*). Section 9.3 details implementations on GPU and FPGA. Section 9.4 compares performances of both technologies. Section 9.5 concludes the chapter.

9.2 Principles of the Algorithm

9.2.1 Overview

For detecting similarities, PLAST assumes that two protein sequences sharing sufficient similarities include, at least, one common word of W residues. From these specific words, larger similarities can be computed by extending the search on their left and right hand sides. These words are called *seeds* because they are the starting point of the alignment procedure.

To anchor two sequences with common seeds, the two banks are first indexed into two separate index-tables having exactly the same structure. The number of entries represents the number of all possible seeds (20^W). Content of one specific entry memorizes all the positions where the associated seed appears in the bank. As an example, suppose a bank composed of the two following sequences, s1 and s2:

```
s1 = AGGTGCTAGCTCT  s2 = TCTGCATCTGCAT
```

The content of the entry associated to the seed TGC will be (s1,4); (s2,3); (s2,9) because the word TGC appears in position 4 in sequence s1 and in positions 3 and 9 in sequence s2.

Taking the same entry of the two index-tables immediately gives the positions where the sequences have a common word (a hit) and, thus, potential local similarity. The next step is then to extend the similarity search in the hit neighborhood. This is done within two distinct phases: the first phase performs a simple extension by considering only substitution errors (ungap extension). A score is calculated regarding the number of matches and mismatches in the immediate neighborhood. If the score exceeds a threshold value, then the second phase is activated. This phase is more complex and considers insertion and deletion errors (gap extension). Again, a score is computed. If it exceeds a threshold value, the alignment is reported as a significant one.

Practically, the PLAST algorithm can be described as follows:

```
Algorithm 1 PLAST principle
 0: GapAlignList = Ø
 1: IndexTable1 = index_bank(Bank1)
 2: IndexTable2 = index_bank(Bank2)
```

```
 3: for all possible seed sk
 4:   AAStringList1 = make_string(IndexTable1[sk])
 5:   AAStringList2 = make_string(IndexTable2[sk])
 6:   for all s1 in AAStringList1
 7:     for all s2 in AAStringList2
 8:       UngapAlign = ungap_extension(s1,s2)
 9:       if UngapAlign.score > T1 and UngapAlign not in
          GapAlignList
10:         then GapAlign = gap_extension(UngapAlign)
11:              if GapAlign.score > T2
12:                 then GapAlignList.add_and_sort(GapAlign)
```

Lines 1 and 2 build the two bank indexes. Line 3 iterates on all possible seeds. Then, for each seed, two lists of short amino acid strings are constructed (lines 4, 5). These strings are made from the left and right neighborhoods of the seeds, and have a fixed length. Pairwise extensions of all elements of the two lists are performed (lines 6, 7, 8). If the ungap alignment resulting from the ungap extension procedure has a score greater than a threshold value (T1) and if it is not included in an alignment already computed (line 9), then the gap procedure is launched. If the score of this new alignment exceeds a new threshold value (T2) then it is added and sorted in the final list of alignments.

The test checking if an ungap alignment is included in the final list of alignments (line 9) is essential: usually, significant alignments contain several anchoring sites from where final alignments can be generated. This test avoids the duplication (and the computation) of gap alignments. To speedup the inclusion search (line 9) the final list of alignments is sorted by their diagonal number (line 12).

Actually, this algorithm has great potentiality for parallelism because the 3 `for all` nested loops are independent. Basically, each seed extension can be performed concurrently. A first medium-grained parallelism, oriented to multicore architecture, is thus to consider a multithreading programming model for the outer `for all` loop (line 3). *N* Threads can thus be associated to *N* different seed extensions. The parallel multithreaded version of the algorithm is the following:

```
Algorithm 2 Parallel scheme
Main Thread
GapAlignList = Ø
IndexTable1 = index_bank(bank1)
IndexTable2 = index_bank(bank2)
create N extension threads
SK = 0
wait until SK >= MAX_SK
Extension Thread
while (SK<MAX_SK)
   sk = SK++
```

```
3:    AAStringList1 = make_string(IndexTable1[sk])
4:    AAStringList2 = make_string(IndexTable2[sk])
5:    for all s1 in AAStringList1
6:      for all s2 in AAStringList2
7:        UngapAlign = ungap_extension(s1,s2)
8:        if UngapAlign.score > T1 and UngapAlign not in
          GapAlignList
9:           then GapAlign = gap_extension(UngapAlign)
10:               if GapAlign.score > T2
11:                then GapAlignList.add_and_sort(GapAlign)
```

The main thread constructs 2 index tables before creating *N* threads dedicated to the computation of the alignments. It sets a share variable *SK* to 0 (line 4) representing the first seed and wait until all the seeds have been processed. The extension threads increment the variable *SK* and compute the alignments associated to this specific seed. The instruction *sk* = *SK*++ is an atomic operation to avoid two threads from getting the same *SK* value.

A second level of parallelism is brought by the two inner `for all` loops (lines 5 and 6). If `i` is the number of elements of IndesxList1 and `j` the number of elements of IndexList2, then there are systematically `i × j` ungap independent extensions to compute. To exploit the regularity of the computation, the lines 5–11 can be decomposed as follows:

```
5: for all s1 in AAStringList1
6:   for all s2 in AAStringList2
7:     UngapAlign = ungap_extension(s1,s2)
8:     if UngapAlign.score > T1
9:      then UngapAlignList.add(UngapAlign)
10: for all x in UngapAlignList
11:   if x not in GapAlignList
12:     then GapAlign = gap_extension(UngapAlign)
13:          if GapAlign.score > T2
14:            then GapAlignList.add_and_sort(GapAlign)
```

The computation is split into two distinct parts: lines 5 to 9 compute ungap extensions and store the successful ungap alignments into the ungap alignment list. This list is then scanned for the gap extension procedure (line 10 to 14). For large databases, it appears that most of the computation time is spent in the first part. Hence, the computation performed by these 2 nested loops (lines 5 to 9) can be deported on specific hardware that is able to support a very high parallelization of this task.

9.2.2 Bank Indexing

The bank indexing process consists of modifying raw genomic data structures (sequence of characters) into more complex structures favoring the fast location of hits between sequences. The protein indexing scheme is based on the concept of subset seeds [14–15]. A subset seed is a word of *W* characters over an extended alphabet: the extra characters represent a specific set of amino acids. Below, a subset seed of size 4 is presented:

- character 1: A,C,D,E,F,G,H,I,K,L,M,N,P,Q,R,S,T,V,W,Y
- character 2: c={C,F,Y,W,M,L,I,V}, g={G,P,A,T,S,N,H,Q,E,D,R,K}
- character 3: A,C,f={F,Y,E}, G, i={I,V}, m={M,L}, n={N,H}, P, q={Q,E,D}, r={R,K}, t={T,S}
- character 4: A,C,D,E,F,G,H,I,K,L,M,N,P,Q,R,S,T,V,W,Y

As an example, the subset seed AcGL represents the words ACGL, AFGL, AYGL, AWGL, AMGL, ALGL, AIGL, and AVGL in the amino acid alphabet. Compared to the BLAST algorithm that requires two neighboring seeds of three amino acids to start the computation of an alignment, we use only one subset seed of four characters. The great advantage is that the computation is highly simplified by eliminating data dependencies and making it much more suitable for parallelism. An extension immediately starts when two identical subset seeds are found in two different protein sequences, avoiding extra computation for managing a couple of seeds. In [16], it is shown that this subset seed structure and the BLAST approach have comparable sensitivity.

The principle of the bank indexing with subset seed is illustrated in Figure 9.1. Each entry of an index table points to a list of subset seed positions. Each word of four amino acids in the bank needs to be translated into its equivalent subset seed. For example, the words AYIL, AMVL, and AVVL, respectively, at positions 0, 12, and 20 are translated into the subset seed word AciL. The entry AciL points to a list of integers where such words occur in the bank. Actually,

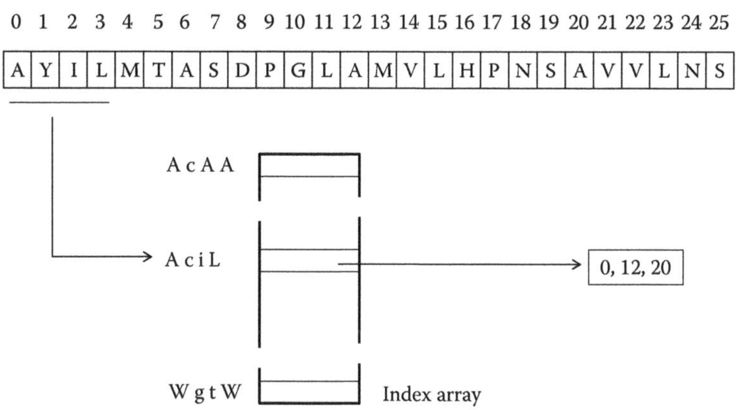

FIGURE 9.1
Principle of the indexing with subset seeds.

for memory optimization purpose, the positions are encoded in a relative way: only the difference between two consecutive positions is reported, leading to a 16-bit encoding. For comparing a protein bank and a DNA bank, the DNA bank is translated into its six reading frames and then indexed in the same way.

The advantage of this index structure is that it immediately provides all the hits between two protein sequences. Coming back to Algorithm 1, line 3, it can be seen that building the two index lists is straightforward.

9.2.3 Ungap Extension

The ungap extension procedure aims to rapidly check if a hit can give raise to a significant alignment. Thus, starting from the hit, left and right investigations are done to measure the similarity of the close neighborhood. In our approach, the neighborhood is fixed to a predefined length of *L*1 amino acids in both directions, and the score of an ungap alignment is only computed on this restricted area as follows:

```
Algorithm 3 ungap extension
1: ungap_extension(s1,s2)
2:    score = 0
3:    max_score = 0
4:    for x = 1 to L1+W
5:      score = score + SUB(s1[x],s2[x])
6:      max_score = max(score,max_score)
 7     score = max_score
 8:    for x = L1+W+1 to 2*L1+W
 9:      score = score + SUB(s1[x],s2[x])
10:      max_score = max(score,max_score)
11:    return score
```

The ungap extension procedure takes as input two strings of amino acids. Their sizes are equal to 2**L*1 + *W* with *L*1 the length of the neighborhood and *W* the length of the seed. The first *W* characters represent the seed, the *L*1 following ones represent the right neighborhood, and the last *L*1 characters represent the left neighborhood. Hence, lines 4 to 6 compute the right extension (including the seed) and lines 8 to 10 compute the left extension. After various tests, *L*1 has been set to 22 (1) for practical implementation issues and (2) because it provides satisfactory results, but the size of the neighborhood could be set to any other values.

This computation is very regular (no if statement) and, consequently, well suited for an implementation on highly parallel hardware.

9.2.4 Gap Extension

The gap extension procedure increases the search space by allowing gaps to be included in the final alignment. It is launched only if the previous step detects enough similarity near the hits. Algorithms for computing alignments with gaps are based on dynamic programming techniques, which are time-consuming procedures. And, as shown in [17], in some cases, this step may represent up to 30% of the total computation time. Parallelizing this procedure is thus sometimes interesting to minimize the overall computation time.

Again, with the objective of making the computation as regular as possible, this step is split into two phases. The first phase, called *small gap extension*, restricts the search both on a close-hit neighborhood and on a specific number of allowed gaps. A dynamic programming algorithm is run, starting from both sides of the hit, but on a limited number of diagonals, as shown Figure 9.2.

$L2$ is the length of the neighborhood and λ the size of the banded diagonals. The search space is represented by the shadow polygon. If the score of the restricted gap alignment exceeds a threshold value ($T3$), then a full gap extension (second phase) is computed using the standard NCBI-BLAST procedure, leading to similar results with this software.

The main reason to break this step into two phases is that the first step exhibits high potential parallelism. As a matter of fact, the small gap extension can be done concurrently on many different sequences of size $W + 2{*}L2$ because each computation requires identical search space.

Again, this phase can be computed in a parallel way.

9.2.5 Generic Hardware Implementation

The implementation of PLAST combines the multithreaded level approach with the fine-grained FPGA or GPU parallelization. The extension thread of

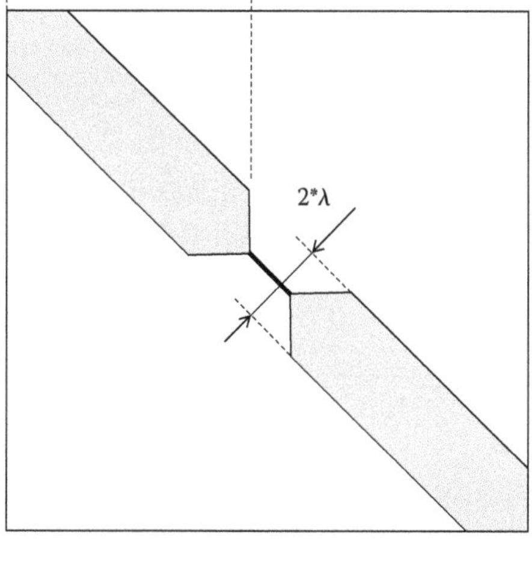

FIGURE 9.2
Search space for the first phase of the gap extension procedure.

Algorithm 2 is modified as shown below, the main thread staying the same. For a given seed, two lists of amino acid strings are built from the index tables (lines 3 and 4). These two lists are processed by the **UNGAP** function, which sent back a list of ungap alignments exceeding a threshold value *T*1 (line 5). The **UNGAP** function can be parallelized using two different technologies: FPGA accelerator or GPU.

Elements of the list that are not included in the list of final alignments are put on a temporary list (lines 6–8). Actually, this list contains couples of sequences of size *W* + 2**L*2 ready for the next step. When the size of this list overcomes its capacity, the first phase of the gap extension is activated. All the ungap alignments inside the temporary list are processed by the **SMALL_GAP** function. Again, this function can be parallelized on specific hardware taking as input a large set of sequences and sending back a list of alignments having a score greater than a threshold value *T*3. These alignments are then extended using the standard NCBI BLAST procedure.

```
Algorithm 4 Multithreaded and fine-grained parallelism
Extension thread
 0: TmpList = Ø
 1: while (SK<MAX_SK)
 2:    sk = SK++
 3:    AAStringList1 = make_string(IndexTable1[sk])
 4:    AAStringList2 = make_string(IndexTable2[sk])
 5:    UngapAlignList = UNGAP (AAStringList1,AAStringList2,T1)
 6:    for all UngapAlign in UngapAlignList
 7:      if UngapAlign not in GapAlignList
 8:       then add UngapAlign in TmpList
 9:            if size(TmpList) >= N
10:              then SmallGapAlignList = SMALL_GAP (TmpList,T3)
11:                   for SmallGapAlign in SmallGapAlignList
12:                   GapAlign = NCBI_BLAST_ALIGN (SmallGapAlign)
13:                    if GapAlign.score > T2
14:                      then GapAlignList.add_and_sort(GapAlign)
13:                 TmpList = Ø
```

This generic hardware implementation allows the PLAST software to adapt itself regarding the available parallel resources.

9.3 Parallelization

Whatever the target technology, the **UNGAP** function takes as input two lists of short amino acid strings and detects the couple of sequences having an ungap alignment score above a threshold value. An all-by-all pairwise comparison is done between all sequences of the two lists, as explained earlier. This function has been parallelized both on GPU and FPGA platforms.

The **SMALL_GAP** function acts as a preprocessing step for computing small gap alignments. It takes as input a list of pairs of sequences and computes a score including gap errors. The search space is however limited by the length of the sequences and by the number of allowed gaps. Parallelization of this function has been done only on GPU.

9.3.1 UNGAP Parallelization on GPU

The parallelization of the **UNGAP** function on GPUs is an adaptation of the matrix multiplication algorithm proposed in the CUDA documentation [18]. Matrices of numbers are simply replaced by blocks of strings of amino acids. More precisely, for each function call, there are two lists of amino acid sequences to process: List1 and List2. Suppose that block $B1[N1, L]$ and block $B2[N2, L]$ correspond, respectively, to List1 and List2, with L the length of the amino acid sequences and $N1$ ($N2$) the number of sequences in List 1 (List2). The result of the computation is a third block $C[N1, N2]$ which stores the scores of all the computation between block B1 and block B2. In other words, $C[i][j]$ hold the score of the ith sequence of List1 and the jth sequence of List2.

The overall treatment is done by partitioning the computation into blocks of threads computing only a subblock of C, called C_{sub}. Each thread within the block processes one element of C_{sub} dimensioned as a 16×16 square matrix. This size has been chosen to optimize the memory accesses, allowing the GPU internal fast memory to store 2×16 amino acid sequences that can simultaneously be shared by 256 threads. Figure 9.3 gives the CUDA kernel code optimized for sequence of length equal to 48 amino acids (`BLOCK_SIZE` = 16).

Each score is computed by first loading the two corresponding 16×16 subblocks from global memory to shared memory with one thread loading one element of each block and by having each thread getting one substitution cost. Each thread accumulates this cost to the current score and performs a maximum operation. When it is done, the result is written to the global memory. By doing the computation in such a way, the shared memory is highly solicited, saving a lot of global memory bandwidth as blocks $B1$ and $B2$ are read from global memory only three times. For a maximal efficiency, the substitution matrix is stored in the texture memory.

Practically, the **UNGAP** function consist in sending to the GPU board two lists of amino acid sequences and getting back an $N1$x$N2$ matrix of scores. A sequential postprocessing is however required to extract significant scores.

9.3.2 UNGAP Parallelization on FPGA

The reconfigurable architecture implementing the computation of the **UNGAP** procedure is a linear array of processing elements (PEs) dedicated to the calculation of a score between two amino acid sequences. If P is the

```
UNGAP _ kernel(char* C, char* B1, char* B2, int N1, int N2)
{
  int bx = blockIdx.x;                    // block index
  int by = blockIdx.y;

  int tx = threadIdx.x;                   // thread index
  int ty = threadIdx.y;

  int Begin1 = N1 * BLOCK_SIZE * by;      // B1 index
  Begin1 += N1 * ty + tx;

  int Step1 = BLOCK_SIZE;                 // B1 iteration step

  int Begin2 = __mul24(BLOCK_SIZE,bx);    // B2 index
  Begin2 += N2 * ty + tx;
  int Step2 = __mul24(BLOCK_SIZE,N2);     // B2 iteration step

  int Csub = 0;                           // initialize results
                                             block
  int CsubMaxi = 0;

  __shared__ int SB1[BLOCK_SIZE][BLOCK_SIZE]; // to store sub-
                                                 block of B1
  __shared__ int SB2[BLOCK_SIZE][BLOCK_SIZE]; // to store sub-
                                                 block of B2

  for (int j=0; j<3; j++)
    {
      SB1(ty, tx) = B1[Begin1 + j*Step1];     // load the
                                                 matrices from
      SB2(ty, tx) = B2[Begin2 + j*Step2];     // device to
                                                 shared memory

      __syncthreads();                    // make sure the
                                             blocks are loaded

      for (int k=0; k<BLOCK_SIZE; k++)    // score computation
        {
          Csub = Csub + texfetch(matrix, SB1(ty, k), SB2(k, tx));
          if(Csub>CsubMaxi) CsubMaxi = Csub;
        }
      __syncthreads();
    }

  int c = Step2 * by + Begin2;           // write the block to
                                            global memory,
  C[c] = CsubMaxi;                       // each thread writes
                                            one element
}
```

FIGURE 9.3
CUDA code for the UNGAP function.

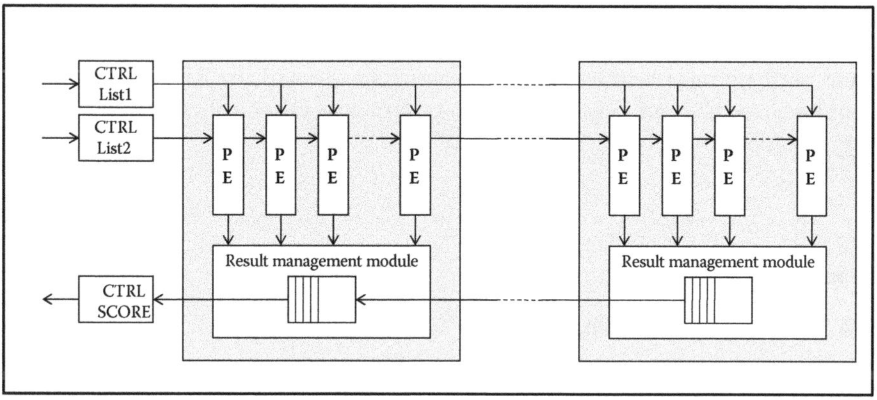

FIGURE 9.4
Principle of the FPGA architecture.

number of PEs, then P scores can be computed simultaneously between one sequence and P sequences. Figure 9.4 depicts the architecture principle of the accelerator. More details can be found in [19]. It works as follows: if $N2$ is the number of sequences of List2, then $N2/P$ iterations are required. One iteration loads P sequences into P different PEs in a systolic way. Then all sequences of List1 are broadcasted to all PEs, character by character, every clock cycle. Each time a PE receives a new amino acid, it updates its score. P scores are then available after L cycles (L is the length of the amino acid string). The scores are sent to a result management module that selects the PEs having scores greater than a predefined threshold value ($T1$). These scores are pushed through a first-in-first-out (FIFO) to the output channel.

For efficiency purpose, the array has been split into subarrays of fixed size that can be pipelined together. The advantages of this structure are twofold: (1) the architecture can be adapted to many FPGA platforms according to the available reconfigurable resources; (2) the performance of the system only depends of the number of PEs; the frequency remains identical whatever the number of subarrays (Figure 9.4).

Compared to the GPU approaches, the host does not need to extract the highest scores. This is done online by the result management module. Instead, the host receives a couple of integers indicating which pair of amino acid sequence has generated a significant score. This mechanism contributes to significantly decrease the need for a high-data bandwidth as small amount of information needs to be transferred from the FPGA accelerator to the host memory.

9.3.3 SMALL GAP Parallelization on GPU

The parallelization of the **SMALL_GAP** function on GPU is straightforward. The host downloads the GPU memory with couples of strings of identical size (2*$L2$ + W). Then, a thread is devoted to the computation of one score between couples of strings. Each thread performs a dynamic programming treatment using a banded Smith–Waterman algorithm. All the scores are sent back to the host, which needs to postprocess these data before triggering, if necessary, a full gap extension.

9.4 Comparison of the GPU/FPGA Technologies

The aim of this section is to evaluate the different approaches on a common base and to discuss the advantages and the drawbacks for each of them having in mind the minimization of the execution time. Here, the sensitivity aspect will not be discussed. A detailed study showing that PLAST and BLAST have an equivalent sensitivity can be found in [16]. Briefly, both algorithms use the same family of heuristics. They slightly differ on the seed choice and, consequently, do not generate exactly the same list of alignments. A few percentages of alignments are found by BLAST and not by PLAST. Inversely, a few are found by PLAST and not by BLAST. The difference is mainly expressed by alignments of weak similarity and represents less than 2% for an e-value of 10^{-3}.

The reference is the execution time of the BLAST software running on a multithreaded mode. It should allow the readers (1) to measure the contributions of the various technologies over a conventional but highly optimized implementation and (2) to appreciate the difference of performances between GPU and FPGA technologies. Two hardware platforms are considered: a GPU platform and a FPGA platform.

9.4.1 GPU Platform

The GPU platform is a Dell Server, 2.6 GHz Xeon Core 2 Quad processor with 8 GB of RAM running Linux Fedora 7. It is equipped with **two** NVIDIA Tesla C870 boards interconnected through peripheral component interconnect (PCI) express buses. Each board houses 1.5 GB of GDDR3 memory and a graphic chip including 128 multithreaded processors. The programming language is CUDA. The **UNGAP** and **SMALL_GAP** functions are run on the GPU boards.

9.4.2 FPGA Platform

The FPGA platform is a SGI ALTIX 350 machine composed of an Intel Itanium2 Core2 (1.6 GHz) with 4 GB of RAM, running SUSE Linux, and equipped with a RASC-100 accelerator (reconfigurable application-specific computing). This device is interconnected to the host system through a NUMAlink bus and is made of **two** Xilinx Virtex-4 FPGA components. The programming language is VHDL. Only the **UNGAP** function is run on the FPGA board. The SMALL_GAP function is processed by the host.

9.4.3 Software and Dataset

Like BLAST, PLAST is declined into several programs targeting various datasets. Here, the comparison between TBLASTN/TPLASTN is only presented. Reasons are as follows:

- Sequence comparison performed by this program is time consuming because of the translation of the DNA bank into its six reading frames, and consequently very well suited for demonstrating the efficient contributions of GPU or FPGA accelerators.
- The gap extension step in these two programs represents generally a minor percentage of the execution time. As an FPGA implementation does not exist for the **SMALL_GAP** function, the FPGA approach will not be too disadvantaged.

The GPU and FPGA version of TPLASTN are, respectively, referenced as GPU-TPLASTN (GPU) and RCC-TPLASTN (reconfigurable computing).

The dataset is composed as follows:

- The human chromosome 1 (220×10^6 bp): hchr1
- Four protein banks randomly constructed from the GenBank Non-Redundant protein database:
  - P1K : 1,000 protein sequences (0.336×10^6 aa)
  - P3K : 3,000 protein sequences (1.025×10^6 aa)
  - P10K : 10,000 protein sequences (3.433×10^6 aa)
  - P30K : 30,000 protein sequences (10.335×10^6 aa)

The BLAST (release 2.2.18) options have been set as follows:

```
blastall -p tblastn -d hchr1 -i pxxK -o rxxK -m 8 -a 2 -e 0.001
```

The –m 8 option provides a tabulated output synthesizing the features of the alignments. The –a 2 option runs BLAST in a multithreaded mode (two threads). The –e 0.001 option sets the e-value to 10^{-3}.

9.4.4 Comparison of the Execution Times

Tables 9.1 and 9.2 report the execution times of the NCBI TBLASTN, GPU-TPLASTN, and RCC-TPLASTN

Note that the NCBI TBLASTN execution time is different for the two platforms. However, this time serves as a reference to compare the speedup with

TABLE 9.1

Execution Time (in Seconds) of NCBI TBLASTN and GPU-TPLASTN on the GPU Platform (2 × C870 TESLA NVIDIA Boards—128 PEs per Chip)

	NCBI TBLASTN (2 threads)	GPU-TPLASTN (2 boards)	Speedup
P1K	754	140	5.38
P3K	2,172	258	8.41
P10K	7,436	744	9.99
P30K	21,951	2165	10.13

TABLE 9.2

Execution Time (in Seconds) of NCBI TBLASTN and RCC-TPLASTN on the FPGA Platform (SGI RASC-100–2 × Xilinx Virtex 4–192 PEs per Chip)

	NCBI TBLASTN (2 threads)	RCC-TPLASTN (2 FPGA)	Speedup
P1K	1,162	363	3.20
P3K	3,441	398	8.64
P10K	11,733	643	18.29
P30K	37,088	1323	28.03

accelerators (GPU or FPGA). To be fair, for each experiment, NCBI TBLASTN, GPU-TPLASTN, and RCC-TPLASTN have been run in a multithreaded mode with two threads. In the GPU mode, each thread drives a NVIDIA Tesla C870 board and progresses independently as explained in Section 9.2. Similarly, in the RCC mode, each thread controls a separate FPGA Virtex-4 device.

Globally, it can be seen that performances increase with the size of the data, whatever the technology or the platform used. A plateau is, however, reached for huge computations (a few hours). This can be explained by the fact that, in that case, the `UNGAP` and `SMALL_GAP` functions represent a very high percentage of the total execution time, which is efficiently parallelized on the accelerators. On the other hand, when the volume of data is low, the ratio between the sequential part and the parallel part increases and, following the Amdahl's law, limits the potential speedup.

9.4.5 GPU Implementation

Adding two NVIDIA C870 Tesla boards provides a speedup factor of 10 for intensive protein sequence comparison compared to the NCBI BLAST multithreaded version. Each board integrates a GPU chip (G80) housing 128 programmable processing units. A question is: can we do better? A first answer is to take the following generation of graphic boards to test the scalability of this approach. Experimentations with the NVIDIA GTX-280 board (T10 chip - 240 processing units) are reported Table 9.3.

In this experiment, the multithreaded mode is disabled for only highlighting the difference of performance between two successive generations of graphic boards. Several comments can be made as follows:

- The threads overhead is negligible. The speedup with two threads and two boards (Table 9.1) is very close to the speedup with one thread and one board (Table 9.3, column 7).
- Moving to the next board generation provides an immediate increasing of performance (columns 7 and 8) without any modification of the CUDA code.
- The **UNGAP** function represents an important percentage of the total execution time. Thus, there are still some rooms for further speedup improvements. In the P30K configuration, the theoretical maximum speedup compared with the NCBI TBLASTN software is about 40 (col2/(col5–col6)). This value is estimated as the NCBI BLASTN execution time divided by the sequential part of GPU-TPLASTN.

9.4.6 FPGA Implementation

Table 9.2 reports the results for 2 × 192-PE arrays implemented on both FPGA devices. This is the maximum of PEs we were able to fit inside the FPGA device. However, we experiment the performances on various array sizes, as shown in Table 9.4.

Again, the multithreaded mode is disabled to only measure the contribution of the FPGA accelerator. It can be seen that the number of PEs is inversely proportional to the execution time of the **UNGAP** function. For example, if the **UNGAP** speedup is measured relatively to 64 PEs, we get a linear speedup, as shown in Table 9.5.

Larger arrays are thus still possible to decrease significantly the overall execution time. The SGI RASC-100 accelerator houses Virtex-4 Xilinx components of 200 K logic cells with 336 × 18 Kb RAM Blocks (Virtex-4 LX 200). With the next generation of Xilinx components, a faster 384 PE array could be easily implemented in a single FPGA (Virtex6: XC6VLX550T) and would at least provide a speedup ranging from 5 to 6 compared to a 100 MHz 64 PE array. In that case, the overall speedup would be somewhere between 45 and 50.

9.5 Conclusion

PLAST is a parallel software for intensive protein comparison. Unlike BLAST, it does not target the scan of genomic databases. It has been designed for

TABLE 9.3

Performance Comparison between the NVIDIA Tesla C870 Board and the NVIDIA GTX-280 Board (Time Is Given in Seconds)

	NCBI TBLASTN (1 thread)	GPU Tesla C870 TPLASTN		GPU GTX-280 TPLASTN		NCBI Speedup		GTX-280 Speedup	
		Total	**UNGAP**	**Total**	**UNGAP**	**Tesla**	**GTX 280**	**Total**	**UNGAP**
P1K	1,369	250	114	216	80	5.47	6.33	1.15	1.42
P3K	4,009	474	306	383	215	8.45	10.36	1.23	1.42
P10K	13,391	1,341	971	1,053	681	9.98	12.71	1.27	1.42
P30K	40,444	3,932	2,917	3,077	2,057	10.38	13.14	1.27	1.42

TABLE 9.4

Execution Time (in Seconds) of RCC-TPLASTN with Different Numbers of PEs

	NCBI TBLASTN (1 thread)	RCC-TPLASTN 64 PEs			RCC-TPLASTN 128 PEs			RCC-TPLASTN 192 PEs		
		Total	UNGAP	Speed up	Total	UNGAP	Speedup	Total	UNGAP	Speedup
P1K	2,185	476	220	4.59	421	176	5.19	414	169	5.27
P3K	6,448	738	462	8.73	554	280	11.63	496	223	13.00
P10K	21,888	1,763	1,366	12.41	1,104	720	16.02	890	510	24.59
P30K	65,461	4,463	3,932	14.66	2,744	2,015	23.85	2,099	1,373	31.86

TABLE 9.5

Speedup Compared to the Number of PEs

# proc	64 PEs	128 PEs	192 PEs
UNGAP *(second)*	3,992	2,015	1,373
Speedup *(relatively to 64 PEs)*	1	1. 98	2.9

processing two large banks of sequences. Different versions are available depending on the nature of the data: PLASTP (protein/protein), PLASTX (DNA/protein), TPLASTN (protein/DNA), and TPLASTX (DNA/DNA). DNA sequences are translated into six reading frames. The heart of these programs and their parallelization scheme are, however, identical.

Like FASTA and BLAST, PLAST uses the concept of seeds to reduce the search space. The main difference is that two index tables are built, allowing groups of identical hits to be immediately identified. Each group can be processed independently on a multithreaded architecture (first level of parallelism), and the computation of each group can be deported on a GPU or FPGA accelerator (second level of parallelism). The combination of these two levels of parallelism fit well with current machines made of multicore processors and this can easily be enhanced with hardware accelerators connected through fast interfaces, like PCI express buses.

The originality of PLAST is that its design has been thought, in its earlier steps, as a parallel algorithm able to target the current and the next generations of computer systems. To compensate the end of systematical increase of the microprocessor clock frequency, to optimize the electric power consumption, and to continue to follow the Moore's law, the future chips will be highly parallel systems. The GPGPU architectures are probably an intermediate (and necessary) phase before more flexible parallel structures of hundreds of PEs. Bioinformatics algorithms need to be revisited to benefit from maximal efficiency provided by these new architectures to face the exponential demand in terms of genomic data processing.

9.6 References

1. UniProt Consortium, The Universal Protein Resource (UniProt), *Nucleic Acids Research* 37 (Database issue): D169–D174, 2009.
2. Benson DA, Karsch-Mizrachi I, Lipman DJ, Ostell J, Wheeler DL, GenBank, *Nucleic Acids Research*, 36 (Database issue): D25–D30, 2008.
3. Shendure J, Hanlee J, Next-generation DNA sequencing, *Nature Biotechnology*, 26(10): 1135–1145, 2008

4. Lavenier D, Giraud M, Bioinformatics applications. In *Reconfigurable Computing: Accelerating Computation with Field-Programmable Gate Arrays*, MB Gokhale, PS Graham, editors, chapter 9, Springer, 2005.
5. Pearson W, Lipman D, Improved tools for biological sequence comparison, *Proceedings of the National Academy of Science*, 85(8): 2444–2448, 1988.
6. Altschul S, Gish W, Miller W, Myers E, Lipman D, Basic local alignment search tool, *Journal of Molecular Biology*, 215(3): 403–410, 1990.
7. Altschul S, Madden T, Schäffer A, Zhang J, Zhang Z, Miller W, Lipman D, Gapped BLAST and PSI-BLAST: a new generation of protein database search programs, *Nucleic Acids Research*, 25: 3389–3402, 1997
8. Muriki K, Underwood K, Sass R, RC-BLAST: towards a portable, cost-effective open source hardware implementation. In *19th International Parallel and Distributed Processing Symposium* IPDPS05, 2005.
9. Lancaster J, Jacob A, Buhler J, Harris B, Chamberlain R, Mercury BLASTP: accelerating protein sequence alignment, *ACM Transactions on Reconfigurable Technology and Systems*, 1(2), 2008.
10. Lavenier D, Gille Georges G, Xinchu L, A reconfigurable index flash memory tailored to seed-based genomic sequence comparison algorithms, *VLSI Signal Processing*, 48(3): 255–269, 2007.
11. Fei X, Yong D, Jinbo X, Hardware BLAST algorithms with multi-seeds detection and parallel extension. In Reconfigurable computing: Architectures, tools and applications, *4th International Workshop*, ARC 2008, 39–50, 2008.
12. Kasap S, Ying L, Benkrid K, High performance FPGA-based core for BLAST sequence alignment with the two-hit method. In *8th IEEE International Conference on BioInformatics and BioEngineering*, 1–7, 2008.
13 Herbordt M, Model J, Gu Y, Sukhwani B, Van-Court T, Single pass, BLAST-like, approximate string matching on FPGAs. In *IEEE Symposium on Field-Programmable Custom Computing Machines*, 217–226, 2006.
14. Roytberg M, Gambin A, Noé L, Lasota S, Furletova E, Szczurek E, Kucherov G, On subset seeds for protein alignment, *IEEE/ACM Transactions on Computational Biology and Bioinformatics*, 99(1): 2009.
15 Peterlongo P, Noe L, Lavenier D, Georges G, Jacques J, Kucherov G, Giraud M, Protein similarity search with subset seeds on a dedicated reconfigurable hardware. In *Proceedings of the 2nd Workshop on Parallel Bio-Computing Workshop* (PBC'07), Springer, 2007, 1240–1248.
16. Nguyen V, Lavenier D, PLAST: Parallel local alignment search tool for database comparison, *BMC Bioinformatics*, to appear, 2009.
17. Nguyen V, Lavenier D, Fine-grained parallelization of similarity search between protein sequences, INRIA Report, RR-6513, 2008.
18. NVIDIA CUDA compute unified device architecture, Programming guide, version 1.0, June 23, 2007.
19. Nguyen V, Cornu A, Lavenier D. Implementing protein seed-based comparison algorithm on the SGI RASC-100 platform, *16th Reconfigurable Architectures Workshop*, May 25–26, Rome, Italy, 2009.

10

Database Searching with Profile-Hidden Markov Models on Reconfigurable and Many-Core Architectures

John Paul Walters, Vipin Chaudhary, and Bertil Schmidt

10.1 Introduction

Hidden Markov models (HMMs) have been applied as statistical models in the area of speech recognition since the early 1970s. The use of HMMs for protein modeling was introduced in the early 1990s by Haussler et al. [1] and Krogh et al. [2]. The general HMM structure to model protein sequence families is known as *profile HMM*. A possible way to construct (or learn) such profile HMMs is by using a multiple sequence alignment (MSA) of proteins belonging to the same family as an input (which is explained in more details in Section 10.2).

A profile HMM M can emit any given protein sequence x with a certain probability $P(x|M)$. A common way to define $P(x|M)$ is by the probability of a path of highest likelihood through M emitting x, which can be computed using the well-known Viterbi algorithm [3]. An alternative way to define $P(x|M)$ is by summing up the probabilities of all possible paths through M emitting x, which can be computed by the Forward algorithm. In any case, the probability $P(x|M)$ can be used as a basic building block for two common database search tasks:

1. *Searching a profile HMM database with a query protein sequence:* significant matches to profile HMMs can identify the query as a member of the modeled protein families. This search procedure is hence frequently used for annotating the protein sequences.
2. *Searching a protein sequence database with a query profile HMM:* significant matches to the query HMM can identify additional homologous proteins of the family modeled by the query HMM.

Both search tasks frequently employ the Viterbi or Forward algorithm to compare (or align) each sequence/HMM to the query HMM/sequence. Owing to the quadratic time complexity of both algorithms the search procedure is therefore highly time consuming. Actual runtime of course depends on the actual database/query sizes.

Prime examples where database searching with profile HMMs requires acceleration are metagenomic sequencing studies such as the global ocean sampling (GOS) expedition [4]. By aligning all generated GOS protein sequences to the Pfam [5] and TIGRFAM [6] profile HMM databases Yooseph et al. [4] were able to identify and annotate a large amount of new proteins and protein families. However, this procedure required 327 hours on a hardware system with multiple field-programmable gate array (FPGA) accelerators.

In this chapter we will show how the Viterbi algorithm for profile HMM database searching can be efficiently parallelized on reconfigurable hardware (FPGAs) as well as on many-core architectures (GPUs) with the compute unified device architecture (CUDA) programming model. The remainder is organized as follows: Section 10.2 provides more detailed background on profile HMMs and the associated Viterbi algorithm. The reconfigurable hardware design and the CUDA implementation are presented and evaluated in Sections 10.3 and 10.4, respectively. Finally, Section 10.5 concludes the chapter.

10.2 Background

In this section we briefly explain how a profile HMM relates to an *MSA*. Each HMM state captures position-specific information about the likelihood of each residue in the corresponding MSA column. Figure 10.1 illustrates this in the case where gaps are not considered: an *ungapped profile* of length four is derived from an ungapped MSA of length four.

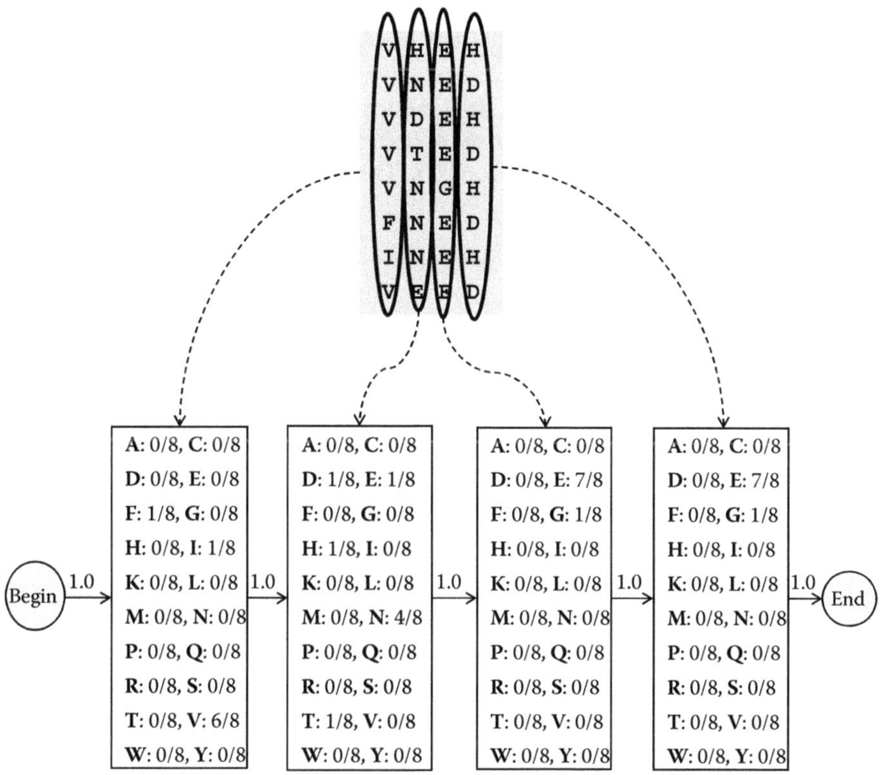

FIGURE 10.1
An ungapped profile of length four derived from an ungapped MSA with four columns.

The derived simple ungapped profile consists of linear sequences of states. States correspond to columns in the associated MSA. Also note that there are two types of probabilities associated with each state: *transition probabilities* and *emission probabilities*. All transition probabilities in ungapped profiles are one (as there is only one possible path). Emission probabilities are based on the probability of an amino acid occurring in the corresponding column in the multiple alignment. Pseudocounts are usually used to avoid over-fitting; that is, the determination of emission probabilities adds pseudocounts to the distribution of the observed amino acids.

The extension of the ungapped profile to a profile HMM needs to model gaps. This extension can be best explained by looking at the alignment of a protein sequence to a profile HMM as a basic operation. This alignment considers gaps in the two following ways:

- *Insertions:* correspond to regions of the sequence that are not present in the profile.

- *Deletions*: correspond to states in the profile that do not correspond to amino acids in the sequence.

Thus, the simple structure of the ungapped profile is extended as follows: three states at each position k (called *node* k) are used: a match (M_k), insert (I_k), and delete state (D_k). These states have the following functionalities.

- *M-states:* emit a single residue (correspond to the states described in Figure 10.1).
- *I-states*: emit a single amino acid (correspond to columns with a lot of gaps in the associated multiple alignment).
- *D-states*: do not emit (i.e., they are silent).

Furthermore, transitions are included so that at each node either the M-state or the D-state is traversed exactly once. I-states also have a self-transition, allowing one or more inserted residues to occur between consensus columns. The general transition structure of a profile HMM with four nodes is shown in Figure 10.2. The transition structure in Figure 10.2 displays profile HMM for global alignments. To allow for other types of alignments (most notably local and multihit alignments) a more flexible HMM structure is required. Therefore, the popular HMMER software [7] uses the so-called *Plan7* architecture [8, 9] (see Figure 10.3). Plan7 extends the general profile HMM structure shown in Figure 10.2 as follows.

- Flanking of the linear sequence of nodes by a begin state (B) and an end state (E)
- Inclusion of the special states: S, N, C, T, and J

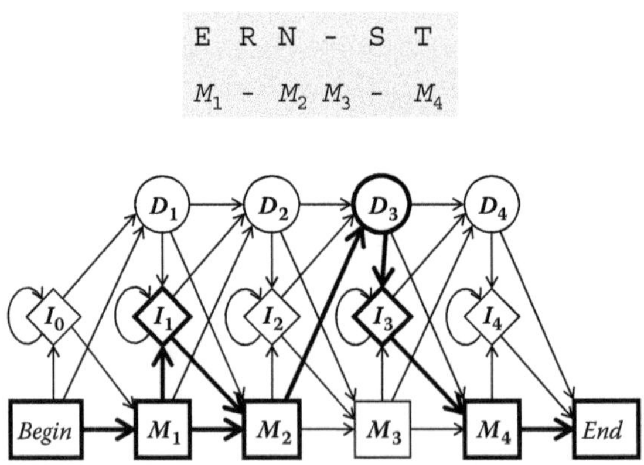

FIGURE 10.2
General transition structure of a profile HMM and a possible alignment of the protein sequence ERNST to the model.

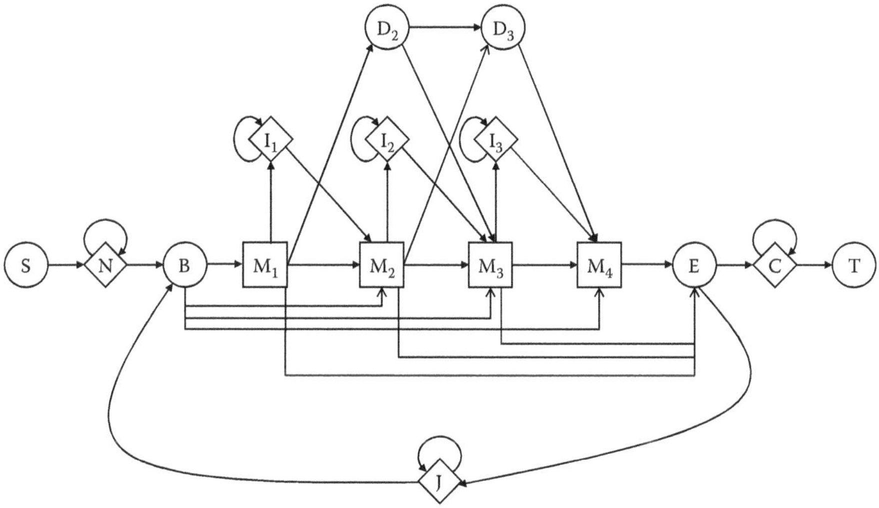

FIGURE 10.3
The Plan7 architecture for a profile HMM of length four.

These additions allow the control of alignment-specific features; for example, how likely the model is to generate various sorts of global, local, or even multihit alignments.

Many alignment algorithms (e.g., Smith–Waterman [10], basic local alignment search tool [BLAST] [11, 12], Needleman–Wunsch [13]) only use *position-independent* scoring parameters; that is, substitution matrix and gap penalties are fixed for all positions. Profile HMMs on the other hand capture *position-dependent* information; that is, amino-acid score and gap penalties can vary depending on the position in the associated multiple alignments. Consequently, databases containing a large number of profile HMMs are available that are applied extensively for genome analysis. The most popular database is Pfam [5], which covers common protein domains and families. The latest version at the time of writing (Pfam 24.0, October, 2009) contains 11,912 profile HMMs in Plan7 format. Construction and usage of Pfam is tightly coupled to the HMMER software package [7].

Profile HMMs can be used for two types of database search tasks. One task is to search a database of profile HMMs against a set of input query sequences. The other one is to search a sequence database for matches to an input profile HMM. For both cases, the similarity score *sim*(*H*,*S*) of a profile HMM *H* and a protein sequence *S* is used to rank all sequences/HMMs in the queried database. The highest ranked sequences/HMMs are then returned as the hits identified by the corresponding database search task.

The key to effective database searching is the accuracy of the similarity score *sim*(*H*,*S*). The similarity score can therefore be recast into finding the Viterbi score of the profile HMM *H* and the protein sequence *S*. The Viterbi score is defined as the most probable path through *H* that generates a sequence equal to *S*. The Viterbi dynamic programming (DP) algorithm for Plan7 profile HMMs is shown in Algorithm 1.

Algorithm 1 Plan7 Viterbi algorithm
Input: A profile HMM *H* of length *k* in Plan7 format (see Figure 10.3) and a protein sequence *S* of length *n*. We describe the profile HMM in terms of transitions between two states and emissions of amino acids at particular states. For example, *tr*(*State1*,*State2*) implies the transition score from *State1* to *State2*. Similarly, *e*(*State1*,*s*) implies the emission score from emitting *s* at *State1*.
Output: The similarity score, *sim*(H,S).

for $j = 1, k$ **do** $M(0,j) = I(0,j) = D(0,j) = -\infty$ **end for**
for $i = 1, n$ **do** $M(i,0) = I(i,0) = D(i,0) = -\infty$ **end for**
$XN(0) = 0$
$XB(0) = tr(N,B)$ {See Figure 10.3 for $tr(N,B)$}
$XE(0) = XJ(0) = XC(0) = -\infty$
for $i = 1, n$ **do**
for $j = 1, k$ **do**

$$M(i,j) = e(M_j, \mathbf{S}[i]) + \max \begin{cases} M(i-1, j-1) + tr(M_{j-1}, M_j) \\ I(i-1, j-1) + tr(I_{j-1}, M_j) \\ D(i-1, j-1) + tr(D_{j-1}, M_j) \\ XB(i-1) + tr(B, M_j) \end{cases}$$

$$I(i,j) = e(I_j, \mathbf{S}[i]) + \max \begin{cases} M(i-1, j) + tr(M_j, I_j) \\ I(i-1, j) + tr(I_j, I_j) \end{cases}$$

$$D(i,j) = \max \begin{cases} M(i, j-1) + tr(M_{j-1}, D_j) \\ D(i, j-1) + tr(D_{j-1}, D_j) \end{cases}$$

end for

$$XN(i) = XN(i-1) + tr(N, N)$$

$$XE(i) = \max_{1 \le j \le k}\{M(i,j) + tr(M_j, E)\}$$

$$XJ(i) = \max \begin{cases} XJ(i-1) + tr(J, J) \\ XE(i) + tr(E, J) \end{cases}$$

$$XB(i) = \max \begin{cases} XN(i) + tr(N, B) \\ XJ(i) + tr(J, B) \end{cases}$$

$$XC(i) = \max \begin{cases} XC(i-1) + tr(C, C) \\ XE(i) \end{cases}$$

end for
Return Final Score: $sim(H,S) = XC(n) + tr(C,T)$

In Algorithm 1, there are three two-dimensional matrices: *M*, *I*, and *D*. *M(i,j)* denotes the score of the best path emitting the subsequence *S*[1...*i*] of *S* ending with S[*i*] being emitted in state M_j. Similarly, *I(i,j)* is the score of the best path ending with S[*i*] being emitted in state I_j, and *D(i,j)* for the best path ending in state D_j. Furthermore, there are five one-dimensional matrices: *XN*, *XE*, *XJ*, *XB*, and *XC*. *XN(i)*, *XJ(i)*, and *XC(i)* denote the score of the best path emitting S[1...*i*] ending with S[*i*] being emitted in special state *N*, *J*, and *C*, respectively. *XE(i)* and *XB(i)* denote the score of the best path emitting S[1...*i*] ending in *E* and *B*, respectively. Finally, the score of the best path emitting the complete sequence *S* is determined by *XC(n)* + *tr(C,T)*. These matrices and their corresponding dependencies are also used for our parallel implementations in the following sections.

10.3 FPGA Parallelization and Results

10.3.1 System Design

Our first step in designing an FPGA system for databases searching with profile HMMs has been to analyze the data dependencies in the recurrence relations presented in the previous section. Direct and indirect data dependencies for computing the cell (*i,j*) in DP matrices *M*, *I*, and *D* are shown in Figure 10.4. The direct dependences for this cell requires the left, upper, and upper-left neighbor as well as *XB*(*i*–1). This leads to an indirect dependency on *XJ*(*i*–1), which in turn depends on *XE*(*i*–1). *XE*(*i*–1) then depends on

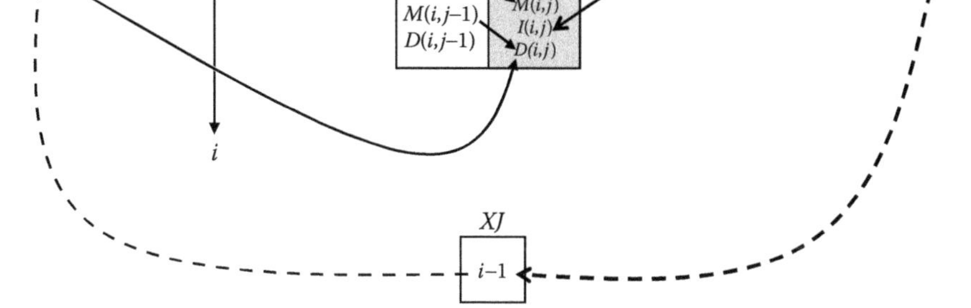

FIGURE 10.4
Data dependencies for computing the values *M(i,j)*, *I(i,j)*, and *D(i,j)* solid lines are used for direct (indirect) dependencies are represented by solid (dashed) lines.

all cells in row i–1 in matrix M. Thus, to satisfy all dependencies the two-dimensional matrices M, I, and D must be filled one cell at a time, in row-major order because of the feedback loop induced by the J state.

A typical strategy used when implementing the Viterbi algorithm in hardware (see e.g., [14, 15]) has therefore been to eliminate the J state. The advantage of this approach is that efficient parallelism can be achieved with an FPGA using a linear systolic array of identical simple processing elements (PEs). Unfortunately, this approach comes at the cost of the implementation's inability to find multihit alignments such as repeat matches of subsequences of S to subsections of H, which in turn can result in a severe loss of sensitivity (in particular for proteins with multiple domains). Therefore, our FPGA solution implements a full Plan7 model. The individual PE design is shown in Figure 10.5. It contains the following features:

- Registers to store the temporary DP matrix values $M(i-1,j-1)$, $I(i-1,j-1)$, $D(i-1,j-1)$, $M(i,j-1)$, $I(i,j-1)$, $D(i,j-1)$.
- $M(i,j)$, $I(i,j)$, and $D(i,j)$ are not stored explicitly, instead they are the inputs to the $M(i,j-1)$, $I(i,j-1)$, and $D(i,j-1)$ registers, respectively.
- Emission ($e(M_j,s_i)$ and $e(I_j,s_i)$) and transition probabilities ($tr(M_{j-1},M_j)$, $tr(I_{j-1},M_j)$, $tr(D_{j-1},M_j)$, $tr(I_j,M_j)$, $tr(I_j,M_j)$, $tr(M_{j-1},D_j)$, $tr(D_{j-1},M_j)$, and $tr(M_j,E)$) are read from the internal FPGA RAM (Block RAM).
- Transition probabilities ($tr(B,M_j)$, $tr(N,N)$, $tr(E,J)$, $tr(J,J)$, $tr(J,B)$, $tr(N,B)$, $tr(C,C)$, and $tr(C,T)$) are stored in registers.
- The PE has a four stage pipeline: *Fetch*, *Comp1*, *Comp2*, and *Store*. In *Fetch*, transition, emissions, and intermediate DP matrix values are read from the Block RAM. All necessary computations are performed in the two compute stages *Comp1* and *Comp2*. Results are written to the Block RAM in *Store*. Computation of the special state matrices uses intermediate values for $XE(i)$ that are computed according to Equation 10.1.

$$XE(i,j) = \max\{XE(i,j-1), M(i,j) + tr(M_j,E))\} \tag{10.1}$$

- Updating of XN, XJ, XB, and XC is only performed at the end of the DP matrix row; that is, if $j = k$.

The PE design has the following implementation details:

- Numbers are represented in 2's complement form.
- Adders use saturation arithmetic.
- Number representation uses two tags to encode special cases: number (00), +max (01), −max (10), and not-a-number (NaN) (11). Adders and max-circuits take advantage of theses tags to compute special

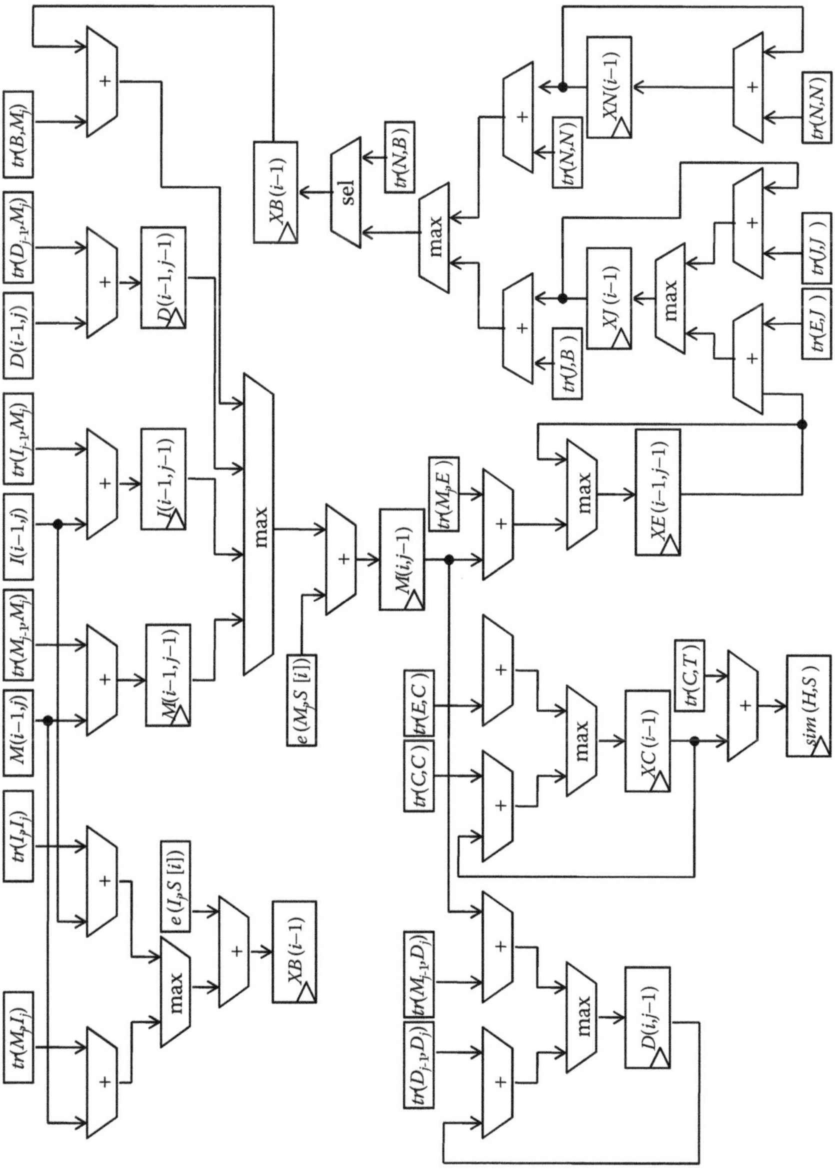

FIGURE 10.5
HMM processing element (PE) design.

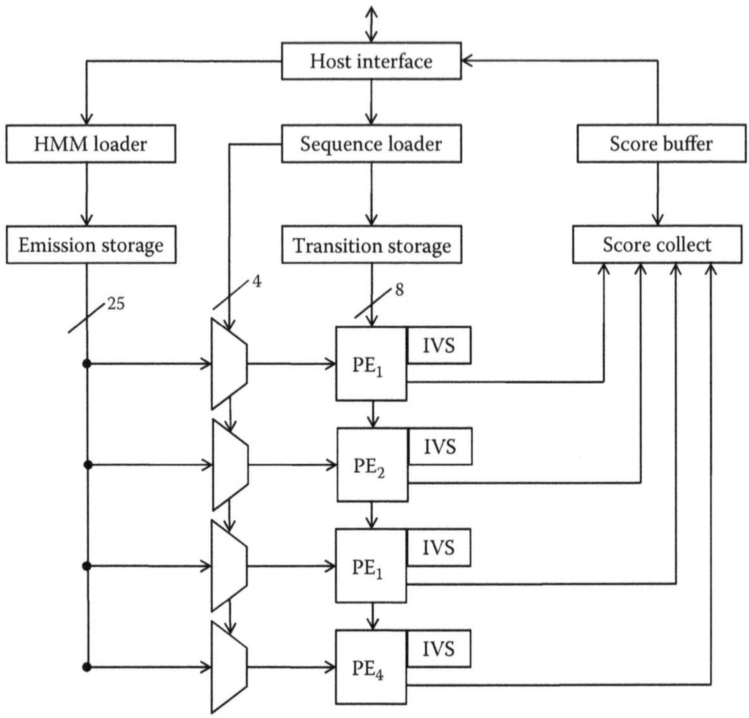

FIGURE 10.6
HMM system design for four PEs.

cases in a very simple and efficient way (e.g., if any of the operand's tags are set in an addition, a simple bit-wise OR operation suffices to compute the result).

Our system design exploits parallelism by aligning different query/subject pairs with the Plan7 Viterbi algorithm independently in separate PEs. Figure 10.6 shows the design for four PEs.

The following features are used in the design:

- *Intermediate value storage* (IVS): Each PE has an IVS that stores one row of previously computed results of the matrices *M*, *I*, and *D*.
- *Emission and transition storage*: We assume that the same profile HMM has to be aligned to different protein sequences. Therefore, PEs are synchronized to process the same HMM state in each cycle, which reduces the bandwidth requirement to access the transition storage to a single state.
- *Score collect and score buffer*: These units are designed to handle cases where PEs produce results in the same clock cycle.

- *HMM loader*: The task of this unit is to transfer emission and transition values into their respective storage.
- *Sequence loader*: Sequence elements are fetched from external memory to the sequence loader. It then forwards them to the emission selection multiplexers.
- *Host interface*: The system is connected to a host via an universal serial bus (USB) interface.

Loading and storing of data to/from the FPGA and postprocessing of relevant hits is performed by the host software. The host portion is described in Algorithm 2. The FPGA is used as a first-pass filter; that is, large database chunks are quickly scanned and narrowed down to a few interesting hits. These hits are then processed on the central processing unit (CPU) host.

```
Algorithm 2 FPGA integration for HMM-based database searching.
Input: T (threshold), E (cutoff value), HMM array hmm[],
Protein sequence array seq[]
Output: Top matches
for all Current HMMs, hmm[j] do
        repeat
       Pack_and_score()
       for all Current sequences, seq[i] do
               FPGA_score = score[i]
               if (FPGA_score ≥ T) and (e-value ≤ E) then
                    Software_score = P7Viterbi(seq[i], H)
               if (Software_score ≥ T)and (e-value ≤ E)then
                               PostprocessSignificantHit(seq[i])
       until No More Sequences
until No More HMMs
```

10.3.2 Performance Evaluation

Our PE design has been described in the Verilog HDL. To investigate the effect of the amount of logic slices and memory on the scalability of our design, we have targeted it to two members of the Xilinx Spartan-3 family: XC3S1500 and XC3S4000. The amount of available logic slices and Block RAMs are 13,312 and 32 for the XC3S1500 and 27,648 and 96 for the XC3S4000, respectively. The achieved size of a single PE is 451 logic slices using Xilinx ISE tools for synthesis, mapping, placement, and routing.

Rather than the amount of logic slices, memory is the crucial resource determining the number of PEs. The amount of memory required is

- Fifty RAM entries per HMM state, comprising 42 emissions and 8 transitions
- Three entries per HMM state for each PE's IVS

Thus, the overall amount of Block RAM entries required is $50 \cdot k + 3 \cdot k \cdot N$, where k is the HMM length and N is the number of PEs. Therefore, the maximum number of PEs that we are able to fit onto an FPGA depends on the HMM lengths. The largest power-of-two HMM lengths we are able to support on an XC3S1500 and XC3S400 are $k = 256$ and $k = 1{,}024$, respectively. In both cases the number of PEs is limited by the number of Block RAM in the targeted FPGA. The number of PEs can therefore be increased for shorter HMM lengths; for example, for $k = 512$ it is possible to fit 30 PEs on an XC3S4000. Further improvement over the 512-state version is then limited by logic slices on the XC3S4000; for example, for 256 states the maximal PE number is still 30. The results can be summarized as follows:

- *XC3S1500:* clock frequency = 70 MHz; number of PEs = 10; maximal supported HMM = 256; theoretical peak performance = 10 PEs × 70 MHz = 700 MCUPS (million cell updates per second)
- *XC3S4000:* clock frequency = 70 MHz; cumber of PEs = 30; maximal supported HMM length = 512; theoretical peak performance = 30 PEs × 70 MHz = 2.100 GCUPS (billion cell updates per second)

We have implemented the FPGA integration as described in Algorithm 2. The acceleration board used for this study is a very low-cost Spartan-3 XC3S1500 board with 64 MB SDRAM and USB 2.0 interface. Figure 10.7 shows the achieved speedups for searching a sequence database with a query profile HMM on an FPGA. Furthermore, Figure 10.8 shows the speedups

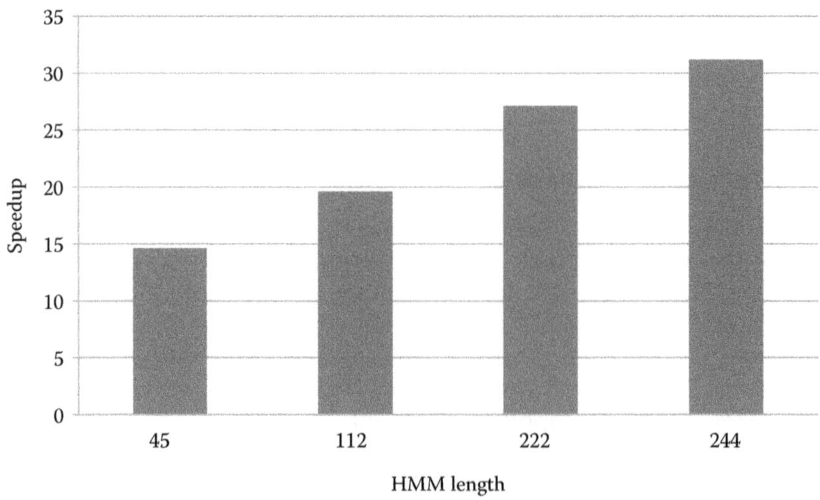

FIGURE 10.7
Speedups for searching a sequence database with profile HMMs of different lengths on a Spartan-3 XC3S1500 board compared to the sequential HMMER 2.3.2 software (*hmmsearch*). The utilized database contains 643,552 protein sequences.

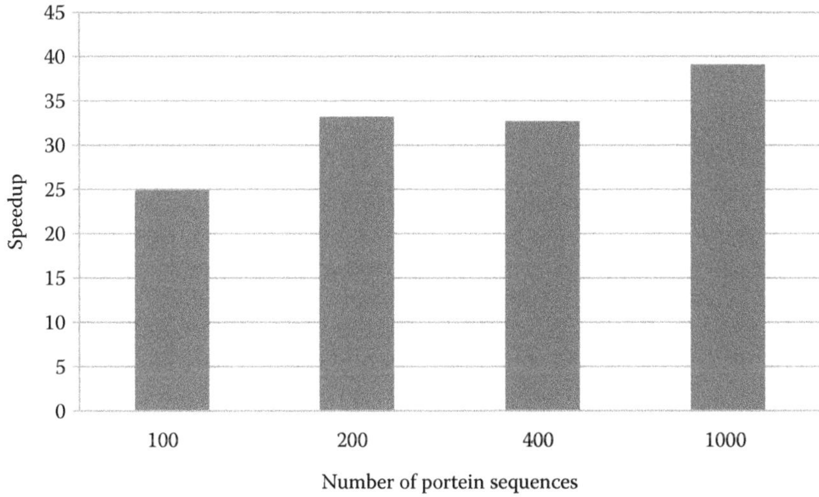

FIGURE 10.8
Speedups for searching an HMM database with a varying number of query protein sequences on a Spartan-3 XC3S1500 board compared to the sequential HMMER 2.3.2 software (*hmmpfam*). A subset of the superfamily database consisting of 1,554 HMMs is used as HMM database.

for searching a profile HMM database with a number of query sequences. Measured timings include data transfer, initialization, and pre- and postprocessing. An AMD Athlon 64 3500+ is used as a host machine. The speedups of the FPGA are compared with the nonaccelerated sequential version of the HMMER 2.3.2 package running on the same PC.

Examining the speedups we can see the effect of the number of states within an HMM and number of protein sequences on the FPGA implementation as compared with the software-only implementation. The speedup generally increases with a larger number of states and sequences. This is to be expected as the software implementation of the Viterbi algorithm does not improve efficiency with a larger number of states or larger number of sequences. However, in case of the FPGA, the greater number of states results in more effective use of the resources, while the larger number of sequences reduces the impact of data transfer overheads. Thus, the FPGA is able to reach an efficiency of up to 94% of the theoretical peak performance stated earlier for large HMMs (Figure 10.8).

10.4 GPU Parallelization and Results

General-purpose programmable GPUs have recently become popular targets for highly parallel applications, including HMM database searching.

In this section we describe our implementation and the performance of a GPU-enabled Viterbi algorithm for HMM database searches. We begin with a description of the GPU hardware that our solution is built on.

10.4.1 CUDA Hardware

Computing with GPUs presents unique challenges and limitations that must be addressed to achieve high performance. Here we describe the NVIDIA 8,800-based GPU that is used in our tests and also explain the unique features of the GPU that presents challenges to efficient programming.

The graphics processors used in our tests are NVIDIA 8,800 GTX Ultra GPUs with 768 MB RAM. The 8,800 GTX Ultra is composed of 16 stream multiprocessors, each of which is itself composed of 8 stream processors for a total of 128 stream processors. Each multiprocessor has 8,192 32-bit registers, which in practice limits the number of threads (and therefore, performance) of the GPU kernel. The GPU is programmed using NVIDIA's CUDA programming model [16]. Each multiprocessor can manage 768 active threads. Threads are partitioned into thread blocks of up to 512 threads each, and thread blocks are further partitioned into groups of 32 threads (called a *warp*). Each warp is executed by a single multiprocessor. Warps are not user controlled or assignable, but rather are automatically partitioned from user-defined blocks. At any given clock cycle, an individual multiprocessor (and its stream processors) executes the same instruction on all threads of a warp. Consequently, each multiprocessor should most accurately be thought of as a single-instruction multiple-data (SIMD) processor.

Programming the GPU is not a matter of simply mapping a single thread to a single stream processor. Rather, with 8,192 registers per multiprocessor, hundreds of threads per multiprocessor and thousands of threads per board should be used to fully utilize the GPU. Memory access patterns, in particular, must be carefully studied to minimize the number of global memory reads. Where possible, an application should make use of the 16 KB of shared memory per multiprocessor, as well as the texture and 64-KB constant memory, to minimize GPU kernel access to global memory. When global memory must be accessed, it is essential that memory be both properly aligned and laid out such that each SIMD thread accesses consecutive array elements to combine memory reads into larger 384-bit reads.

10.4.2 Results

The C code of HMMER's Viterbi algorithm was ported to CUDA with a variety of performance optimizations. The kernel operates on multiple sequences simultaneously, with each thread operating on a unique sequence. The number of threads that can be executed in parallel is limited by two factors: (1) GPU memory will limit the number of sequences that can be stored, and (2) the number of registers used by each thread will limit the number of threads that can run in parallel. In our implementation, register use is the most prohibitive resource.

In the remainder of this section we describe the optimizations made to the GPU kernel. We consider a variety of optimizations in our implementation including database-level load balancing, memory layout and coalescing, loop unrolling, and shared/constant memory use. Results are shown with a variety of HMMs of increasing length.

We test HMMs of length 77, 209, 456, 789, and 1,431 states. All HMMs except the 77-state HMM were taken directly from the Pfam database, while the 77-state HMM is distributed with the HMMER source. We note that the average length of an HMM within the Pfam database is 209 states. All tests are taken against the publicly available NCBI nonredundant database NCBI NR [17]. The 3-GByte NR database used in these tests consists of more than 5.5 million sequences with sequence lengths varying from 6 to 37,000 amino acids.

10.4.2.1 Database Sorting

HMMER's Viterbi function is sensitive to both the length of the query HMM and the length of an individual sequence. CUDA provides limited support for thread synchronization; a barrier synchronization function is provided that returns only when all threads have finished executing `cudaThreadSynchronize()`. In our implementation, 3,072 threads are run in parallel on a single GPU, with each thread operating on its own sequence. A typical database is unordered, placing short sequences in close vicinity to long sequences. On a CUDA-enabled GPU this results in threads operating on the shorter sequences completing early and being forced to wait for the thread computing the longest sequence in the current batch before the barrier synchronization completes. The solution is to presort the sequence database by length, thereby balancing a similar load over all 3,072 threads participating in the computation. This has the advantage of being both effective and quite straightforward as we are able to achieve a nearly 7× performance improvement over the unsorted database without changing the GPU kernel in any way. For the database used in these experiments, only 262.36 seconds were required for sorting. Further, the sorted database can be reused for the entire useful life of the database, making the one-time cost of sorting it negligible.

10.4.2.2 Memory Layout Optimizations

The most effective optimization to the Viterbi is from optimizing memory layout and usage patterns within HMMER's Viterbi algorithm. Because the CUDA environment does not allow threads to dynamically allocate GPU memory, all memory allocations (even those allocating the GPU's on-board memory) must be performed by the host system and copied to the GPU before instantiating the kernel. By default, the Viterbi function requires integer arrays of size $3 \cdot m \cdot l + 5 \cdot l$, where m and l are the length of the HMM and sequence, respectively. For large HMMs and large sequences, this can easily result in several megabytes of data per thread. With only 768 MB memory for 4,096 threads, this can quickly exhaust the GPU's memory.

Through careful optimization we are able to reduce the memory requirements of the Viterbi scoring computation to $6 \cdot m + 10$ integer array elements. This was accomplished by noting that the Viterbi algorithm described in Section 10.2 largely requires only the current and previous rows of the DP matrices M, I, and D over the length of the inner-most loop, m (the number of HMM states). Thus, M, I, and D contribute $6 \cdot m$ array elements.

Reducing the memory footprint means that we can no longer perform the Viterbi trace-back procedure. Fortunately, the trace-back is only needed when a database hit is made. In our tests less than 2% of the database entries result in hits, so we simply perform a full software Viterbi, including trace-back, on all database hits. We also exploit three opportunities within the core Viterbi loop to reuse intermediate values within registers, rather than repeatedly writing/reading from the GPU's DRAM. Specifically, the current and previous values of the M, I, and D matrices may be reused in subsequent iterations through the Viterbi loop.

10.4.2.3 Memory Hierarchy Optimizations

Finally, we also include the use of the shared and constant memories. We note that the HMM stays constant throughout the entire computation and is used by each thread for each sequence. In most cases we can fit the entirety of the core transition matrices (denoted as $tr(M_{j-1}, M_j)$, $tr(I_{j-1}, M_j)$, $tr(D_{j-1}, M_j)$, and $tr(B, M_j)$ in Section 10.2) into the 64-KB constant memory. In cases where the size of the HMM exceeds the amount of constant memory, we utilize the full constant memory before switching over to texture memory for the remaining portions of the HMM.

Further, we use shared memory to temporarily store our index into each thread's digitized sequence that is referenced repeatedly throughout the core Viterbi loop. As a consequence, we are able to reduce the number of texture reads to two per iteration (four if the loop is unrolled).

In Figure 10.9 we present the results of our final GPU kernel. As Figure 10.9 shows, we are able to achieve between 12× and 37× speedup, depending on the size of the HMM. We note that the largest HMM (size 1,431) runs for more than 1 day before completion (serial time). This results in a much higher speedup as the vast majority of the CUDA runtime is spent on the GPU. For the same reason, the 77-state HMM results in much lower speedup as more of its time is spent reading from the sequence database and postprocessing. In fact, between postprocessing, database reading, and DMA transfers to the GPU, the 77-state HMM spends twice as much time outside of the GPU kernel as within the GPU kernel.

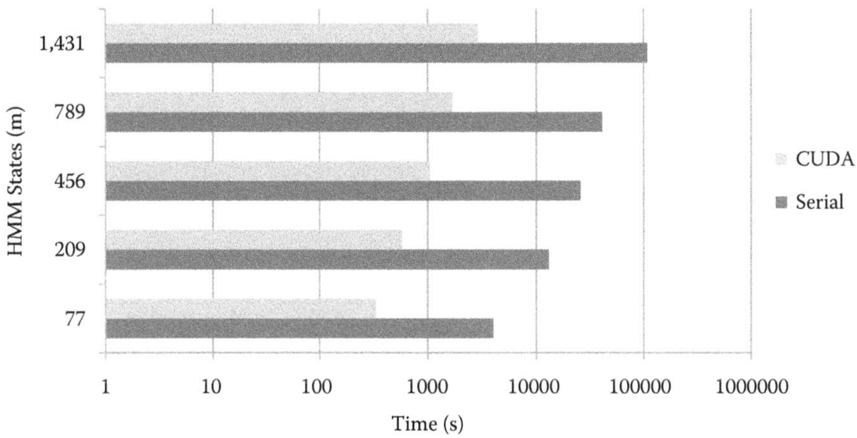

FIGURE 10.9
Runtime of final Viterbi kernel without host optimizations.

10.4.2.4 Host Optimizations

To compensate for such overhead, particularly for small or average-sized HMMs, we performed several host-side optimizations. Specifically, the serial overhead of reading the database and postprocessing the database hits between GPU kernel invocations was addressed. To do so we created two threads: the first for database reading, and the second to postprocess database hits. We noted that the 8,800 GTX GPU did not permit us to overlap DMA operations to the GPU during kernel executions. However, current hardware is capable of such optimizations.

In Figure 10.10 we compare our final implementation to the existing ClawHMMER GPU implementation by Horn et al. [18]. Because ClawHMMER runs within Windows XP, we were forced to use a smaller version of the NR database as well as smaller HMMs to stay within the Windows XP 2GB memory limit. Nevertheless, as we show, our implementation substantially outperformed the ClawHMMER implementation for every tested HMM. Moreover, as the size of the HMM increased, the performance of our CUDA implementation increased relative to the ClawHMMER implementation.

In Figure 10.11 we present our final performance results for the full NR database and a variety of HMMs of increasing size. Here the benefits of the host-side optimizations become clear—the 77-state HMM, for example, now achieved a performance of 19×, compared to the 12× previously achieved. Further, both the 209-state and 456-state HMMs also improved in performance, from 22.5× and 24.6× to 28× and 27× for the 209- and 456-state HMMs, respectively. The 789- and 1,431-state HMMs improved slightly, from 24× to 26×, and from 37× to 38.6×. Their improvement was less dramatic as their runtimes dwarfed the serial portions of the computation.

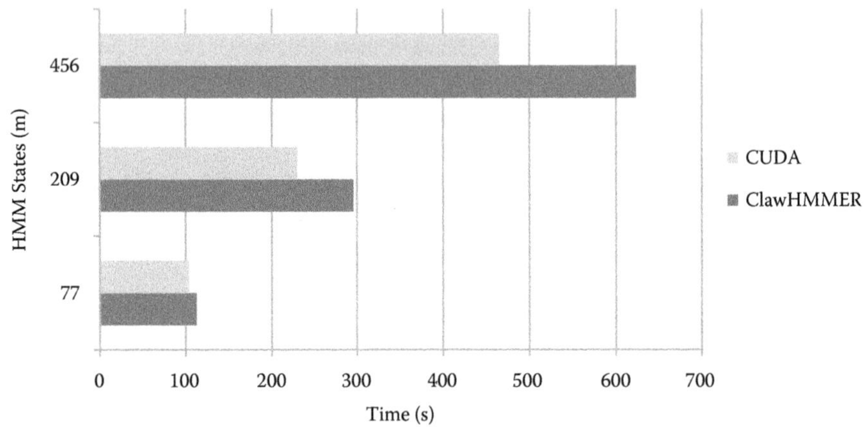

FIGURE 10.10
Comparison of our final implementation to ClawHMMER.

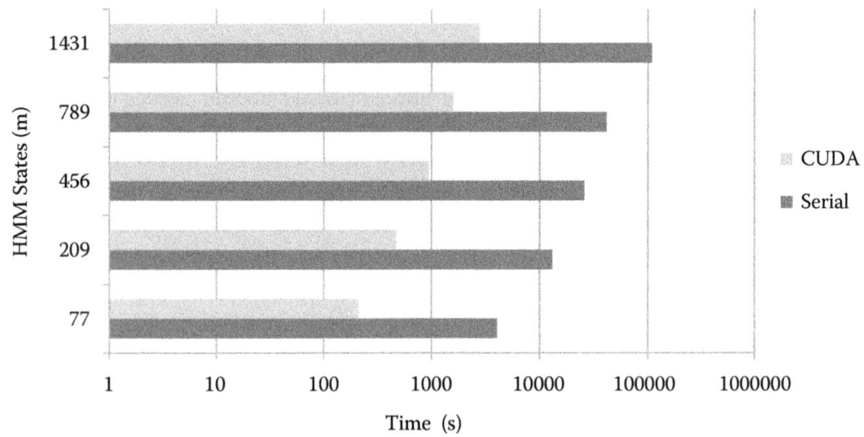

FIGURE 10.11
Runtime and speedup of final Viterbi kernel including host optimizations.

10.5 Discussion

In this chapter we have shown how GPUs and FPGAs can be efficiently used to accelerate HMM-based database searching. Both architectures used the same parallelization approach (i.e., running different HMM/sequence comparisons in parallel) and achieved similar speedups of around 30× on low-end hardware platforms.

An advantage of the GPU-based solution is the more convenient programmability with CUDA as well as the likely portability to newer runtime environments such as OpenCL. Further, the widespread availability of CUDA-enabled GPUs makes the adoption of GPU-based solutions easy and exceptionally low cost.

On the downside, the utilized graphics card has significantly higher power consumption than the USB-based FPGA solution. Moreover, USB-based solutions are easily scalable within a single machine. Indeed modern desktop PCs are able to easily accommodate many such USB devices. GPUs, however, are currently limited by the number of PCIEx16 slots that are available on the host system's motherboard. Finally, the GPU-HMMER solution does not currently implement the Pfam search functionality included within the HMMER distribution. The main obstacle to its implementation is that hmmpfam is known to I/O-bound rather than compute-bound. This, along with the added memory consumption of thousands of HMMs executing in parallel on the GPU, make hmmfpam a challenge. We are currently investigating alternative strategies to enable Pfam searches while keeping memory consumption low.

Therefore, it would be interesting to compare both approaches on higher-end systems. Extending these single node solutions to multiple independently executing nodes (via MPI, for example) would also prove instructive. Not only would this expose issues such as host bandwidth, but it also would likely expose load-balancing issues that may not otherwise be visible. Finally we intend to examine how these parallel architectures may be used for accelerating other HMM-based search methods, such as the already announced HMMER3 tool [19].

10.6 References

1. Haussler, D., Krogh, A., Mian, I. S., Sjölander, K.: Protein modeling using hidden Markov models: Analysis of globins. In: *Proceedings of the Hawaii International Conference on System Sciences,* volume 1 pp. 792–802, Los Alamitos, CA: IEEE Computer Society Press (1993).
2. Krogh, A., Brown, M., Mian, S., Sjolander, K., Hausler, D.: Hidden Markov Models in computational biology: Applications to protein modeling, *Journal of Molecular Biology* 235, 1501–1531 (1994).
3. Viterbi, A.J.: Error bounds for convolutional codes and an asymptotically optimum decoding algorithm, *IEEE Transactions on Information Theory* 13, 2, 260–269 (1967).
4. Yooseph, S., et al.: The *Sorcerer II* global ocean sampling expedition: Expanding the universe of protein families, *PLoS Biology* 5(3), e16 March (2007).
5. Finn, R.D., et al.: The PFAM protein families database, *Nucleic Acid Research* 36, D281–D288 (2008).
6. Haft, D.H., Selengut, J.D., White, O.: The TIGRFAMs database of protein families, *Nucleic Acids Research* 31, 371–373 (2003).
7. Eddy, S.R.: HMMER: Profile HMMs for protein sequence analysis. http://hmmer.janelia.org, Accessed February 16, 2010 (2009).

8. Eddy, S.R.: Profile hidden Markov models, *Bioinformatics* 14, 755–763 (1998).
9 Durbin, R., Eddy, S., Krogh, A., Mitchison, G.: *Biological Sequence Analysis, Probabilistic Models of Proteins and Nucleic Acids,* Cambridge University Press, (1998).
10. Smith, T.F., Waterman, M.S.: Identification of common molecular subsequences, *Journal of Molecular Biology* 147, 195–197 (1981).
11. Altschul, S.F., Gish, W., Miller, W., Myers, E.W., Lipman, D.J.: Basic local alignment search tool, *Journal of Molecular Biology* 215, 403–410 (1990).
12. Altschul, S.F., Madden, T.L., Schaffer, A.A., Zhang, J., Zhang, Z., Miller, W., Lipman, D.J.: Gapped BLAST and PSI-BLAST: A new generation of protein database search programs, *Nucleic Acids Research* 25, 17, 3389–3402 (1997).
13. Needleman, S., Wunsch, C.: A general method applicable to the search for similarities in the amino acid sequence of two sequences, *Journal of Molecular Biology* 48, 3, (1970).
14. Maddimsetty, R.P., Buhler, J., Chamberlain, R., Franklin, M., Harris, B.: Accelerator design for protein sequence HMM search, *Proc. 20th ACM International Conference on Supercomputing (ICS06),* 288–296, (2006).
15. Oliver, T.F., Schmidt, B., Yanto, J., Maskell, D.L.: Accelerating the Viterbi algorithm for profile hidden Markov models using reconfigurable hardware, *Lecture Notes in Computer Science*, Springer, Vol. 3991, 522–529 (2006).
16. NVIDIA Corporation: Compute Unified Device Architecture (CUDA) Programming Guide. NVIDIA, 1.0 edition (2007).
17. NCBI. The NR (non-redundant) database. ftp://ftp.ncbi.nih.gov/blast/db/FASTA/nr.gz, Accessed February 16, 2010 (2009).
18. Horn, D.R., Houston, M., Hanrahan, P.: ClawHMMER: A streaming HMMer-Search implementation, *ACM/IEEE Conference on Supercomputing* (2005).
19. Eddy, S.R.: A probabilistic model of local sequence alignment that simplifies statistical significance estimation, *PLOS Computational Biology* 4, e1000069 (2008).

11

COPACOBANA: A Massively Parallel FPGA-Based Computer Architecture

Manfred Schimmler, Lars Wienbrandt, Tim Güneysu, and Jost Bissel

11.1 Introduction

11.1.1 History of Complexity

Several complexity measures have been used to evaluate the quality of algorithms running of different computer architectures. The dominating measures for sequential computers used to be the required computing time T and memory space S [1]. With the idea of parallel computations the number N of processors became an additional important complexity measure. Over the years, since about 1980, compute-intensive parallel algorithms have been implemented as full-custom or semicustom very-large-scale integration (VLSI) chips. Since then the chip area A became one of the important measures for such implementations. Combinations like AT and AT^2 were widely used [2] to evaluate the quality of VLSI algorithms. Lower bounds for the time complexity of parallel algorithms could be proven by means of these combinations [3]. In the last 10 years another complexity measure became dominant: power consumption P. As heat dissipation is one of the major problems of modern high-performance computer systems, the power is often the limiting parameter for computational performance [4].

One conclusion from this historical observation could be drawn. Any time computations were evaluated by the computing time T and some second measure that was equivalent to an amount of money: the memory space could be increased by money, additional processors were a matter of investment, and the cost of chip area and power is also a question of how much money one is willing to spend. Therefore, a useful complexity measure for the performance of compute-intensive algorithms is the product of time T and cost C [5]. It can be discussed whether T should be weighted more than C or if the T^2C measure should be taken straight [6]. Of course, a parallel system reduces T. Thus, it would benefit from such a modified measure. For the purpose of this chapter we keep in mind that we are heading for solutions optimized with respect to the TC measure and among those we prefer the faster ones.

If we consider compute-intensive applications that can be partitioned into a large number of parallel threads, it is obvious to think about a computer architecture optimizing the TC product. This approach leads to a field-programmable gate array (FPGA)-based parallel system. FPGAs are chips that are configurable to meet the requirements for the desired application. Here we are interested in reconfigurable FPGAs; that is, chips whose configuration is performed by writing into static random access memory (SRAM) components integrated on the chip. By rewriting the SRAM, the FPGA is reconfigured; that is, it gets a different functionality.

FPGAs are very well suitable for implementation of fine-grained parallel algorithms because a large number of processing elements (PEs) that are tailored toward the desired application can be fitted onto one single FPGA [7–10]. By using FPGAs the hardware costs can be optimized. In contrast to standard processor-based systems there are no (costs for) components being inactive during the whole computation. Furthermore, FPGAs are extremely power efficient. Current has to flow only when the computation requires switching events.

11.1.2 Basic Idea of the COPACOBANA Series

The initial idea to develop cost optimal parallel code breaker (COPACOBANA) goes back to 2004 with the intent to build some low-cost hardware (less than US$ 10,000) that is able to break the 56-bit data encryption standard (DES) within 30 days. The choice was an FPGA-based parallel architecture optimized with respect to cost and time performance. It was to consist of a large number of low-cost FPGAs with some interconnection network able to deliver operand data and to transfer result data to the host system. The best individual price-performance ratio in 2004 had been provided by the contemporary Spartan3-1000 of Xilinx. It is an integrated circuit with a number of around one million system gates, 17,280 equivalent logic cells, and 1,920 configurable logic blocks (CLBs) equivalent to 7,680 slices with a total 15,360 4-input-lookup tables (LUTs) for the price of € 40.

For the interconnection and for data paths from and to the FPGAs the dominating requirements were simplicity and low costs. The same held for power supply and global signals for all FPGAs. Thus, as the desired application did not need any interprocess communication, the choice was to design a simple single-master multiple-slave backplane bus with a shared medium and broadcast ability. Furthermore, the speed of the bus did not need to be high due to the fact that computations heavily dominate communication requirements. Hence, a single conventional host computer was sufficient to provide the required data packets and fetch the results.

The final decision on the numbers of FPGAs to be taken was dependent on the time boundary of 30 days for DES cracking. The solution was to take about 120 FPGAs. For the purpose of maintenance and occupancy of space, six FPGAs each were placed on an own small printed circuit board

(PCB), leading to 20 boards in total. All FPGAs were connected through an 64-bit shared bus with additional 8 bits to address an FPGA, 2 bits to signalize read-write operations or configuration and 5 bits for optional register addressing. One extra bit was needed for the bus clock and 8 bits of the 64 data bits were double bound for the configuration data. The bus was clocked with 33 MHz leading to a theoretical bandwidth of 2 Gbit/s, which in fact was never reached, but definitely more than enough. The bottleneck was a simple USB1.0 controller module acting as bus master and interface to the host computer. Later, the frequency was even set to 20 MHz and the controller replaced with a Transmission control protocol/ Internet protocol (TCP/IP) interface via Ethernet to gain more stability and flexibility.

For testing the FPGAs, the controller, and the bus a simple "MemoryTest" application was developed. The configuration code for the FPGAs contains an implementation of only 32 registers, accessed by the five register addressing bits. On a write operation the FPGA took the data from the bus and stored it in the addressed register while it responded to a read operation with the content of the addressed register. The host application now sent some data to every register first. Afterward it fetched the data and compared it to what it sent before.

After many hardware patches and extensive testing finally the first prototype of the COPACOBANA 1000 was running with the DES breaking application in 2006.

11.2 COPACOBANA 1000

COPACOBANA 1000 is a massively parallel reconfigurable architecture. As cryptanalytical applications, which COPACOBANA was intentionally developed for, need plenty of computing power, a total of 120 low-cost FPGAs are installed. COPACOBANA 1000 fits in standard 19 inches server racks using only 2 height units (2HE). The measures are 13 × 45 × 84 cm, and it weighs less than 18 kg. For rack enclosure COPACOBANA 1000 has front-to-back cooling. Power consumption is only 600 W on full load. The main components of COPACOBANA are the FPGA modules, the backplane providing the interconnection of the modules with a shared bus medium, and the controller card acting as interface to a host computer.

The first prototype of COPACOBANA 1000 was taken into service in 2006. Already in 2007, a small first series of 15 pieces was fabricated and delivered to research institutions all over the world to find out what kind of applications could be implemented on such a machine. Furthermore, this was to find out if VLSI designers were able to easily run configurations on COPACOBANA.

FIGURE 11.1
COPACOBANA 1000.

Although the next version of the COPACOBANA series, the COPACOBANA 5000, is already available, COPACOBANA 1000 is still successfully sold by SciEngines [11] because of its robustness, reliability, and easiness of use. Figure 11.1 depicts a photo of the COPACOBANA 1000 machine.

11.2.1 FPGA Module

There is a tradeoff between performance and price in the design of these PCBs: For higher performance, the number of FPGAs per PCB should be maximized. But larger PCBs are more expensive:

- The number of layers of conductors increases with the number of FPGAs.
- The price for the PCB rises approximately quadratic with the size.
- More FPGAs per PCB require more expensive power modules per PCB.
- It is cheaper to replace a small PCB in case of fabrication faults.
- The width of the PCB determines the size of the box of the whole machine.

As regards these assumptions, it turned out that a small PCB with six FPGAs provided the best price-performance ratio.

For an easy and stable mounting having enough pins available, but occupying as little space as possible, the dual in-line memory module (DIMM) connection seems to be the most appropriate. This is also working very well in standard PCs for mounting random access memory (RAM) modules on the mainboard. The backplane has to provide the DIMM slots whereas no additional connector is required on the FPGA modules. The PCB simply has to fit the form of the connector. With this type of connection it is easy to remove or exchange FPGA modules from the backplane, for example, for maintenance purposes.

In addition to the FPGAs the module needs bidirectional bus transceivers to forward data, address, and clock signals. The address decoding has already been done on the backplane, so every FPGA now needs only a selection signal instead of slot and FPGA address. The register address still needs to be completely available.

To get a system clock on the FPGA no oscillators are found on the FPGA module. It is the bus clock that simply acts as input for the digital clock manager (DCM). This has to be configured together with the FPGA to modify the bus clock for appropriate system clocks.

The whole system is powered by 3.3 V. However, the FPGA core needs 1.2 V. Hence, a power converter to supply the core voltage to all FPGAs on the PCB is also mounted. Each FPGA-card has a maximal power dissipation of 20 W. For cooling purposes the FPGA-cards are adjusted in a vertical way such that the air can be transported through every two adjacent FPGA-cards. Nine fans are responsible for this air movement. Five of them are located in the front panel. They blow the air into the box. The remaining four are in the back wall of the box. They suck the warm air out. The inside temperature is thus kept below 68°C. A picture of the FPGA module is shown in Figure 11.2.

11.2.2 Backplane

The functional behavior of the backplane is the interconnection between the FPGA modules and the interface controller card. It provides the shared medium of the bus as simple connections of the corresponding data pins. One exception is made for the controller slot. The bidirectional bus transceivers are mounted on the backplane instead of the module to obtain flexibility in easier substitution of the controller card.

The backplane provides 20 DIMM slots for FPGA modules and one slot for the controller card. For historical reasons, the slot for the controller card is not a DIMM slot, like the one for the FPGA modules, but a simple 96-pin DIN 41612 connector. This decision was made after choosing the CESYS USB2FPGA development board for the first version of the controller. It comes with an easy-to-use universal serial bus (USB) interface, fits in size, and

FIGURE 11.2
FPGA module of COPACOBANA 1000.

already has the DIN 41612 plug for general-purpose input/output (IO). The backplane has no oscillators to provide the bus clock. This has to be done by the bus master to assure a minimal clock skew. On the backplane, like on the FPGA modules, the clock is handled like any other bus signal and also simply forwarded by the bus transceivers.

One more important task of the backplane is the decoding of the slot and the FPGA address. A decoder GAL is located next to every module slot. Each decoder allocates six selection signals, one for each FPGA on the corresponding module. A selection signal is set to high only if the appropriate address is latched at the input of the decoder.

Power supply has been one of the major problems in the design of the backplane of COPACOBANA 1000. Because each of the 20 FPGA modules has a peak requirement of 20 W there is an overall power consumption of 400 W. With a voltage of 3.3 V it means the maximal current is 120 A. It is impossible to lead currents of that magnitude through one layer of a PCB. Therefore, four extra power rails have been installed for the supply voltage of the FPGA modules.

11.2.3 Interface Controller

The interconnection between the host system and the backplane is an interface controller board. The first version of the COPACOBANA 1000 used the already mentioned CESYS USB2FPGA development board providing an interface via USB. The programming interface on the host side was a simple application programming interface (API) for using the USB driver of the CESYS board written in the language C. Later, this was made more comfortable. The USB board vanished for a TCP/IP interface via Ethernet. COPACOBANA could now be located somewhere in the local network and connected to/from every PC in this network. In addition, the new API is written in Java, providing an easy-to-use and platform-independent interface for communicating with COPACOBANA.

The interface module is a Memec Virtex-4 FX12 Mini Module by Avnet [12] connected to an adapter card for the DIN 41612 connector. This way the controller module can simply be replaced in every COPACOBANA without changing other parts of the hardware. The Memec module contains a Virtex-4 FPGA with an integrated PowerPC processor running a TCP/IP server. Data packets from a TCP/IP connection are translated for an inbuilt hardware entity representing the bus master that serves the COPACOBANA bus. Even the bus clock is generated here.

Another benefit from the new controller is the ability to store some data. It provides a little memory of approximately 64 MB of which 32 MB could be freely accessed. This can speedup applications where the same data is needed very often by the FPGAs. It is accessible very fast by the bus master and bypasses the bottleneck of the slower TCP/IP connection.

Although the connection can be made with 100 Mbit/s or even Gigabit Ethernet the achieved bandwidth is far below these terms. This is due to the slow inbuilt PowerPC processor, the little memory available and thus a nonefficient implementation of the TCP/IP stack. Hence, the next version of a controller board is currently in the last phase of development. This time a PicoITX PC board in combination with an FTDI USB2.0 controller card will act as controller module. This design will reach a bandwidth sufficient for 100 Mbit/s Ethernet and will still keep the flexibility for the location of COPACOBANA.

11.2.4 Application Development

Because FPGAs are not ready-to-use processors like in conventional PCs but have to be configured in its functionality, a developer needs extra skills in hardware design using a hardware description language like VHDL. Thus, the functionality of an FPGA has to be designed in creating several PEs and control structure in hardware. Most FPGAs, as well as the Spartan3-1000, have already built-in components like block RAM, multipliers, DSPs, or, in some cases, even CPU cores. They could easily be integrated in an individual design to make it even more efficient for special requirements and help speed up the design process.

Thus, the first and most important layer to develop an application for COPACOBANA 1000 is to create a hardware design for the FPGAs with a hardware description language like VHDL. So, the problem instance should typically not be too complex for hardware implementation. Furthermore, to achieve massive parallelism by using all FPGAs efficiently at the same time, and every FPGA even with several PEs, it is beneficial to have a problem that is easy to parallelize without much interprocess data dependency.

The benefit of a real hardware design in comparison to a program running on a standard processor is the area and speed efficiency of the problem implementation. While a standard processor is developed to solve a large variety of problem instances and therefore needs a whole lot of resources, a problem-specific hardware design can exactly be equipped with only the resources needed. For simple problem instances, a PE can be designed that are small in terms of area usage such that several PEs fit into a single FPGA and thus working in real parallelism.

In addition, a standard processor runs with very high frequency to process a lot of basic commands representing a problem instance. In hardware such a problem instance could be a simple step in an automation depending on a Boolean function that could be solved in one clock cycle. Thus, the frequency in an FPGA design has limited significance and so a high frequency is not always necessary.

For a little help in VHDL FPGA design for COPACOBANA the little "MemoryTest" application, which checks the functioning of each FPGA and the internal bus, is provided. This gives the developer a clue as to how to use the COPACOBANA bus system on the FPGA side.

The second layer of development is the creation of an application on the host PC. This application generally does not have to run computation-intensive processes because these should be solved by the COPACOBANA FPGAs. Its main purpose should be the data exchange between the host PC and COPACOBANA. Thus, it simply has to fit the communication interface with the provided API. Other tasks should be data preparation and postprocessing.

An optional third layer is the user interface. In most cases it is always some user who simply wants to change some problem parameters and is waiting for a result afterward. He/She does not want to learn complex command line usage reading tons of cryptic output data afterward. In most cases this leads to a graphical user interface (GUI), but will leave us with another topic here.

11.3 Cryptanalysis with COPACOBANA 1000

The security of symmetric and asymmetric ciphers is usually determined by the size of their security parameters, in particular, the key length. Hence, when designing a cryptosystem, these parameters need to be chosen according to the assumed computational capabilities of an attacker. Depending on the chosen security margin, many cryptosystems are potentially vulnerable to attacks when the attacker's computational power increases unexpectedly. In real life, the limiting factor of an attacker is often financial resources. Thus, it is quite crucial from a cryptographic point of view to not only investigate the complexity of an attack, but also to study possibilities to lower the cost-performance ratio of attack hardware. For instance, a cost-performance improvement of an attack machine by a factor of 1,000 effectively reduces the key length of a symmetric cipher by roughly 10 bit (as 1,000 ~ 2^{10}).

Cryptanalysis of modern cryptographic algorithms involves massive and parallel computations, usually requiring more than 2^{40} operations. Many cryptanalytical schemes perform their computations in independent operations, which allows for a high degree of parallelism. Such parallel functionality can be realized by individual hardware blocks that can be operated simultaneously, improving the time complexity of the overall computation by a perfect linear factor. At this point, it should be remarked that the high nonrecurring engineering costs for application-specific integrated circuit (ASICs)—which can often consume more than US$ 100,000 for large projects—have put most projects for building special-purpose hardware for cryptanalysis out of reach for commercial or research institutions. However, with the recent advent of low-cost FPGAs, which host vast amounts of logic resources, special-purpose cryptoanalytical machines have now become a possibility outside government agencies.

In this chapter, we will show how COPACOBANA can be used to break the DES block cipher [13] and a corresponding cryptosystems in real-world products. Though DES was revoked as standard in 2004, it is still a popular choice for low-end security system as well as available in many legacy systems. Still in 2008, we identified a class of cryptotokens that generate one-time-passwords (OTPs) according to the ANSI X9.9 standard in which the DES encryption is still in use. Alarmingly, we are aware of online banking systems in Europe, and North and Central America that still distribute such tokens to users for authenticating their financial transactions. (Because we do not want to support hacking of bank accounts, we will not give further details here.)

Besides DES breaking and other symmetric ciphers, cryptanalysis on asymmetric cryptography can also be supported by COPACOBANA; for example, for solving the Elliptic Curve Discrete Logarithm Problem [14], which is known as the fundamental primitive for cryptosystems based on elliptic curves. Further work employing COPACOBANA for cryptanalysis (which is also not in the scope of this chapter) has been done on breaking the legacy hard disk encryption (*Norton Diskreet*) [7], the GSM A5/1 stream cipher [15], and recent Machine Readable Travel Documents (*ePassport*) [16].

11.3.1 Previous Work on DES Breaking

Since the invention of the computer, a continuous effort has been taken to build clusters providing the recent maximum of computing power. For our purpose, we focus on the cost-efficient COPACOBANA system instead of reviewing all recent variants of such high-performance clusters or supercomputers for investments of several millions of dollars. Thus, we now shortly survey the history of breaking the most popular block cipher of the last decades—the DES. After that, we will come up with an evaluation of the security margin provided by DES nowadays with respect to available machines like COPACOBANA.

Although the DES was reaffirmed for use in (U.S. government) security systems several times until 1999, the worries about the inherent threat of its short key space was already raised in 1977 when it was first proposed. The first estimates were proposed by Diffie and Hellman [17] for a brute force machine that could find the key within a day at a cost of US$ 20 million. Some years after that, a first detailed hardware design description for a brute force attack was presented by Michael Wiener at the rump session of CRYPTO'93; a printed version is available in [18]. It was estimated that the machine could be built for less than a million U.S. dollars. The proposed machine consists of 57,000 DES chips that could recover a key every 3½ hours. In 1997, a detailed cost estimate for three different approaches for DES key search, distributed computing, FPGAs, and custom ASIC designs, was presented by Blaze et al. [19]. In 1998, the Electronic Frontier Foundation (EFF) finally built a DES hardware cracker called *Deep Crack,* which could perform an exhaustive key search within 56 hours [20]. Their DES cracker consisted of 1,536 custom-designed ASIC chips at a cost of material of around US$ 250,000 and could search 88 billion keys per second. To our knowledge, the latest step in the history of DES brute force attacks took place in 2006, when the COPACOBANA was built for less than US$ 10,000 [8]. COPACOBANA is capable of breaking DES in less than 1 week on average as shown in the next sections. We would like to note that software-only attacks against DES still take more than 1,000 PC-years (based on Intel Pentium-4@3GHz) in worst case.

11.3.2 Exhaustive Key Search on DES

The DES with a 56-bit key size was chosen as the first commercial cryptographic standard by NIST in 1977 [13]. A key size of 56-bits was considered to be a good choice considering the huge development costs for computing power in the late 1970s, which made a search over all the possible 2^{56} keys appear impractical. As DES was designed to be extremely efficient in terms of area and speed for hardware, an FPGA implementation of DES can be orders of magnitude faster than an implementation on a conventional PC at much lower costs [8]. This allows a hardware-based engine for a DES key search to be much faster and efficient compared to a software-based approach.

Our attack is based on simple known-plaintext scenario; that is, we assume to have knowledge of a single pair of ciphertext and its corresponding plaintext. Although it might seem to be a strong assumption in the first place that the attacker has access to a piece of unencrypted information, it is indeed valid in many scenarios. Whenever data in protocols or files are encrypted, there are requirements on the structure and formatting of data. The information

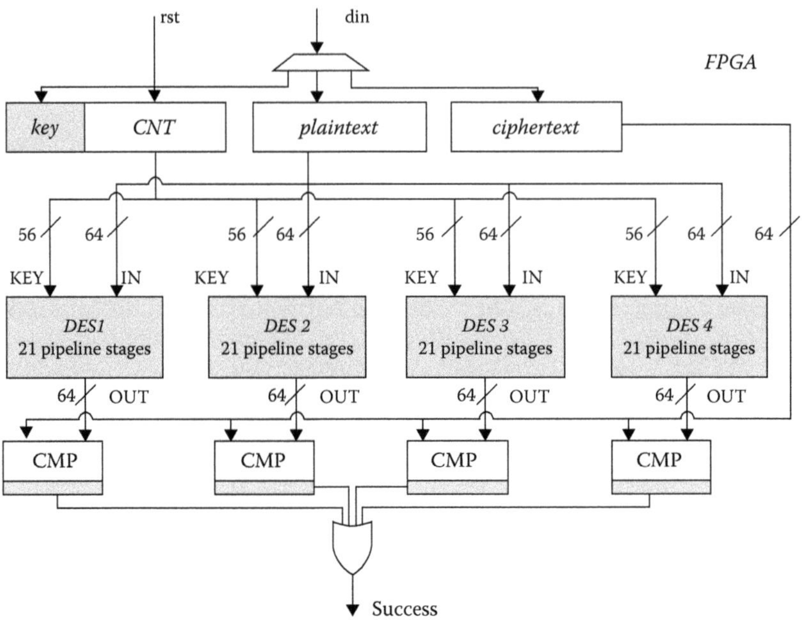

FIGURE 11.3
Architecture for exhaustive key search with four DES key search units.

about formats and data structures is usually publicly known, also to a potential attacker. As an example, all pictures in the graphics interchange format (GIF) start with the fixed token `GIF89a`; many other formats include (redundant) length information of the document. Whenever the type of encrypted data is known, we can exploit these and similar bits of information to construct a valid plaintext–ciphertext pair from the document.

Then, the approach to use such a plaintext–ciphertext (p,c) pair in an exhaustive key search is simple: we sequentially encrypt the given plaintext p with all possible key candidates of the cipher's key space K (i.e., $\tilde{c} = enc_k(p)$ with $k \in K$) and compare each returned ciphertext $\tilde{c}$ to the given ciphertext c. As soon a match is found (i.e., $\tilde{c} = c$), we can identify the correct key candidate.

Our core component to perform the key search is an improved version of the efficient DES engine developed by the Université Catholique de Louvain's Crypto Group [21] based on 21 pipeline steps. Our design can test one key per clock cycle and engine. On the COPACOBANA, we can fit four such DES engines inside a single FPGA, which allows for sharing plaintext–ciphertext input pairs and the key space as shown in Figure 11.3.

We can operate each of the machine's FPGAs at a clock rate of 136 MHz, which is an improvement in performance by 36% compared to our original design in [8]. Consequently, a partial key space of 2^{42} keys can completely be checked in $2^{40} \times 7.35$ ns by a single FPGA, which is approximately 135 minutes. As COPACOBANA hosts 120 of these low-cost FPGAs, the key search machine can check $4 \times 120 = 480$ keys every 7.35 ns; that is, 65.28 billion keys per second. To find the correct key, COPACOBANA has to search through an average of 2^{55} different keys. Thus, COPACOBANA can find the right key in approximately $T = 6.4$ days on average. As more than one COPACOBANA can be attached to a single host and the key space can be shared among all machines, the search time then reduces to T/l, where l denotes the number of machines.

11.3.3 Breaking DES-Based Crypto Tokens

In this section, we employ COPACOBANA for an attack on a real-world system. More precisely, we attack cryptographic tokens that are used for user authentication and identification according to FIPS 113 and ANSI X9.9, respectively. Their authentication method is based on one-time passwords (OTP) generated using the DES algorithm. Unfortunately, such devices are still used in many security relevant applications. (We are aware of online banking systems in some places of the world still relying on ANSI X9.9-based tokens for authorization of financial transactions. We prefer not to give any details at this point.) Hence, the attack presented in the following still has an impact on affected online banking systems used worldwide.

11.3.3.1 Basics of Token-Based Data Authentication

We will now describe an OTP token-based data protocol according to FIPS 113 or ANSI X9.9, which is used for authentication in some real-world online banking systems. Please note that we assume that OTP tokens have a fixed, securely integrated static key inside and do not use additional entropy sources like time or events for computing the passwords. Indeed, there are tokens available that generate new passwords after a dedicated time interval (e.g., products like the RSA SecurID solution [22]) but those will not be the focus of this chapter. This type of tokens require additional assumptions concerning the unknown plaintext, and thus are harder to attack. More precisely, our contribution assumes fixed-key OTP tokens that can be used in combination with a challenge–response protocol. In such protocols, a decimal-digit challenge is manually entered into the token via an integrated keypad. The token in turn computes the corresponding response according to the ANSI X9.9 standard. Tokens implementing this standardized authentication scheme (incorporating ANSI 3.92 DES encryption) often have a fixed-size liquid crystal display (LCD) allowing for displaying eight-decimal digits for input and output.

After the user has typed in eight-decimal digits as input (challenge), the value is converted to binary representation using standard ASCII code notation according to the ANSI X9.9 standard. For instance, the typed number "12345678" is converted into the 64-bit challenge value in hexadecimal representation $c = (0 \times 31, 0 \times 32, 0 \times 33, 0 \times 34, 0 \times 35, 0 \times 36, 0 \times 37, 0 \times 38)$. After recoding, c is used as plaintext to the DES encryption function $r = e_k(c)$ with the static key k stored securely in the token. The output of the encryption function is the 64-bit ciphertext $r = (r_1, r_0)$ where each r_i denotes a 32-bit word to be transformed using a mapping μ to fit the eight-digit display of the token. The mapping μ takes the eight hexadecimal digits of r_1 (32 bits) of the DES encryption as input and converts each digit individually from hexadecimal (binary) notation to decimal representation. Let $H = \{0,\ldots,9,A,\ldots,F\}$ and $D = \{0,\ldots,9\}$ be the alphabets of hexadecimal and decimal digits, respectively. Then μ is defined as

$$\mu : H \rightarrow D : \{0_H \mapsto 0_D;\ldots;9_H \mapsto 9_D; A_H \mapsto 0_D; F_H \mapsto 5_D\}$$

Hence, the output after the mapping μ is an eight-decimal digit value that is displayed on the LCD of the token. Figure 11.4 shows how the response is generated on the token according to a given challenge. In several countries, this authentication method is used in banking applications whenever a customer needs to authenticate financial transactions. For this, each user of such an online banking system owns a personal token used to respond to challenges presented by the banking system to authorize every security-critical operation. In this context, for example, a security-critical operation can be the login to the banking system as well as the authorization of a money transfer. Figure 11.5 depicts a token-based challenge–response protocol interaction

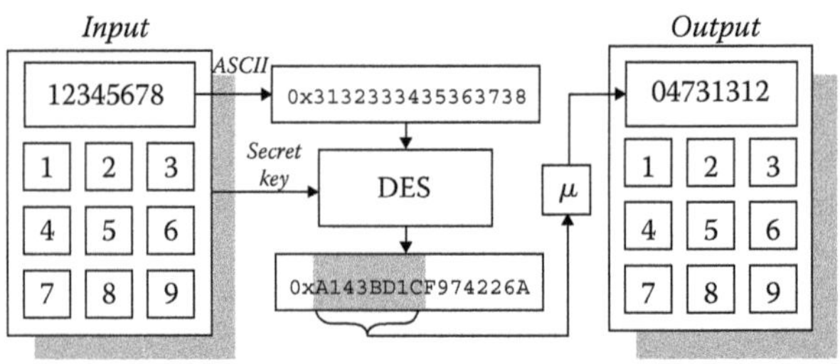

FIGURE 11.4
Principle of response generation with ANSI X9.9-based crypto tokens.

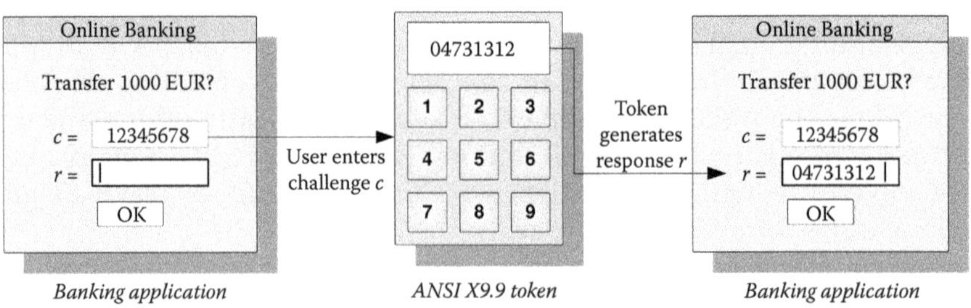

FIGURE 11.5
Token-based challenge–response protocol for online banking.

with an online banking system from a user's perspective. The central role in such a security-related application makes the secret token an interesting target for an attack.

11.3.3.2 Cryptanalysis of the ANSI X9.9-Based Challenge–Response Authentication

With the knowledge of how an authenticator is computed in the challenge–response protocol, we will continue with identifying weaknesses to attack this authentication scheme. Firstly, ANSI X9.9 relies on the DES algorithm for which we built a low-cost special-purpose hardware machine and this can perform an exhaustive key search in less than a week. Second, the output r of the DES encryption is only slightly modified. Note that a more complex scrambling with additional dynamic input, like hash functions with salt, would make the attack considerably more complex or even impossible. The output r is only truncated to 32 bits and modified using the mapping μ to convert c_1 from hexadecimal to decimal notation. Owing to the truncation to 32 bits, we need to acquire knowledge of at least two plaintext–ciphertext pairs when mounting an exhaustive key search to return a single key candidate only. The digit conversion μ additionally reduces the information leaked by a single pair of plaintext–ciphertext, which is addressed by Observation 1.

Observation 1: Let $D=\{0,\ldots,9\}$ be the alphabet of decimal digits. With a single challenge–response pair (c,r) of an ANSI X9.9-based authentication scheme with $c,r \in D^8$, on average 26 bits of a DES key can be determined (24 bits in the worst case, 32 bits in the best case).

As only 32 bits of the output for a given challenge c are exposed, this is a trivial upper bound for the information leakage for a single pair. Assuming the DES encryption function to be a pseudorandom function with appropriate statistical properties, the 32 most-significant bits of c form eight hexadecimal digits that are uniformly distributed over $H^8=\{0,\ldots,9,A,\ldots,F\}^8$. The surjective mapping μ has the domain $F=\{0,\ldots,9\}$ of which $T=\{0,\ldots,5\}$ are doubly assigned. Hence, we know that $\Delta = F\backslash T = \{6,\ldots,9\}$ are four fixed points that directly correspond to output digits of c yielding four bit of key information (I). The six remaining decimal digits $\Omega = F \cap T$ can have two potential origins allowing for a potential deviation of one bit (II). According to a uniform distribution of the eight hexadecimal output digits, the probability that (I) is given for an arbitrary digit i of c is $\Pr(i \in \Delta) = 1/4$. Thus, on average we can expect two out of eight hexadecimal digits of c to be in Δ revealing four bits of the key, whereas the remaining six digits introduce a possible variance of one unknown bit per digit. Averaged, this leads to knowledge of $R = 2 \cdot 4 + 6 \cdot 3 = 26$ bits of DES key material. Obviously, the best case with all eight digits in Δ and worst case with no digits out of the set Δ provide 32 and 24 key bits, respectively.

According to Observation 1, we can develop two distinguished attacks based on the knowledge of two and three known challenge–response pairs:

Observation 2: Given two known challenge–response pairs (c_i, r_i) for $i \in \{0,1\}$ of the ANSI X9.9 authentication scheme, an exhaustive key search using both pairs will reveal $2^4 = 16$ potential key candidates on average (256 candidates in the worst case, and in the best case the actual key is returned).

Assuming independence of two different encrypted blocks related to the same key in block ciphers, we can use accumulated results from Observation 2 for key determination using multiple pairs (p_i, c_i). Hence, on average we can expect to determine 52 bits of the key where each c_i has two digits from the set Δ. Given a full DES key of 56-bit size, the results are 2^4 possible variations for key candidates. Having at least four digits from Δ for each c_i, we can determine the best case resulting in a single key candidate. In the worst case and with no Δ digits in any c_i, we will end up with 48 bits of determined key material and $2^8 = 256$ possible remaining key candidates. As a consequence, the number of potential key candidates is directly dependent on how many digits of a c_i are fixed points from the set Δ.

Observation 3: Given three known challenge–response pairs of the ANSI X9.9 authentication scheme, an exhaustive key search based on this information will uniquely reveal the DES key.

This directly follows from Observation 2. For this attack, $3 \cdot 24 = 72 > 56$ bits of key material can directly determined (even in the worst case) resulting in the correct key to be definitely identified.

11.3.3.3 Possible Attack Scenarios on Banking Systems

With these fundamental observations at hand, we can begin to develop two attack variants for two and three plaintext–ciphertext pairs. As we need only few pairs of information, an attack is feasible in a real-world scenario. For instance, if we consider a phishing attack on an online banking system, we can easily imagine that two or three (faked) challenges are presented to the user, who is likely to respond with the appropriate values generated by his token. Alternatively, spying techniques, for example, based on malicious software like key-loggers or hidden cameras, can be used to observe the user while responding to a challenge. Note that the freshness of these values do not play a role as we use the information only for computing the secret key and not for an unauthorized login attempt. Figure 11.6 shows a possible attack scenario on ANSI X9.9 tokens and associated banking applications based on phishing of challenge–response pairs c, r. With at least two pairs of challenge–response data, we can perform an exhaustive key search on the DES key space implementing the specific features of ANSI X9.9 authentication. To cope with the DES key space of 2^{56} potential key candidates we will propose an implementation based on dedicated special-purpose hardware.

In case that three challenge–responses pairs are given, we are definitely able to uniquely determine the key of the secret token using a single exhaustive

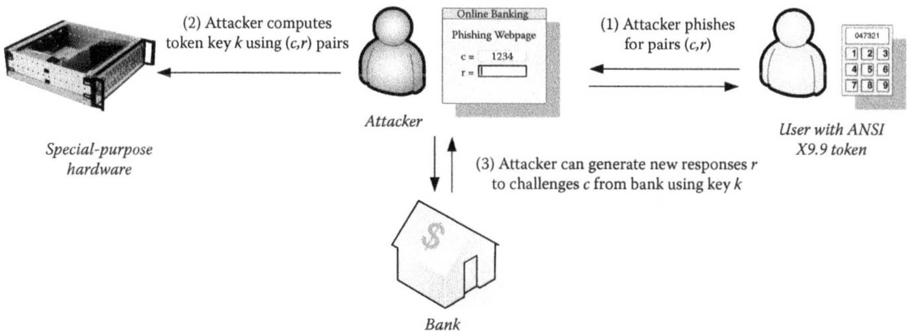

FIGURE 11.6
Attack scenario for token-based banking applications using phishing techniques.

key search. When only two pairs (c_i, r_i) are available to the attacker, then it is likely that several potential key candidates are returned from the key search (cf. Observation 2). With 16 potential solutions on average, the attacker can attempt to guess the right solution by trial and error. As most banking systems allow the user to enter up to three erroneous responds to a challenge in a row, two key candidates can be tried by the attacker at a time. Then, after a period of inactivity, the authorized user has probably logged into the banking application that resets the error counter and allows the attacker to start another trial session with further key candidates. On average, the attacker can expect to be successful after about four trial-and-error sessions, testing 8 out of the 16 keys from the candidate list. Hence, an attack on an ANSI X9.9-based token is very likely to be successful even with knowledge of only two given challenge–response pairs.

11.3.3.4 Implementing the Token Attack on COPACOBANA

As before, the main goal of our hardware design is a key search of the token to be done in a highly parallelized fashion by partitioning the key space among the available FPGAs on the COPACOBANA. This requires hardly any interprocess communication, as each of the DES engines can search for the right key within its allocated key subspace.

Within the FPGAs, we use again a slightly modified version of the highly pipelined DES implementation of the Université Catholique de Louvain's Crypto Group [21], which computes one encryption per clock per engine. As with the brute force attack, we can fit four such DES engines inside a single FPGA, and therefore allow for sharing of control circuitry and the key space as shown in Figure 11.7. The FPGA architecture comprises two 64-bit *plaintext* registers for the challenges and two 32-bit *ciphertext* registers for storing the corresponding responses that can be acquired from the OTP token. The key space to be searched is allocated to each chip as the most significant 14 bits of the key that is stored in the *key* register. The counter (*CNT 1*) is used

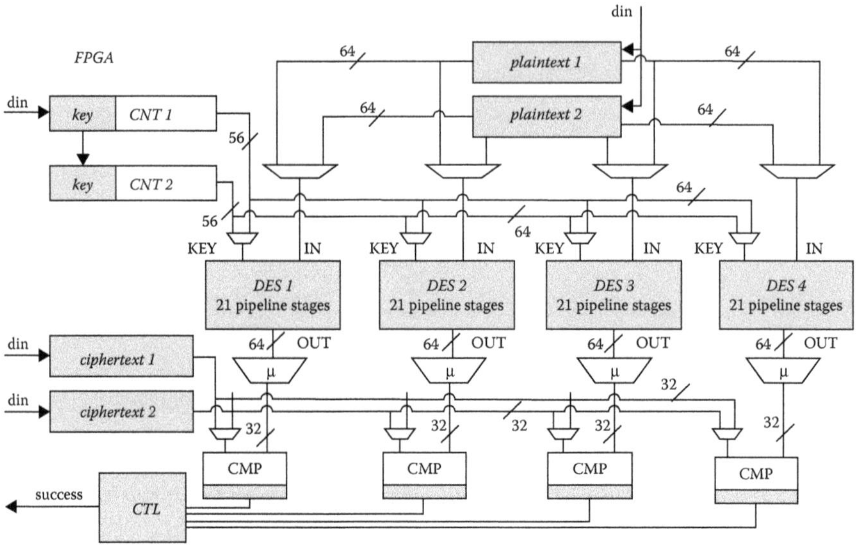

FIGURE 11.7
Four ANSI X9.9 key search units based on fully pipelined DES cores in a Xilinx Spartan-3 FPGA.

to run through the least significant 40 bits of the key. The remaining two bits of the 56-bit key for each of the DES engines are hardwired and dedicated to each of them. Thus, for every such FPGA, a task is assigned to search through all the keys with the 16 most-significant bits fixed, in total 2^{40} different keys. The key space is partitioned by a connected host PC so that each chip takes around 150 minutes (at 120 MHz) to test all ANSI X9.9 authenticators in its allocated key subspace. During a single check of an authenticator, the DES engines use the first challenge (*plaintext 1*) as a primary input to the encryption function. Then, the upper 32-bits of the generated ciphertext are mapped digit per digit by the function μ and compared with the value of the response stored in the register *ciphertext 1*.

If any of the DES engines provides a positive match, the corresponding engine switches its input to the second challenge encrypting it with the same key. To match the pipelined design of the DES engine, we are using a shadow counter (*CNT 2*) tracking the key position at the beginning of the pipeline. In case that the derived authenticator from the second encryption compares successfully to the second response, the controller *CTL* reports the counter value to the host PC as a potential key candidate. The host PC keeps track of the key range that is assigned to each of the FPGAs and, hence, can match the right key from a given counter value. If no match is found until the counter overflows, the FPGA reports completion of the task and remains idle until a new key space is assigned.

In case that a third challenge–response pair is specified, the host PC performs a verification operation of the reported key candidate in software. In case the verification is successful, the search is aborted and the key returned as a result of the search.

We have implemented the FPGA architecture shown in Figure 11.7 using the Xilinx ISE 9.1 development platform. After synthesis of the design incorporating four DES engines and the additional logic for the derivation of the ANSI X9.9 authenticator, the usage of 8,729 flip-flops (FF) and 12,813 LUT was reported by the tools (56% FF and 83% LUT utilization of the Spartan3-1000 device, respectively). As discussed in Section 1.3, we included specific optimizations like pipelined comparators because n-bit comparators are likely to introduce a long signal propagation path reducing the maximum clock frequency significantly. By removing these potential bottlenecks, the design can be clocked at 120 MHz after place and route resulting in a throughput of 480 million keys per FPGA and second. In total, a fully equipped COPACOBANA with 120 FPGAs can compute 57.6 billion ANSI X9.9 authenticators per second. On the basis of this, we can present time estimates for an attack provided that two challenge–response pairs are given. Recall that in this scenario we will be faced with several potential key candidates per run so that we have to search the entire key space of 2^{56} to build a list with all of them. This ensures that we are able to identify the actual key in a separate step.

Similarly, we can present figures for an attack scenario where three challenge–response pairs are available. In this attack, we must test 2^{55} ANSI X9.9 authenticators on average to find the corresponding key that is half the time complexity of an attack having two known pairs of data. Note that all presented figures of Table 11.1 include material costs only (not taking energy and development costs into account).

For comparison with our hardware-based cluster, we have included estimations for an Intel Pentium 4 processor operating at 2.4 GHz. This microprocessor allows for a throughput of about 2 million DES computations a second, what we also assume as appropriate throughput estimate for generating ANSI X9.9 authenticators.

TABLE 11.1

Cost-Performance Figures for Attacking the ANSI X9.9 Scheme with Two and Three Known Challenge–Response Pairs (c_i, r_i)

Hardware System	Material Cost	Two Pairs (c_i, r_i)	Three Pairs (c_i, r_i)
1 Pentium 4 @ 2.4 GHz	US$ 50	1,170 years	585 years
1 FPGA XC3S1000	US$ 50	4.8 years	2.4 years
1 COPACOBANA	US$ 10,000	14.5 days	7.2 days
100 COPACOBANAs	US$ 1 million	3.5 hours	104 minutes

11.4 COPACOBANA 5000

11.4.1 Direction toward New Applications

When the architecture of COPACOBANA 1000 was published [7–8, 23–26] it was mainly seen as a special-purpose device for cryptoanalysis. But the scientific discussion [27, 28] brought up the idea to make some minor changes to the architectural concept to be able to cover a much wider application spectrum. With the new upcoming questions in scientific computing [29] and bioinformatics, like short-read genome assembly [30–32], it turned out that these changes were feasible and it was obvious that such a machine would have a significant impact on research advancement in these areas.

11.4.2 Requirements

The bottleneck of COPACOBANA 1000 for I/O-dominated applications was the small data rate of the global bus. This was the case due to two reasons: first, the bus speed is bounded to the bandwidth of TCP/IP over Ethernet (at most 100 Mbit/s or 1 Gbit/s). Second, the capability of the internal controller module (the older Memec controller module as well as the newer FTDI USB bridge) decreases the bandwidth again. In addition, the fan out of one to twenty from the controller to the FPGA modules limited the clock rate for the global bus, and the physical distance between the controller board and the last FPGA module gave a lower bound for the speed of the communication over the bus.

For applications like error correction in genome assembly, motif search, or alignment a data rate of at least 2 Gbit/s between controller and FPGA would be desirable. This created the need of some more sophisticated interconnection network. The LUTs and the block RAM of the Spartan FPGAs allowed a fast access to a small amount of memory with some limited degree of parallelism. This was not sufficient if genome or amino acid sequences or large parts of them had to be locally stored for access of the individual processing units on the FPGAs. Therefore, a second level of memory hierarchy had to be introduced into the COPACOBANA concept.

The problems in cryptoanalysis are typically of the form that identical, compute-intensive operations have to be applied to a large number of distributed data. Therefore, the algorithms consist of a first phase, where these data are distributed (or generated in the FPGAs); a second phase of computation; and a third phase of collecting the results. The dominating part is the second. This means there was no need to cope with complicated communication patterns between the active units of COPACOBANA 1000. As the requirements for bioinformatics applications were much more demanding in terms of communication it was necessary to develop a communication network that allowed efficient point-to-multipoint communication and parallel point-to-point communication as well. A set of different communication patterns were developed to specify the needs of the bus and communication system for the new version of COPACOBANA.

One obvious weakness of COPACOBANA 1000 was the computing power of the Spartan3-1000. One of the requirements for the new system was therefore to find a more powerful FPGA with comparable values in power consumption and price.

As COPACOBANA 1000 was designed as a monolithic system for cryptoanalysis, there had been no standardization of its modules. This fact made it difficult to keep track with the performance development of newly upcoming chips. As an example, it was impossible to adapt the FPGA-cards to make use of the DSP-units of the Virtex-4 series of Xilinx. The new system therefore has been designed in classes of modules where the modules in each class could be exchanged among each other.

11.4.3 Architecture of COPACOBANA 5000

The new architecture COPACOBANA 5000 consists of an 18-slot backplane equipped with 16 FPGA modules and two controller modules interconnected with a high-performance bus system. This massively parallel FPGA-computer is connected to an in-system off-the-shelf PC via the two controller modules and the PCIe interface. The embedded PC is capable of running computation-intensive applications and/or acting as storage database due to at least a quad-core CPU, 8 GByte RAM, and 1 TByte SATA hard disk. To preserve some backward compatibility and usability COPACOBANA 5000 provides nearly all interfaces a standard PC also does, for example, Gigabit Ethernet.

Each of the replaceable FPGA modules carry eight high-performance FPGAs (e.g., the Spartan3-5000, not fixed in advance) interconnected in a one-dimensional array. The 1.8 kW main power supply with 12 V and 125 A on the output site provides enough energy to run all 128 FPGAs, the embedded PC, controllers, and six high-performance fans for front-to-back cooling. An optional standard power supply for the embedded PC can be mounted to discharge the main supply on extreme power-consuming applications.

Like in COPACOBANA 1000 the COPACOBANA 5000 comes in a case mountable in standard 19-inch racks, now still occupying only 3 height units (3HE). A photo of the new machine is depicted in Figure 11.8.

11.4.3.1 Bus Concept and Backplane

The COPACOBANA 5000 backplane provides 18 slots for communication modules. In general, these are 16 FPGA modules and two controller modules. The backplane does not contain any further important electronic components as only the interconnection of the slots and the supplying of power is done here.

FIGURE 11.8
COPACOBANA 5000 front view.

Owing to the requirements of communication speed, the interconnection of all individual FPGAs and the two controller modules is organized as a systolic chain: There are fast point-to-point connections between every two neighbors instead of one physical global bus. The first controller communicates with the first FPGA on the first module and the last FPGA on the last module communicates with the second controller. This implies a long queue of all FPGAs putting the embedded PC at the beginning and at the end of this queue. To speedup certain communication profiles, for example, global broadcasts, there are shortcuts in this chain between adjacent FPGA modules.

The point-to-point interconnections consist of eight pairs of wires in each direction. Each pair is driven by low-voltage differential signaling (LVDS) with a speed of 250 MHz. All FPGAs are clocked globally and synchronously. Each clock cycle data is transferred from one FPGA to the adjacent one forming a huge communication pipeline. Thus, inside the system this technique allows a systolic data flow of 2 Mbit/s in each direction. Between the controller modules and the embedded PC the maximum data rate is also limited to 250 MByte/s (2 Gbit/s) due to the PCIe connection.

As all communication is done serially, there are no extra pins for addressing. All relevant information is bound in a communication protocol requiring an overhead of bandwidth of about 20%. Only the configuration is done over special data paths leading to each FPGA.

The interconnection network can be used flexibly either for local exchange of data between every two adjacent FPGA modules or as a systolic point-to-multipoint bus. The idea of systolic communication is not new: it is simply a pipelined bus, where in every clock cycle one data item is propagated form one unit to the next. As all units operate in parallel, the throughput is maximized at the prize of a considerable latency of several clock cycles between two distant communication points. Former architectures have exploited this concept of systolic communication as well [33]. Figure 11.9 shows the overall bus architecture of COPACOBANA 5000.

11.4.3.2 FPGA Module

To overcome the limits of the Spartan3-1000 different FPGA chips have been taken into account. Virtex-4 and Virtex-5 turned out to be too expensive with respect to their performance. The best price-performance results were

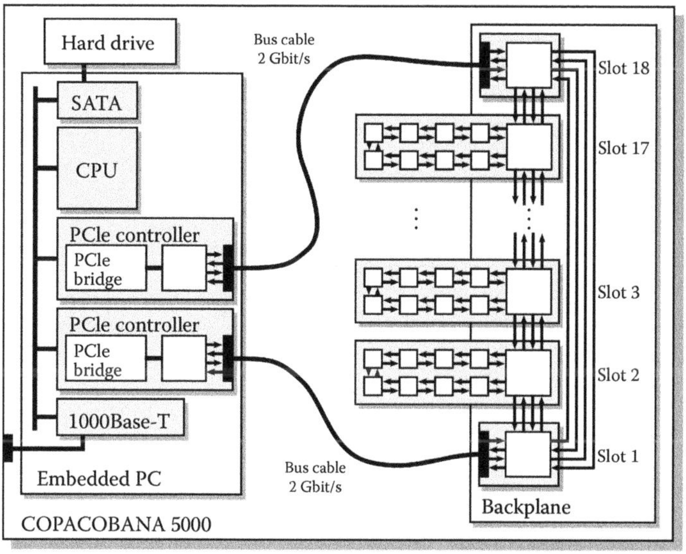

FIGURE 11.9
Architecture of the COPACOBANA 5000.

provided by the Xilinx Spartan3-5000. It comes with different packages, whereas the chosen one is compatible with many other Xilinx FPGAs. Hence, it is possible to easily build FPGA modules mounted with different FPGAs.

Thus, in the standard version of COPACOBANA 5000 each FPGA module contains eight freely configurable Spartan3-5000 FPGAs. One additional FPGA with a fixed configuration is mounted on each controller module to do intelligent routing of the systolic data streams. The simple communication protocol provides address information in the header fields of the data stream, so this FPGA can easily filter packets out.

As the new COPACOBANA architecture is intended to be used for applications with dependence on large amounts of data and FPGAs generally do not provide the ability to store more than a few kilobytes in their block RAMs, each FPGA can optionally be equipped with a 256-Mbit DRAM module. This provides a lot more flexibility to an application developer and can release the bus traffic significantly. Figure 11.10 shows the components and the data path of the FPGA module.

In comparison to the COPACOBANA 1000 the connection mechanism for the FPGA modules to the backplane has changed. Now, because high-performance FPGAs have higher heat dissipation, some more space between two adjacent FPGA modules is required for cooling mechanisms. This implies the possibility of taking a connection mechanism which is not that small and better to handle than the DIMM connectors in COPACOBANA 1000. Hence, the FPGA modules are connected to the backplane via PCI connectors, providing good electrical characteristics, stability, enough pins, and ease of use.

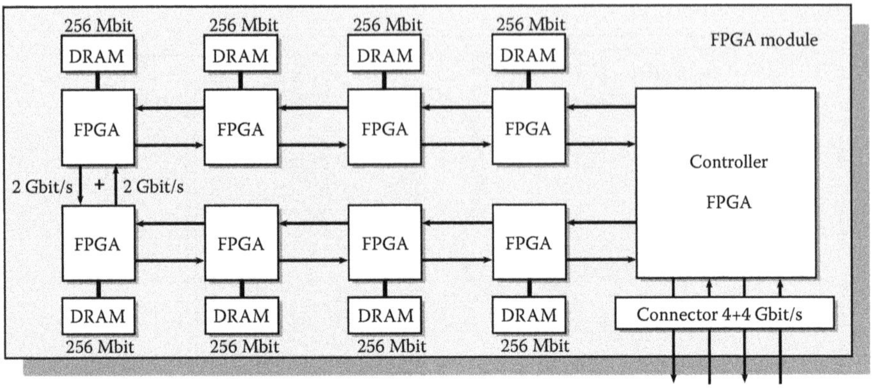

FIGURE 11.10
Design of the COPACOBANA 5000 FPGA module.

11.4.3.3 Interface Controller

The root entity of control is normally located on some host PC outside the COPACOBANA 5000 machine. However, COPACOBANA 5000 contains a fully featured embedded PC as front end directly connected to the two controller modules for the COPACOBANA bus. The connection to this PC can be done in several ways. The easiest way is to integrate the COPACOBANA machine into a local area network (LAN) and connect via TCP/IP. The usage of this connection is then very flexible.

Two scenarios are considered here: the first scenario is to let the host computer control the FPGAs and the internal bus directly via a special API. This is the way it was already done on COPACOBANA 1000. For this purpose COPACOBANA 5000 provides two Gigabit Ethernet interfaces to assure enough bandwidth for communication.

The second scenario is the preferable one in most cases: A database of input data is physically stored inside the COPACOBANA machine, for example, on the 1TB SATA hard drive attached to the embedded PC. The hard disk provides a last stage of memory hierarchy for the parallel system as well as for the interchange of data between COPACOBANA and the outside world. The Gigabit Ethernet connection can then be used for preparative data transfers. In this case, the controlling application is running remotely on COPACOBANA and the host computer acts as user terminal only. As the embedded PC is running a conventional operating system, the user can thus access COPACOBANA in a familiar way.

11.4.3.4 Power Supply and Cooling Mechanism

COPACOBANA 5000 comes in a box fitting in standard 19-inch racks using 3HE. As a lot of power-consuming components were to be placed in comparatively small space the designers were confronted with heat dissipation—one of the major problems in the design of COPACOBANA 5000. As the FPGAs tend to be configured such that they exploit the majority of their CLBs, each FPGA will come to its limits with respect to power consumption. Eight computing FPGAs plus one routing FPGA on each board reach a maximal power of about 75 W. In addition, the ATX-board with its main processors and the hard disks contribute to the heat that has to be dissipated. The cooling air is blown through the box of COPACOBANA by six high-performance fans. Three of them force the intake of cold air; three others transport the heated air out. The FPGA modules are plugged in vertical direction to let the heated air rise up and to maximize the air flow.

Closely related to the cooling system is the power supply. Under maximal load, COPACOBANA takes 1.8 kW. The FPGAs are driven with a core voltage of 1.2 V (2.5 V for signals). Therefore, the current to each FPGA can reach a value up to 7 A, which makes a total of 63 A on each FPGA module. To discharge the backplane from transporting current in this order of magnitude, power converters are mounted on each FPGA module converting the supply voltage of 12 V to the core voltage of 1.2 V. This decreases the amount of current by a factor of 10, but still, the total current to all FPGA modules is greater than 100 A. Specific power lines have been implemented into the backplane to be able to cope with currents of this order of magnitude.

11.4.3.5 Application Development

As COPACOBANA 5000 also benefits from the massive parallelization of PEs implemented on FPGAs, the way to develop applications is not much different from the way for COPACOBANA 1000. The benefit of the new machine is the expansion of suitable application types. Now, applications can be considered requiring more interprocess communication and/or high data throughput.

The first layer of programming is still the design of the FPGA configuration. But now on the one hand, because the bus system is more complicated, the developer has to involve a bus controller in his/her design and use another API for accessing data of it. On the other hand, this controller releases the developer from taking care of data packets that simply have to run through the FPGA.

For developing the control software, the second layer of COPACOBANA application development, the amount of required resources has to be considered. Does the application depend on fast access of huge amounts of data or does data dependency have less priority? This is important for the decision if this application should run either on the host computer or directly on the embedded PC. An application running directly on the embedded PC has easier and faster access to the onboard resources. The developer can benefit from anything the operating system provides like simple file I/O on the hard disk. The communication with the FPGAs is done directly via an API operating the COPACOBANA bus controllers via the PCIe interface. This

API provides functions to easily manage the whole systolic data flow on the COPACOBANA bus. Hence, there is only a user interface to implement on the host computer, communicating with the COPACOBANA application. This usually does not require much bandwidth and can be done via Ethernet for example. In addition the interface can be operating system independent. Only the client for the host computer has to be implemented for the desired operating system, which implies an easier portability.

The implementation of an application to run on the host computer depends on a special server running on the embedded PC. This server provides another API to control the COPACOBANA bus, which is similar to the one operating directly on the PCIe interface. There are some drawbacks to consider in this scenario. The first to mention is the loss of flexibility as the usage of onboard resources is limited and depends on the API. Second, data has to be streamed directly from the host, which can be a lot slower depending on the network infrastructure and third-party traffic. And last but not least, portability is more difficult because the whole application including the user interface is running on a specific operating system. But because there are many applications already running for COPACOBANA 1000, porting them to COPACOBANA 5000 for speedup is easy using this type of application implementation.

As a helper tool again the little "MemoryTest" application is provided. It checks the functioning of each FPGA, the applied memory, and the internal bus. This tool acts as an example of how to use the different APIs for hardware and software on both sides, the FPGA side and the host side.

11.5 Applications in Bioinformatics

The new hardware architecture COPACOBANA 5000 is designed especially for streaming algorithms that need a high throughput and that profit easily by massive parallelization. It aims for many algorithms in bioinformatics as they have to cope with large amounts of genome data. In most cases these datasets could easily be analyzed in streams.

The greatest difficulty software designers have to face when porting algorithms to the COPACOBANA 5000 architecture is the hardware design of the PEs for the COPACOBANA FPGAs. As FPGAs are no real processors like the CPU in a standard PC running a simple program, the functional behavior depends on a hardware description written in a hardware description language like VHDL. Fundamentals of the design are its speed and area usage. Therefore, it is often beneficial not to implement existing algorithms directly in hardware, but to alter these algorithms or even create new ones to fit the requirements of the new architecture and to gain better results in a more efficient way.

Two of the most challenging problems in bioinformatics are motif finding and alignment. As there is still no universal motif finding algorithm that satisfies the needs of every biologist, for optimal global and local alignments the algorithms by Needleman and Wunsch [34] and Smith–Waterman [35] are widely accepted. Though it is hard to implement them directly in hardware for high efficiency, the next section describes a very powerful solution. Afterward the motif finding problem is addressed by an efficient hardware solution searching for unknown regulatory elements of fixed size but allowing variations in the instances of the located strings.

11.5.1 Sequence Alignment

The alignment of nucleotide sequences deals with the problem of finding the best fitting alignment of two nucleotide sequences against each other. Algorithms used to handle this problem may be classified to either heuristic or nonheuristic ones. The heuristic alignment algorithms such as basic local alignment search tool (BLAST) [36] have become the common tools to search for alignments as they are much faster than nonheuristic ones. Although they produce a large amount of false results and may not succeed in finding all the correct solutions, in terms of computing time they outperform nonheuristic algorithms by far and have therefore gained broad acceptance within the group of molecular biologists.

In recent years, the advancing hardware technology has allowed the revival of nonheuristic alignment algorithms. In this section we want to demonstrate how COPACOBANA 5000 can be used for this purpose.

11.5.1.1 Smith–Waterman Alignment

To demonstrate the applicability of nonheuristic alignment algorithms the so-called *Smith–Waterman algorithm* [35] is introduced as an example. This algorithm is capable of finding the best fitting alignment of one sequence inside of another, meaning it searches for occurrences of one sequence. For convenience the sequence that is searched for is called "query sequence" while the one that is searched in is called "database sequence" To find the best of every possible alignment a score is generated. These scores are calculated by the simple scoring function:

$$H(i,j) = \max \begin{Bmatrix} 0 \\ H(i-1,j) + GapPenalty \\ H(i,j-1) + GapPenalty \\ H(i-1,j-1) + Match/Mismatch \end{Bmatrix}$$

Here, H is a matrix and the values for *GapPenalty*, *Match*, and *Mismatch* can be user defined. H is the so-called *scoring matrix*.

A software implementation of the Smith–Waterman algorithm usually calculates the entries of the scoring matrix one after the other using one or more processes. Typically, this is done line by line or column by column from left to right as the scoring function only uses the values of the top, the left, and the top-left neighbors. Once the matrix calculation has finished, the actual alignment is found by starting a backtracking through the matrix at the cell with the highest value. This technique is not practical to be implemented in hardware because the matrix can easily get too large to store it in the RAM of an FPGA. So a work around has to be found to solve this kind of problem in hardware.

Example: The scoring matrix for a Smith–Waterman alignment of the query "CGGA" on "ACGAT" can look like the matrix shown in Table 11.2.

This scoring matrix was generated with a score of 2 for *Match* and –1 for *Mismatch* and *GapPenalty*. The backtracking results in the alignment shown in Table 11.3.

11.5.1.2 Hardware Implementation

To cope with the memory limitations mentioned earlier it is essential to know that in a biologist's workflow it is very likely to align thousands of query sequences at a time with only looking at the maximal scores of the alignments. These scores are analyzed and the actual alignment of a very little selection of query sequences with the highest scores may easily be computed on a standard PC. Hence, it is not important to store the whole matrix, but only the maximum and the values needed for computation. Then, the output is simply the maximum cell value.

Like in a software implementation, parallelization can be done line by line. Thus, every PE calculates one line, meaning that it is responsible for exactly one character of the query sequence. Therefore, referencing the value of the

TABLE 11.2

Scoring Matrix for Smith–Waterman Alignment

	A	C	G	A	T
C	0	2	1	0	0
G	0	1	4	3	2
G	0	0	3	3	2
A	2	1	2	5	4

TABLE 11.3

Alignment Example ("-" Denotes a Gap)

Base:	A	C	G	-	A	T
Query:	-	C	G	G	A	-

left neighbor is simply accessing its own value of the previous clock cycle. Like this, referencing the value from the top or top-left neighbor means referencing the value of the preceding PE one or two clock cycles before, respectively. In addition, the update of the maximum value that was seen so far has to be done.

With a parallelization scheme like this, the algorithm processes the matrix diagonally with the database sequence streaming through the chain of PEs. The processing of a matrix is seen in the following example while the illustration of an FPGA implementation is shown in Figure 11.11.

Example: The processing scheme of the scoring matrix from the previous example is depicted in Table 11.4.

11.5.1.3 Performance on COPACOBANA 5000

Common FPGA implementations of the Smith–Waterman algorithm only use one single FPGA. Hence, they either are limited in the length of the query sequence or need to reinitialize their chain of PEs. Implementing the

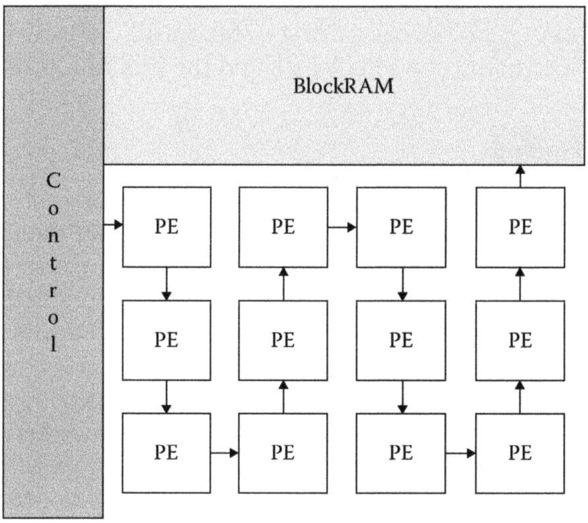

FIGURE 11.11
Implementation of the Smith–Waterman algorithm on an FPGA.

TABLE 11.4
Processing Scheme for Smith–Waterman Alignment

	A	C	G	A	T
C	0	2	1		
G	0	1			
G	0				
A					

TABLE 11.5

Performance Results on Smith–Waterman Alignment

	COPACOBANA 5000	Cray XD-1	AMD Opteron (2.2 GHz)
Runtime	0.1 hours	7.39 hours	75 hours
Speedup (vs. PC)	750	10.15	1
Energy consumption	0.06 kWh	2.7 kWh	22.5 kWh
System cost	200,000 $	100,000 $	500 $

Smith–Waterman algorithm on COPACOBANA 5000 offers new ways to think about scaling. Now, there are 128 instead of 1 FPGA available. In addition, in contrast to the interprocess communication ability of COPACOBANA 1000, all FPGAs can communicate in a systolic manner. Thus, it is possible to simply align 128 small query sequences at once, or, even better, to align a query sequence that is 128 times the size of what was possible before the COPACOBANA 5000 was available. It is even able to mix different query lengths and process them altogether in parallel.

Table 11.5 shows the performance results of the alignment of 3,685 20-mers against the human genome, comparing a standard PC (AMD Opteron at 2.2 GHz), a Cray XD-1 using one FPGA [37], and the COPACOBANA 5000.

11.5.2 Motif Finding

The biological background for motif finding is the discovery of regulatory sequences in DNA. It is often described as the "needle-in-the-haystack problem" [38]. Algorithms search for overrepresented occurrences of unknown short sequences that in addition could have slight variations in their instances. This makes it one of the most challenging problems in bioinformatics. In fact there are problem instances of motif finding that are unsolvable by current techniques. There are two reasons that make this problem so difficult. First, the parameters of a given problem instance (like sequence and motif length, grade of mutation) can make it impossible to identify motifs due to background noise. Second, it is computationally expensive as nearly nothing is known on the instances intended to be found.

Decades have already been spent on this issue but no single algorithm meets all the demands of the biologists. Every scientist has his/her own special needs on special ways of the problem with modified criteria for the algorithms. Thus most algorithms developed as yet only perform well on special problem instances, but deliver worse results on general problem instances [39, 40]. Although the MEME algorithm [41, 42] reached a broad acceptance in academic circles, like BLAST [36] for alignment problems, its results often only give a clue about what was expected for real. Important instances simply stay hidden although additional criteria could help to discover them.

The MEME approach and the similar Gibbs sampler [43, 44] develop matrices as representation for motif candidates using an expectation maximization method. Other algorithms like PROJECTION [45] only precompute start values for the expectation maximization method to filter out false-positive results. The greedy approaches like the original CONSENSUS [46] simply find the most likely instances by aligning only small parts of the genome at a time, but have a high risk of missing other important instances.

The approach iterative generation of matrices (IGOM), later enhanced to Boolean matrix algorithm (BMA) [47], uses position frequency matrices as representation of motif instances. It iteratively generates these matrices by measuring a value similar to signal-to-noise ratio. This is done in correlation to signal theory where the signal strength opposing to the background noise of the medium is maximized. The restrictions of the position frequency matrices are weakened to gain the highest signal-to-noise ratio, thus generating a candidate for a motif. The step from IGOM to BMA is due to the awareness that in general the exact distribution of the different bases on several positions in the motif instances is not necessarily needed for further analysis. It is only important to know which bases are permitted to exist on a weakened position. So, a Boolean representation of the position frequency matrices is sufficient, leading to the algorithm BMA.

11.5.2.1 The BMA Algorithm

In the following section the simple algorithm BMA [47] is described. This algorithm is highlighted by the success in discovering a new stress factor in *Bacillus subtilis* (in cooperation with the Institute of Virology, Free University of Berlin). Given clean input data, that is, for example a preanalyzed sequence with repeat regions already cut out, not much effort has to be made for filtering the result data again. The results already provide clean indications for what is a real motif candidate and what is not.

In addition, BMA is highly customizable for special criteria; for example, the searching for underrepresented sequences, or motifs where instances with slight modifications do not occur, could easily be adapted.

11.5.2.2 Implementation of BMA

The basic implementation of BMA uses Boolean matrices as representation for a motif. The Boolean matrices used here have similarities to commonly used position frequency or position weight matrices. They consist of four rows, each for one of the four possible bases A, C, G, and T. The number of columns is equal to the size of the expected motif and represents a position in the motif. Thus exactly one "true"-value in a column forces the corresponding base to be present in every motif instance at the column's position. This is called a "preserved" position. More than one "true"-value in a column imply a weakened preservation of bases in this position. BMA

TABLE 11.6

An Example of a Boolean Matrix

A	1	1	1	0	0	1
C	0	0	0	0	1	0
G	0	1	1	0	0	0
T	0	0	0	1	0	0

only allows maximal two "true"-values in one column, which is called a "semi-preserved" position. The example in Table 11.6 shows a Boolean matrix with semipreserved positions at the second and third base, thus representing the sequences "AAGTCA," "AGATCA," "AGGTCA," and "AAATCA."

The first step in the BMA algorithm is the initialization of the Boolean matrix representing a possible candidate for a motif. The initialization value could either be taken from a substring of the genome data to analyze or be taken out of all possible initialization values. The preferable way depends on the size of the genome data and the expected motif size. As there are four different bases the number of possible initialization values is limited to $4^{(\text{motif size})}$. If this number exceeds the number of subsequences of the same length in the genome data, the first way should be preferred.

After initialization the number of occurrences of the actual motif candidate in the genome data is simply counted and interpreted as score. In the same time variances of the actual candidate are registered by counting the differences in each position in some scoring matrix. The registration is done only in cases where exactly one base differs from the candidate. This way the most popular variation of the actual motif instance is voted and the corresponding position will be weakened to "semipreserved" after analyzing the genome data once. In the following example an analyzing run is illustrated. A Boolean matrix and a sequence mismatching the matrix in exactly one position are shown. The corresponding counter to this mismatching position in an example scoring matrix is emphasized, leading to the ensuing weakened Boolean matrix.

Example 11.1: An Analysis Run with BMA

The Boolean matrix is taken from the example in Table 11.6. An example sequence ("AGACCA") with exactly one mismatch according to this matrix is stated in Table 11.7. The fields corresponding to the characters in the sequence are highlighted.

An example of a scoring matrix emphasizing the corresponding counter is stated in Table 11.8.

The weakened Boolean matrix resulting from the scores of this scoring matrix can be seen in Table 11.9.

TABLE 11.7

Matching a Sequence with a Boolean Matrix

A	1	1	1	0	0	1
C	0	0	0	0	1	0
G	0	1	1	0	0	0
T	0	0	0	1	0	0
	A	G	A	C	C	A
	✓	✓	✓	✗	✓	✓

TABLE 11.8

An Example of a Scoring Matrix, Corresponding to the Previous Example

A	–	–	–	6	0	–
C	0	5	0	*12*	–	0
G	7	–	–	0	4	4
T	0	0	2	–	0	1

TABLE 11.9

The Resulting Weakened Boolean Matrix from the Previous Example

A	1	1	1	0	0	1
C	0	0	0	*1*	1	0
G	0	1	1	0	0	0
T	0	0	0	1	0	0

This procedure is repeated until the desired degree of weakened positions is reached. The score of each matrix then represents the degree of overrepresentation of the motif candidate in the genome.

11.5.2.3 Parallelization of BMA in Hardware

Because of their Boolean representation the matrices easily fit in a hardware design, occupying less space than matrices containing integer or floating point numbers. The simplicity of the algorithm itself, the independence of several PEs, and the accessing of the genome data as stream make it perfectly well suited for massive parallelization on COPACOBANA [10].

The parallelization is straightforward. As every actual motif candidate has to be analyzed with the genome data, the best way is to take as many PEs as possible, each handling one motif candidate, and analyze the streamed genome data in parallel. Each PE has to hold the Boolean matrix for the candidate, some counters for its score and several counters in a scoring matrix. Thus, the most expensive component concerning area usage is the scoring matrix, but it is rarely accessed beneficially throughout the analysis. So, the scoring matrix could be placed in the block RAM structure abundant on every COPACOBANA FPGA. This way the resources of logic cells are saved and it is possible to place more PEs on each FPGA. To ensure the access to the genome data at high clock rates, the PEs on the FPGA are arranged in several queues along the location of the block RAM. Each element provides the data to its neighbor after one clock cycle. The same happens for the results whose data flow is just the other way round. In addition a preselection is done by providing only the best results per chip. Figures 11.12 and 11.13 show an illustration for the arrangement of the components on a COPACOBANA 1000 FPGA and the data flow for a single run.

There is not much difference in the implementations for the COPACOBANA 1000 and the COPACOBANA 5000 architectures. The basic PEs are the same, but it is apparently the count that differs. While there are 32 elements fitting on a COPACOBANA 1000 FPGA, it is 128 elements on a COPACOBANA 5000 FPGA. In total 3,840 versus 16,384 PEs are working in parallel.

The second main difference between the architectures is the communication structure and thus the speed of a single analysis run. The COPACOBANA 1000 provides the genome data to the PEs by broadcasting on the internal single-master multiple-slave bus. COPACOBANA 5000 communicates on a

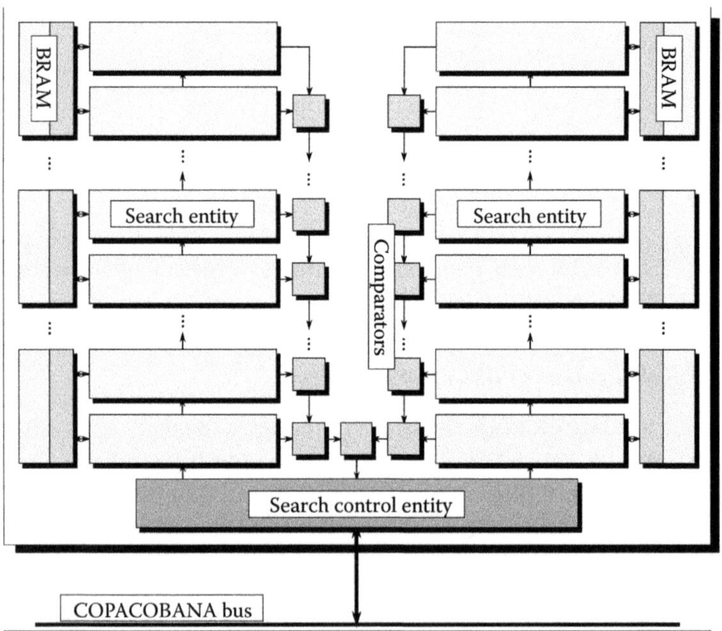

FIGURE 11.12
Hardware design overview of the parallelized BMA algorithm. The processing elements are arranged in a queue on each FPGA.

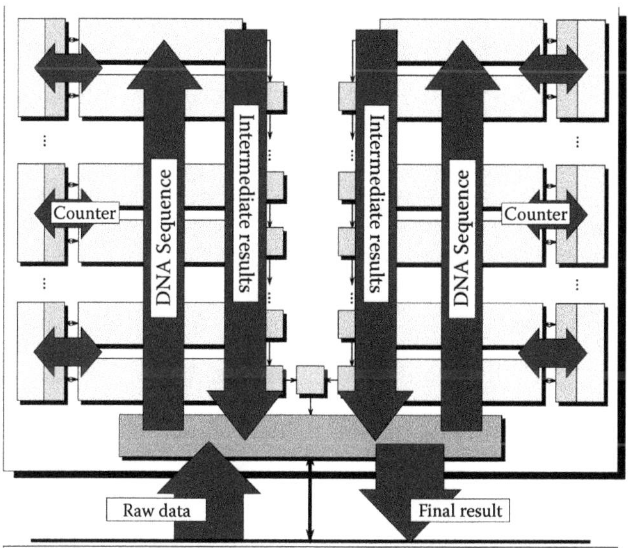

FIGURE 11.13
The data flow of a parallelized BMA run on a single FPGA.

communication ring basis. Broadcasting is done by sending chunks of data one by one to the first client in the ring, which repeats it in sending it to the next client, and so on. The benefit of this behavior is that communication is done over short distances. Hence, it could be very fast, but with the little drawback of a relative high delay for the least clients. Actually communication between two clients reaches the data rate of 2 Gbit/s on COPACOBANA 5000. This is definitely enough to make the runtime of the whole algorithm solely depend on the speed of the internal clock while the critical factor on COPACOBANA 1000 is the speed of the bus.

To gain high clock frequency in a BMA PE for the COPACOBANA 5000 the design was being optimized by introducing pipeline stages while comparing the motif candidate with the actual subsequence of the input data. This raised the clock speed from only 40 MHz in the COPACOBANA 1000 design to a possible frequency of about 100 MHz for the COPACOBANA 5000 design. For comparison with a standard PC the algorithm was accordingly adapted, implemented in C++, and compiled with highest processor specific optimization.

11.5.2.4 Performance Results of BMA

Table 11.10 shows the performance results of some motif-finding runs on several genomes for the COPACOBANA 5000, COPACOBANA 1000, and a standard PC architecture. The PC features the following properties: an Intel Core2 Quad CPU at 2.4 GHz, 8 GB DDR-II RAM, and SATA hard disk

TABLE 11.10

Performance Results on Motif Finding with BMA

	COPACOBANA 5000 (1,000W)	COPACOBANA 1000 (600W)	Intel Core2 Quad 2.4 GHz (300W)
Cowpox virus (230 kbp):			
Runtime	~2 seconds	3 minutes	3 hours
Speedup (vs. PC)	5,400	60	1
Energy consumption	0.0006 kWh	0.03 kWh	0.9 kWh
Energy costs (0.20 €/kWh)	€ 0.0001	€ 0.006	€ 0.18
Rickettsia can. (1.2 Mbp):			
Runtime	21 seconds	21 minutes	4 days 19 hours
Speedup (vs. PC)	19,714	329	1
Energy consumption	0.006 kWh	0.21 kWh	34.5 kWh
Energy costs (0.20 €/kWh)	€ 0.0012	€ 0.042	€ 6.90
Bacillus subtilis (5.9 Mbp):			
Runtime	7 minutes	6 hours 10 minutes	122 days
Speedup (vs. PC)	25,097	475	1
Energy consumption	0.117 kWh	3.7 kWh	878.4 kWh
Energy costs (0.20 €/kWh)	€ 0.023	€ 0.74	€ 176

running a Linux operating system. The algorithmic parameters were a motif size of 12, a number of six runs for each initialized candidate, and for every genome their subsequences of the same length as the motif were used for initialization instead of using all possible sequences. In the case of taking all possible initialization sequences, the runtime would increase linearly for all architectures. The measurement of the power consumption of the three architectures resulted in about 1000 W for COPACOBANA 5000, 600 W for COPACOBANA 1000, and 300 W for the standard PC.

The results show that the COPACOBANA architectures outperform a standard PC in an order of magnitude while taking the computation time back to rational means. The efficiency factor calculated by dividing the hardware costs of one PC (€ 300) with the costs of one COPACOBANA 5000 machine (€ 150,000) and multiplied with the speedup of about 25,000, reaches 50. This means COPACOBANA is 50 times more efficient than using a PC or PC cluster for solving this kind of problems.

Further, according to the energy consumption, COPACOBANA even provides energy savings in several orders of magnitude. Imagine, a PC cluster working in parallel to decrease the runtime will not decrease the amount of energy needed! Thus, given the costs for a COPACOBANA 5000 machine of about €150,000 and the energy costs for a PC cluster of about €180 just for solving only one problem of size *B. subtilis* (s. table), it will take only 830 runs to reach energy costs the size of the hardware costs for one COPACOBANA 5000 machine. While scaling the size of the problem to higher amounts, this factor gets even worse on side of the PC. That makes the acquisition of COPACOBANA even cheaper than maintaining a cluster of standard PCs.

11.5.3 Future Work

As COPACOBANA 5000 is suitable for all kinds of algorithms processing huge amounts of data preferably in streams without difficult computations, the research area of bioinformatics has a lot of topics adequate for this massively parallel architecture. For instance, further research is done on genome assembly with reassembly as well as de novo assembly on COPACOBANA. These kinds of algorithms require the processing of several ten thousand megabytes of data in rational means of time. The capability of the COPACOBANA 5000 to handle these requirements has already been shown in the examples earlier. The machine outperforms a standard PC by an order of magnitude while saving energy costs in several orders of magnitude. Hence, the commercial availability makes it indeed an interesting opportunity for researchers in molecular biology to replace their existing PC clusters.

11.6 References

1. Aho A.V., Ullman J.D., Hopcroft J.E.: *The Design and Analysis of Computer Algorithms*, Addison Wesley, 1973.
2. Chazelle B., Monier L.: Unbounded hardware is equivalent to deterministic Turing machines, *Theoretical Computer Science* 24, 123–130, 1983.
3. Hromkovic J.: *Communication Complexity and Parallel Computing*, Springer, 1997.
4. Floyd T.L.: *Digital Fundamentals*, Pearson International Education, 2006.
5. Robert Y., Quinton P., Tchuente M.: *Parallel Algorithms and Architectures*, North Holland, Amsterdam, 1986.
6. Schröder H.: VLSI-sorting evaluated under the linear model, *Journal of Complexity* 4, 330–355, 1988
7. Kumar S., Paar C., Pelzl J., Pfeiffer G., Rupp A., Schimmler M.: How to break DES for € 8,980. Presented at the Workshop on Special Purpose Hardware for Attacking Cryptographic Systems (SHARCS'06), 2006.
8. Kumar S., Paar C., Pelzl J., Pfeiffer G., Schimmler M.: Breaking Ciphers with COPACOBANA—A cost-optimized parallel code breaker. In L. Goubin and M. Matsui, editors, *Proceedings of the Workshop on Cryptograpic Hardware and Embedded Systems* (CHES 2006), LNCS 4249, pp. 101–118. Springer-Verlag, 2006.
9. Pfeiffer G., Baumgart S., Schröder J., Schimmler M.: A massively parallel architecture for bioinformatics, *Computational Science—ICCS 2009* International Conference on Computational Science, LNCS 5544, pp. 994–1003, 2009.

10. Schröder J., Wienbrandt L., Pfeiffer G., Schimmler M.: Massively parallelized DNA motif search on the reconfigurable hardware platform COPACOBANA. International Conference on Pattern Recognition in Bioinformatics, Oct 2008.
11. SciEngines GmbH. http://www.sciengines.com Accessed February 18, 2010.
12. Avnet EMG GmbH: Memec Virtex-4 FX12 Mini Module User Guide, 2005. http://www.avnet-memec.eu Accessed February 18, 2010.
13. National Institute for Standards and Technology (NIST). Data Encryption Standard, FIPS PUB 46–3, January 1977.
14. Güneysu T., Paar C., Pelzl J.: Attacking elliptic curve cryptosystems with special purpose hardware. In *Proceedings of the International Symposium on Field Programmable Gate Arrays* (FPGA 2007), pp. 207–215. ACM Press, 2007.
15. Gendrullis T., Novotný M., Rupp A.: A real-world attack breaking A5/1 within hours. In E. Oswald and P. Rohatgi, editors, *Proceedings of the Workshop on Cryptographic Hardware and Embedded Systems* (CHES 2008), LNCS 5154, pp. 266–282. Springer-Verlag, 2008.
16. Liu Y., Kasper T., Lemke-Rust K., Paar C.: E-passport: Cracking basic access control keys. In R. Meersman and Z. Tari, editors, *Proceedings of On the Move to Meaningful Internet Systems Conferences* (OTM 2007), LNCS 4804, pp. 1531–1547. Springer-Verlag, 2007.
17. Diffie W., Hellman M.E.: Exhaustive cryptanalysis of the NBS data encryption standard. *Computer*, 10, 6, 74–84, 1977.
18. Wiener M.J.: Efficient DES key search. In W.R. Stallings, editor, *Practical Cryptography for Data Internetworks*, pp. 31–79. IEEE Computer Society Press, 1996.
19. Blaze M., Diffie W., Rivest R.L., Schneier B., Shimomura T., Thompson E., Wiener M.: Minimal key lengths for symmetric ciphers to provide adequate commercial security. Technical report, Security Protocols Workshop, Cambridge, UK, January 1996. Available at http://www.counterpane.com/keylength.html Accessed February 18, 2010.
20. Electronic Frontier Foundation. *Cracking DES: Secrets of Encryption Research, Wiretap Politics & Chip Design*. O'Reilly & Associates Inc., 1998.
21. Rouvroy G., Standaert F.-X., Quisquater J.-J., Legat J.-D.: Design strategies and modified descriptions to optimize cipher FPGA implementations: Fast and compact results for DES and triple-DES. In *Field-Programmable Logic and Applications—FPL*, pp. 181–193, 2003.
22. RSA. RSA SecurID, 2007. http://www.rsa.com/. Accessed February 18, 2010.
23. Güneysu T., Kasper T., Novotny M., Paar C., Rupp A.: Cryptanalysis with COPACOBANA, *IEEE Transactions on Computers*, 57, 11, 2008.
24. Güneysu T., Paar C., Pelzl J., Pfeiffer G., Schimmler M., Schleiffer C.: Parallel computing with low-cost FPGAs: A framework for COPACOBANA, ParaFPGA Symposium LNI 2007, September 2007.
25. Kumar S., Paar C., Pelzl J., Pfeiffer G., Schimmler M.: A configuration concept for a massively parallel FPGA architecture, International Conference on Computer Design (CDES'06), June 2006.
26. Kumar S., Paar C., Pelzl J., Pfeiffer G., Schimmler M.: COPACOBANA—A cost-optimized special-purpose hardware for code-breaking, IEEE Symposium on Field-Programmable Custom Computing Machines (FCCM'06), April 2006.

27. Oliver T., Schmidt B., Jacop Y., Maskell D.: High-speed biological sequence analysis with hidden Markov models on reconfigurable platforms, *IEEE Transactions on Information Technology in Biomedicine*, 13, 5, 740–746, 2007.
28. Oliver T., Leow Y.Y., Schmidt B.: Integrating FPGA acceleration into HMMer, *Parallel Computing*, 34, 11, 681–691, 2008.
29. Börm S.: Construction of data-sparse H^2-matrices by hierarchical compression, *SIAM Journal of Scientific Computing*, 31, 1820–1839, 2009.
30 Pevzner P.A., Tang H., Waterman M.S.: An Eulerian path approach to DNA fragment assembly, *Proceedings of the National Academy of Science*, 98, 9748–9753, 2001.
31. Simpson J.T., Wong K., Jackman S.D., Schein J.E., Jones S.J.M., Birol I.: ABySS: A parallel assembler for short read sequence data, *Genome Research*, 19, 1117–1123, 2009.
32. Zerbino D.R., Birney E.: Velvet: Algorithms for de novo short read assembly using De Bruin graphs, *Genome Research*, 18, 821–829, 2008.
33. Lang, H.-W.: The instruction systolic array, a parallel architecture for VLSI, *Integration, the VLSI Journal*, 4, 65–74, 1986.
34. Needleman S.B., Wunsch C.D.: A general method applicable to the search for similarities in the amino acid sequence of two proteins, *Journal of Molecular Biology*, 48, 3, 443–453, 1970.
35. Smith T.F., Waterman M.S.: Identification of common molecular subsequences, *Journal of Molecular Biology*, 147, 195–197, 1981.
36. Altschul S.F., Gish W., Miller W., Myers E.W., Lipman D.J.: Basic local alignment search tool, *Journal of Molecular Biology*, 215, 403–410, 1990.
37. Storaasli O., Yu W., Strenski D., Maltby J.: Performance evaluation of FPGA-based biological applications, Cray Users Group Proceedings 2007, Seattle, Washington.
38. Bailey T.L., Williams N., Misleh C., Li W.W.: MEME: Discovering and analyzing DNA and protein sequence motifs, *Nucleic Acids Research*, 34 (WebServer Issue), W369–W373, 2006.
39. Das M.K., Dai H.-K.: A survey of DNA motif finding algorithms, *BMC Bioinformatics*, 8, 7, S21, 2007.
40. Tompa M., Li N., Bailey L. et al.: Assessing computational tools for the discovery of transcription factor binding sites, *Nature Biotechnology*, 23, 1, 137–145, 2005.
41. Bailey T.L., Elkan C.: Fitting a mixture model by expectation maximization to discover motifs in biopolymers, *International Conference on Intelligent Systems for Molecular Biology*, 2, 28–36, 1994.
42. Bailey T.L., Elkan C.: Unsupervised learning of multiple motifs in biopolymers using expectation maximization, *Machine Learning*, 21, 51–80, 1995.
43. Lawrence C.E., Altschul S.F., Boguski M.S., Liu J.S., Neuwald A.F., Wootton J.C.: Detecting subtle sequence signals: A Gibbs sampling strategy for multiple alignment, *Science*, New Series, 262, 5131, 208–214, 1993.
44. Lawrence C.E., Reilly A.A.: An expectation maximization (EM) algorithm for the identification and characterization of common sites in unaligned biopolymer sequences, *Proteins*, 7, 1, 41–51, 1990.
45. Buhler J., Tompa M.: Finding motifs using random projections, *Journal of Computational Biology*, 9, 2, 225–242, 2002.

46. Hertz G.Z., Hatzwell G.W., Stormo G.D.: Identification of consensus patterns in unaligned DNA sequences known to be functionally related, *Computer Applications in the Biosciences*, 6, 2, 81–92, 1990.
47. Schröder J., Schimmler M., Tischer K., Schröder H.: BMA—Boolean matrices as model for motif kernels, *International Conference on Bioinformatics, Computational Biology, Genomics and Chemoinformatics*, 36, 41, 2008.

12

Accelerating String Set Matching for Bioinformatics Using FPGA Hardware

Yoginder S. Dandass

12.1 Introduction

String set matching is an important operation in computational biology. For example, when proteomics data is used for genome annotation in a process called *proteogenomic mapping* [1–5], a set of peptide identifications obtained using mass spectrometry is matched against a target genome translated in all six reading frames. Given a large number of peptides and long translated genome strings, the fundamental problem here is to efficiently search a large text corpus (i.e., the translated genome) to identify the locations of individual strings that belong to a large set of patterns (i.e., the set of peptides).

12.1.1 String Matching Approaches

Efficient substring search algorithms such as Boyer–Moore [6] and Knuth–Morris–Pratt [7] that locate single pattern strings within a larger text string can be used in a multipass manner (i.e., one pass for each string in the set of peptides). However, this approach does not scale well with an increasing number of pattern strings. In particular, assuming p patterns with an average length of n characters and a text corpus of length m characters, naïve, multipass approaches have computational complexity of $O(p \times (m + n))$.

The Aho–Corasick algorithm (ACA) [8] provides a scalable solution to the string set matching problem because it incorporates the search mechanism for the entire set of patterns into a single deterministic finite automaton (DFA). The power of ACA stems from its ability to find the location of the strings belonging to the pattern set in the large text corpus in a single pass. The computational complexity of Aho–Corasick search is $O(m + k)$ where k is the total number of occurrences of the pattern strings in the text. This linear processing time complexity has resulted in the widespread use of ACA in string matching applications.

The performance of the ACA can be further enhanced by implementing it in hardware. Tan and Sherwood [9] were the first to describe an area-efficient hardware approach for implementing the Aho–Corasick for network intrusion detection systems implemented in application-specific integrated circuits (ASICs). However, the cost associated with ASIC development is a significant impediment to their adoption in computational biology. Field-programmable gate array (FPGA) devices, conversely, can be repeatedly reconfigured to create a variety of application-specific processing elements, making FPGAs a popular low-cost alternative for developing low-cost hardware-based processing accelerators.

Although the fundamental ACA is identical for all string set matching applications, optimization for specific applications and target hardware results in significant performance and storage efficiencies. This chapter expands on previous work by Dandass et al. [10], and demonstrates how an Aho–Corasick architecture and DFA organization can be specifically optimized for implementation in FPGA hardware. This chapter also shows how the 18-kbit blocks of random access memory (BRAM) available on Xilinx's Virtex-4 FPGAs and 9-kbit RAM blocks available in Altera's FPGAs can be utilized to create resource-efficient amino acid sequence set matching engines.

12.1.2 Use of the ACA in Computational Biology

The ACA is widely used in computational biology for a variety of pattern matching tasks. For example, Brudno and Morgenstern use a simplified version of ACA to identify anchor points in their CHAOS algorithm for fast alignment of large genomic sequences [11, 12]. The TROLL algorithm of

Castelo, Martins, and Gao uses ACA to locate occurrences of tandem repeats in genomic sequence [13]. Farre et al. use Aho–Corasick as the search algorithm for predicting transcription binding sites in their tool PROMO v.3 [14]. Hyyro et al. demonstrate that Aho–Corasick outperforms other algorithms for locating unique oligonucleotides in the yeast genome [15]. The SITEBLAST algorithm [16] employs the ACA to retrieve all motif anchors for a local alignment procedure for genomic sequences that makes use of prior knowledge. Buhler, Keich, and Sun use an Aho–Corasick DFA to design simultaneous seeds for DNA similarity search [17]. The AhoPro software package adapts the ACA to compute the probability of simultaneous motif occurrences [18].

12.1.3 Use of FPGAs in Computational Biology

There has been significant interest in using FPGAs to address bottlenecks in computational biology pipelines. Examples include the use of FPGAs to improve the speed of homology search [19, 20] for computing phylogenetic trees [21], for the pairwise alignment step in multiple sequence alignment using ClustalW [22], and for acceleration of the Smith–Waterman sequence alignment algorithm [19]. In computational proteomics, Alex et al. have demonstrated the use of FPGAs to accelerate peptide mass fingerprinting [23]. Bogdan et al. have applied FPGAs to the problem of analyzing mass spectrometric data generated by MALDI-ToF instruments by developing hardware implementations of algorithms for denoising, baseline correction, peak identification, and deisotoping [24].

12.1.4 Use of String Set Matching in FPGAs in Other Domains

Hardware implementations of ACA have been developed for applications other than bioinformatics. Snort is a popular computer security program that looks for a set of "signature" patterns corresponding to known intrusion attacks in network packets. Tan and Sherwood split the Aho–Corasick implementation for Snort into four separate FSMs such that each FSM is responsible for two separate bit positions in the signature string set and network packet [9]. This bit-split implementation is more efficient in terms of hardware area. However, their paper does not exploit the availability of specialized hardware resources in FPGAs.

Jung, Baker, and Prasanna describe an implementation of the bit-split ACA for Snort using FPGA technology [25]. They optimize the bit-split implementation of Aho–Corasick for Snort by using RAM blocks available on Xilinx FPGAs. However, in Snort, the input alphabet consists of all 256 distinct symbols that can be represented using 8 bits in a byte. By contrast, in string matching for proteogenomic mapping, the alphabet consists of 20 amino acids and a small number of additional symbols that can be represented in 5 bits. Furthermore, in genomics only 3 bits are required to represent the four nucleic acid bases and any other special symbols. Furthermore, Jung et al.

do not exploit the dual-ported nature of RAM blocks in modern FPGAs that enables more efficient utilization of storage resources. Therefore, the previously described bit-split implementations designed for Snort are not optimal for proteomics processing in FPGAs.

Sidhu and Prasanna describe a technique for constructing nondeterministic finite automata (NFA) from regular expressions that can be used for string matching [26]. Their solution requires $O(n^2)$ space where n is the number of characters in the regular expressions to be searched. Because their NFA is implemented entirely in FPGA logic, this technique requires large FPGAs to implement searches for large-string sets. Lin et al. describe a technique for improving the space efficiency by up to 20% for NFA implementations in FPGA logic fabric [27]. Their architecture optimizes space by sharing common prefixes, infixes, and suffixes between multiple regular expressions. Fide and Jenks provide an extensive survey of string matching techniques and implementations in hardware [28]. The survey focuses on intrusion detection and network router implementation.

12.2 Approach

Utilization of ACA for string set matching requires two phases (a) preprocessing, and (b) searching. In the preprocessing phase, a DFA is constructed from the set of strings to be matched. The DFA is used to perform that matching in the subsequent search phase. The preprocessing phase has a runtime complexity of $O(pn)$ and the search phase has a runtime complexity of $O(m + k)$. Detailed description and analysis of ACA can be found in [8]. A brief description follows below.

12.2.1 The Aho–Corasick Preprocessing Phase

In the preprocessing phase, the DFA is constructed using three steps. In the first step, the set of target strings, P, is organized into a *keyword tree* data structure, $\Gamma = \{N, E\}$ where N is the set of nodes and E is the set of directed edges. A keyword tree is rooted at node $r \in N$, each node $v \in N$ represents a prefix in the set of strings and each edge $e \in E$ is labeled with a single character. In a keyword tree, all edges out of a node have unique labels and the sequence of edges on the path from r to v specifies a prefix of a string in P. Furthermore, all strings in P map to nodes in Γ. Figure 12.1 shows an example keyword tree for the set of strings P = {"ACA," "ACACD," "ACE," "CAC"}. The paths from node 0 to nodes 3, 5, 6, and 9 correspond to strings "ACA," "ACACD," "ACE," and "CAC," respectively.

The keyword tree can be used to search corpus T for strings in P by starting at the root node 0 and following the path indicated by the prefix of T

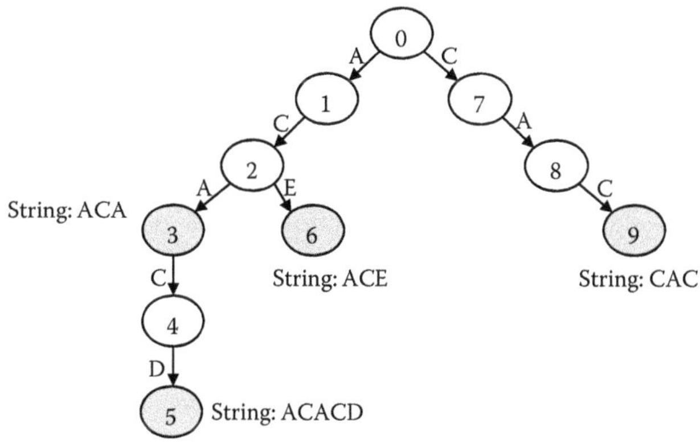

FIGURE 12.1
Keyword tree for P = {"ACA," "ACACD," "ACE," "CAC"}.

starting at position 1 until either a node that maps to a pattern in P is found or no edge leading out of a node matching the corresponding character in the prefix exists. If a node v maps to a pattern, the match is reported and the search can continue along the path to match longer strings if v is a nonleaf node (e.g., node 3 in Figure 12.1). If v is a leaf node or no match is found, then the search starts over again at the root of T with the prefix of T starting at position 2. This process continues until all of T has been searched. However, this naïve search method has a runtime complexity of $O(n \times m)$.

In the second preprocessing step, "failure links" are added to the tree to speedup the search. Failure links are nonlabeled edges that account for overlapping patterns strings in the corpus and can be used to continue the search, without starting at the root of the tree, when the current branch of the tree fails to produce a match because of the current symbol in the text string does not match an edge out of the current node. The algorithm for adding failure links is given below (Algorithm 1):

```
Algorithm 1 Add failure links to keyword tree Γ
The failure link of the root node, r in Γ, leads to r;
The failure links of all immediate child nodes of r also lead
to r;
Repeat for every node v in Γ for which the failure link is not
yet known traversed in a breadth-first manner:
        v' := parent of v;
        x := character that labels edge (v', v);
        w := node that results from following the failure link
             out of v';
        While there is no edge out of w labeled x and w ≠ r:
                w := node that results from following the
                     failure link out of w;
        End while;
```

```
      If there is an edge (w, w') labeled x
               Add failure link from v to w';
      Else
               Add failure link from v to r;
End repeat;
```

Note that the algorithm requires the failure links of all nodes with shorter path lengths than v to be known before v can be processed. This is guaranteed by computing failure links for all nodes in a breadth-first manner. Figure 12.2 shows the result of adding failure links to the keyword tree constructed in Figure 12.1. In Figure 12.2, the failure links from nodes 0, 1, 5, 6, which lead to node 0 and are not shown to preserve clarity.

In the third preprocessing step, a DFA is constructed from the keyword tree by adding labeled edges corresponding to the information encoded within the keyword tree using the following algorithm (Algorithm 2):

```
Algorithm 2 Add edges corresponding to failure links to
            keyword tree Γ
Repeat for every nonroot node v in Γ traversed in a breadth-
first manner:
      w:= node that results from following the failure link
          out of v;
      Add to v any pattern matches indicated by w;
      For every edge (w, w') out of w labeled x:
           If there is not already an edge out of v labeled x
                   Add edge (v, w') out of v labeled x;
      End for;
End repeat;
```

Figure 12.3 illustrates the DFA resulting from applying the algorithm to the example keyword tree. In the figure, state 0 is the start state and the shaded states 3, 4, 5, 6, and 9 match peptides ACA, CAC, ACACD, ACE, and

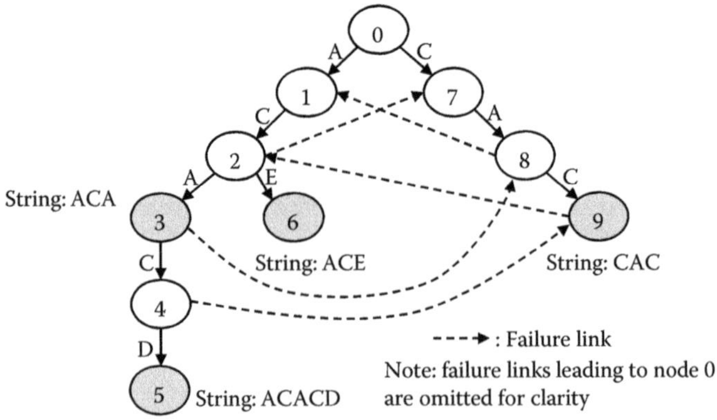

FIGURE 12.2
Keyword tree with added failure links.

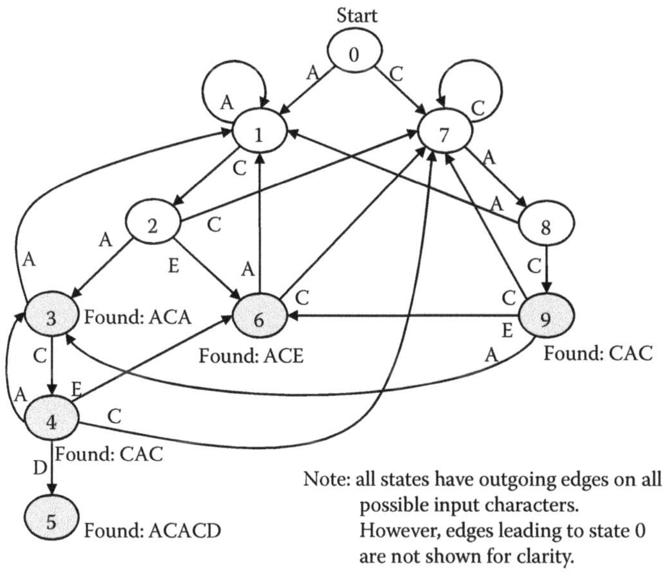

FIGURE 12.3
DFA for matching P = {"ACA," "ACACD," "ACE," "CAC"}.

CAC, respectively. The DFA can be used to match sets of strings in an input corpus as follows (Algorithm 3):

```
Algorithm 3 Search input corpus using a DFA
      State v:= 0;
      While there are characters in the input corpus:
            c:= character at the front of the corpus;
            Remove the character at the front of the corpus;
            If there exists an edge (v, v') out of v labeled c:
                   v:= v';
                  If v indicates a match then report that
                  match;
            Else
                  v:= 0;
End while;
```

In a computer, the DFA state transitions can be represented in the form of a table. Table 12.1 presents a table-oriented representation used for implementation of the DFA in Figure 12.3.

12.3 FPGA Implementation of the String Set Matching DFA

The DFA resulting from the ACA processing phase can be directly used for implementing a string set matching solution in an FPGA. However, the large

TABLE 12.1

A Table-Oriented Representation of the DFA in Figure 12.3

	Input Character				
Current State	A	C	D	E	Match
0	1	7	0	0	Ø
1	1	2	0	0	Ø
2	3	7	0	6	Ø
3	1	4	0	0	ACA
4	3	7	5	6	CAC
5	1	7	0	0	ACACD
6	1	7	0	0	ACE
7	8	7	0	0	Ø
8	1	9	0	0	Ø
9	3	7	0	6	CAC

size of the resulting DFA for any realistic problem in computational biology will require the DFA be stored in external memory (e.g., in the form of external DDR or DDR2 modules). While off-FPGA memory is plentiful and relatively inexpensive, it is also slow and requires several clock cycles of latency for every memory access. Therefore, using external memory for searches is not ideal. Splitting the set of strings to be matched into several small subsets, each encoded into small DFAs that are executed separately, is a straightforward approach to reducing storage requirements. However, using this approach alone will likely require several passes over the input corpus, thereby increasing the total processing time. Therefore, other approaches for reducing storage requirements of string set matching DFAs are described below.

12.3.1 Bit-Split DFA Architecture

For a given DFA, its branching factor and the number of states have a significant impact on its storage requirements. Therefore, both these attributes must be addressed in an FPGA implementation. As observed in [25], the branching factor of the DFA can be reduced by splitting the original DFA, A, into smaller DFAs $A_0, A_1, \ldots, A_B$, where B is the number of bit positions required to encode the characters in the input corpus and automata A_b processes the bit position b in the input corpus. In proteomics, 5-bit positions are sufficient for encoding 32 characters representing all possible amino acids and special characters. In genomics, two-bit positions can encode the four bases. However, three bits are required if additional special characters such as "*N*" are required. A small translation module can convert ASCII-encoded input characters into the binary equivalents, if required.

Table 12.2 shows the table-based structure of the 5-bit-split DFAs corresponding to the DFA in Figure 12.3. The states in the bit-split DFAs are composed from sets of states in the original DFA. Furthermore, each state in the bit-split DFA has exactly two edges labeled 0 or 1. The central idea behind the bit-split DFA construction algorithm shown below (Algorithm 4) is to aggregate states in the original DFA into states in the bit-split DFA based on the identity of the corresponding bit positions of the edge labels. In following discussion, $A{:}p$ refers to state p in the original DFA and $A_b{:}q$ refers to state q in the bit-split DFA for bit position b; bit positions are specified in a right-to-left order starting at 0.

```
Algorithm 4 Construct DFA A_b for bit position b from DFA A.
V:= {v | v is state 0 in A};
Add state V to A_b;
Insert V into Q, a list of newly added states in A_b;
While Q is not empty:
        V:= the state in the front of Q;
        Remove the state in front of Q;
        E_0:= {(v, v') | (v, v') ∈ A ∧ v ∈ V ∧ the label of (v,
             v') at bit position b is 0};
        V_0' := { v' | (v, v') ∈ E_0};
        If V_0' does not already exist in A_b;
                Add V_0' to A_b;
                Insert V_0' to the back of Q;
        Add edge (V, V_0') with a label of 0 to A_b;
        E_1:= {(v, v') | (v, v') ∈ A w v ∈ V ∧ the label of (v,
             v') at bit position b is 1};
        V_1' := { v' | (v, v') ∈ E_1};
        If V_1' does not already exist in A_b;
                Add V_1' to A_b;
                Insert V_1' to the back of Q;
        Add edge (V, V_1') with a label of 1 to A_b;
End while;
```

Consider the construction of bit-split DFA A_0 given the original DFA A in Figure 12.3. Initially, the start state $A_0{:}0 = \{A{:}0\}$ is added to A_0. Next, all states in A that can be reached from the states that comprise $A_0{:}0$ when the bit position 0 in the edge label is 0 are determined and aggregated into a new bit-split node $A_0{:}1$. In the example, $A{:}1$ and $A{:}7$ are aggregated to form $A_0{:}1$. Because $A_0{:}1$ does not already exist in A_0 (i.e., there is no state in A_0 that is aggregated from exactly $A{:}1$ and $A{:}7$), it is added to A_0. Furthermore, an edge $(A_0{:}0, A_0{:}1)$ with a label of 0 is also added to A_0. Next, all states in A that can be reached from the states that comprise $A_0{:}0$ when the bit position 0 in the edge label is 1 are determined and aggregated; in this example, the aggregation results in state $\{A{:}0\}$ that already exists in A_0. Therefore, the edge $(A_0{:}0, A_0{:}0)$

TABLE 12.2

Bit-Split DFAs Equivalent to the DFA in Figure 12.3

(a) A_0

State	0	1	Match
0	1	0	Ø:0000
1	2	0	Ø:0000
2	3	0	Ø:0000
3	4	0	4,3,1:1101
4	4	5	4,3,1:1101
5	1	0	2:0010

(b) A_1

State	0	1	Match
0	1	2	Ø:0000
1	1	3	Ø:0000
2	4	2	Ø:0000
3	5	2	Ø:0000
4	1	6	Ø:0000
5	1	7	3,1:0101
6	5	2	4:1000
7	5	8	4:1000
8	4	2	2:0010

(c) A_2

State	0	1	Match
0	1	0	Ø:0000
1	2	0	Ø:0000
2	3	4	Ø:0000
3	5	4	4,1:1001
4	1	0	3:0100
5	6	4	4,1:1001
6	6	4	4,2,1:1011

(d) A_3

State	0	1	Match
0	1	0	Ø:0000
1	2	0	Ø:0000
2	3	0	Ø:0000
3	4	0	4,3,1:1101
4	5	0	4,3,1:1101
5	5	0	4,3,2,1:1111

(e) A_4

State	0	1	Match
0	1	0	Ø:0000
1	2	0	Ø:0000
2	3	0	Ø:0000
3	4	0	4,3,1:1101
4	5	0	4,3,1:1101
5	5	0	4,3,2,1:1111

with a label of 1 is added to A_0. This process is repeated for all newly added states in A_0 until there are no new unprocessed states in A_0.

A_0:1 was added to A_0 previously and is examined next. Recall that A_0:1 is an aggregate of *A*:1 and *A*:7. Therefore, all states in the original FSM that can be reached from either *A*:1 or *A*:7 when the edge label at bit position 0 is 0 are aggregated into A_0:2. In this example, A_0:2 is created from *A*:2 and *A*:8. Because A_0:2 does not already exist in A_0, it is added to A_0 and the edge (A_0:1, A_0:2) with a label of 0 is also added to A_0. Next, all states in the original DFA that can be reached from either *A*:1 or *A*:7 when the edge label at bit position 0 is 1 are aggregated; in this example, the aggregation results in state {*A*:0} that already exist in A_0. Therefore, the edge (A_0:1, A_0:0) with a label of 1 is added to A_0. This process is continued until there are no new states added to A_0. Note that only unique new nodes are added to A_b. When a new state A_b:*n* is created by aggregation but another node, A_b:*k*, created by aggregating the same set of states in *A* already exists in A_b, then instead of inserting the new node, A_b:*n*, an edge leading to A_b:*k* from the state currently under examination is inserted into A_b.

Indicators for string matches are also handled by aggregation (i.e., state A_b:*k* matches all the strings that are matched by the states in the original DFA that were aggregated into A_b:*k*). Given a set of *p* strings, each state in A_b can match up to *p* strings. Therefore, a sequence of bits is used for indicating the matches represented by state A_b:*k* such that a value of 1 at bit position *p* indicates a match with string *p* in the string set. For the example in Table 12.2(b), a sequence of four bits is used to represent the matches in each state and because state A_1:5 is created by aggregating states *A*:3 and *A*:6, a matching bit sequence of 0101 is used to indicate matches to strings 1 and 3 (note that the right-most bit corresponds to bit position 1).

Table 12.2 illustrates the five bit-split DFAs resulting from the original DFA in Table 12.1 assuming proteomics processing. The characters A, C, D, and E are encoded as 00000, 00010, 00011, and 00100, respectively. Essentially, there is no need to create a more compact encoding that eliminates the codes for letters not used to represent amino acids (e.g., B) because all five bits are needed to encode the 20 amino acids and any other special characters (e.g., Z).

12.3.2 Implementing Bit-Split DFA Tables in FPGAs

Implementing the bit-split DFAs using lookup tables is more resource efficient as compared with encoding sequences of conditional state transitions in the FPGA logic fabric. In modern FPGAs, the DFA tables can be stored using either configurable logic block resources (i.e., distributed RAM) or BRAM. However, BRAM is more efficient when storing large tables because it has higher storage density than distributed RAM [9].

Xilinx FPGAs provide a large number of 18-kbit BRAMs that can be organized into 512 rows of 36-bit wide words [29]. The Xilinx BRAMs are dual ported; therefore, by tying the high-order bit of the 9-bit BRAM address input to 0 on one port and to 1 on the other port, the BRAM can be divided

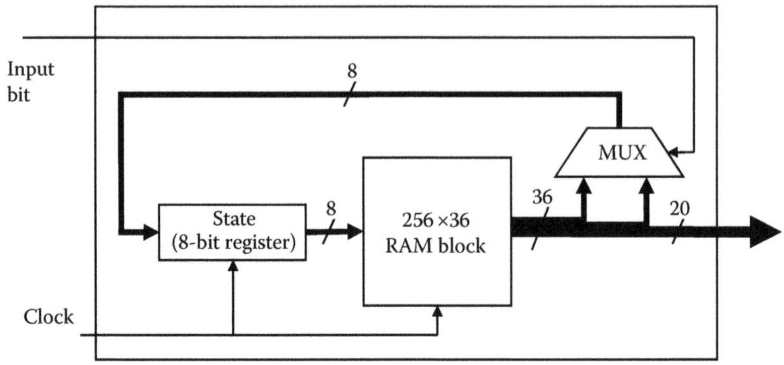

FIGURE 12.4
Architecture of the bit-split DFA implementation.

into two independent 9-kbit RAM blocks containing 256 rows of 36-bit words each. Altera FPGAs also provide a large number of 9-kbit BRAMs that can be organized into 256 rows of 36-bit wide words (other BRAM configurations are also possible but are not used in this application) [30].

A 256×36-bit block of RAM can hold 256 rows of a bit-split Aho–Corasick DFA. Essentially, the bit-split DFA reads the row corresponding to the current state to output the string match bit vector and to determine the next state. The DFA can transition into one of two states (note that the DFA can transition back into the state it is currently in) depending on the input value (i.e., 0 or 1). Because 8 bits are required to represent each of the 256 possible next state values, 16 bits in each 36-bit wide row are used for storing the two possible next state values. The remaining 20 bits in the row are used to store the 20-position string match bit vector.

Figure 12.4 illustrates the architecture of a bit-split DFA. In addition to the 9-kbit RAM block, the implementation requires an 8-bit register to store the current state and a multiplexer to select one of the two next state values based on the value of the input bit. For proteomics applications, five of these bit-split DFA modules are combined to create a complete Aho–Corasick tile as depicted in Figure 12.5. Inside a tile, the 5-bit input to the Aho–Corasick implementation is distributed to the 5 bit-split DFAs. A bit-wise and operator combines the bit-split string match vectors into the consensus 20-bit string match vector output.

To conserve signal routing resources, the consensus string match vector is converted into a five-bit numerical value using a 20-to-5 bit priority encoder (the reason for using a priority encoder is explained toward the end of this section). The encoder essentially scans the consensus string match vector in increasing index order and returns the index of the first bit that has a value of 1. Therefore, strings that appear near the beginning of the list of strings have higher priority than those appearing later. If all consensus string match vector bits are clear (i.e., there is no match), the priority encoder returns an undefined value. Therefore, to indicate that a string has been found, an output

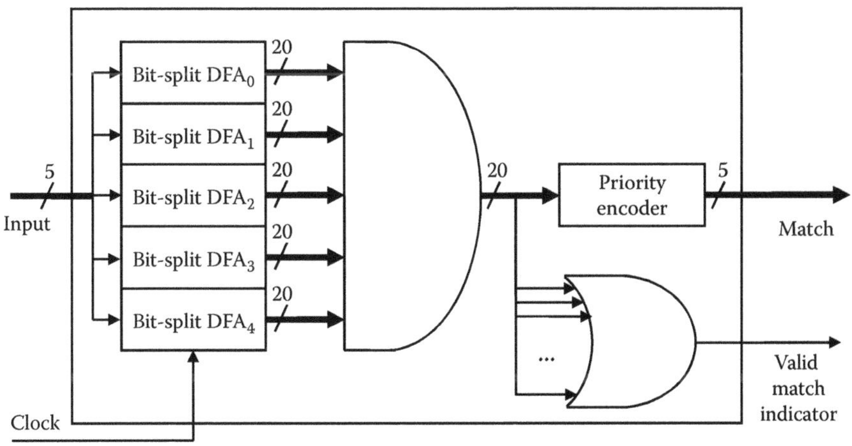

FIGURE 12.5
Architecture of an Aho–Corasick tile.

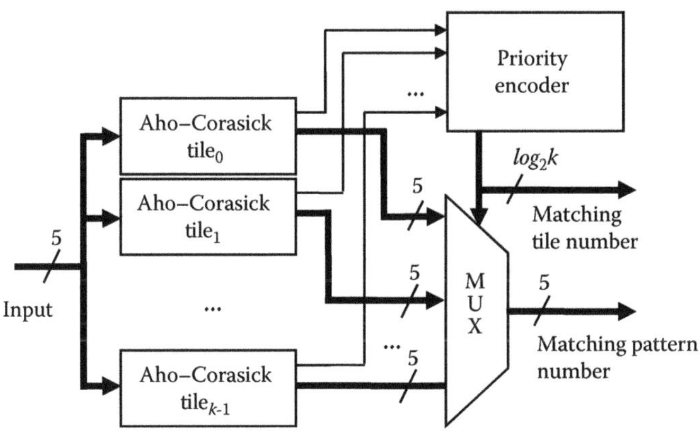

FIGURE 12.6
Aho–Corasick implementation with k tiles.

valid indicator signal is also generated when any of the consensus string match vector bits are set.

Typically, proteomics pipelines require the detection of more than 20 peptides. In this case, several Aho–Corasick tiles can be utilized in parallel as shown in Figure 12.6. The input reading frame is simultaneously streamed to all tiles. The output of the tiles is combined into a single output peptide number using a priority encoder. Because the priority encoder produces an undefined value when no peptide is matched, a match indicator signal is also generated when any of the tiles indicate a valid match.

Using the architecture described earlier, each tile can detect up to 20 peptides in an input stream of reading frame data. However, because the tile has a capacity of only 256 states per bit-split DFA, in some cases, it may be necessary to reduce the number of peptides that can be detected to create a bit-split DFA with no more than 256 states (note that all five peer bit-split DFAs must represent the same reduced number of peptides). A simple iterative greedy algorithm can be employed to allocate peptides to tiles using a trial-and-error approach. Initially, the algorithm assigns a set of 20 peptides to a tile. If the bit-split DFA for the given number of peptides has more than 256 states, the algorithm reduces the number of peptides assigned to the tile and tries again until the bit-split DFA is successfully created.

To minimize the number of states required in the Aho–Corasick implementation, strings beginning with the same sequence of characters should be grouped together into the same tile. This is because the strings will share the same initial states in the DFA. One way to achieve this is to sort the set of peptide strings in ascending alphabetical order before assigning them to the Aho–Corasick tiles. In addition, this bit-split Aho–Corasick implementation architecture can only indicate a single string match at any given time. This is typically not a problem unless one string is a suffix of another peptide. String p' is a suffix of string p if and only if the length of p is greater than or equal to the length of p' and p ends with a substring that is identical to p'. In this case, if the string for p appears in the corpus text, it is sufficient to simply indicate that p has been found because this also implies that p' has been found. The priority encoding architecture ensures that the detection of a match with p receives higher priority than p' as long as p appears before p' in the sorted set of strings. Therefore, the sorting of the peptide string set must account for both alphabetical and suffix-based priority ordering.

12.3.3 Analysis of DFA Storage Utilization Efficiency

Assume that P_i, such that $1 \le P_i \le 20$, is the number of strings that can be detected by tile i. This means that in each bit-split DFA table row, $20 - P_i$ bits are unused for indicating matches. Furthermore, in a majority of cases, a bit-split DFA requires fewer than 256 states to detect P_i strings. This means that when DFA $A_{i,b}$ of tile i requires $S_{i,b}$ states, such that $1 < S_{i,b} \le 256$, then $256 - S_{i,b}$ rows of available storage in the 9-kbit RAM block are unused.

The storage utilization efficiency of a single 256 by 36-bit block used by a single bit-split DFA is computed as follows:

$$\eta_{i,b} = \frac{16S_{i,b} + P_i S_{i,b}}{256 \times 36} = \frac{(16 + P_i)S_{i,b}}{9216} \tag{12.1}$$

The overall storage utilization for an implementation requiring T tiles can be computed using the following expression:

$$\eta = \frac{\sum_{i=1}^{T} \sum_{b=0}^{4} (16 + P_i)S_{i,b}}{5 \times 9216 \times T} = \frac{\sum_{i=1}^{T} \sum_{b=0}^{4} (16 + P_i)S_{i,b}}{46080T} \tag{12.2}$$

12.4 Case Study

The bit-split ACA was implemented for a hypothetical proteomics application requiring the matching of peptides with the human genome using the Virtex-4 family of FPGAs as reported in [10]. For example, the Virtex-4 FX-100 device has 376 18-kbit BRAM blocks, of which 350 were used for implementing Aho–Corasick tiles and the remaining 26 were reserved for meeting the storage requirements of other modules that support the implementation (e.g., input/output [I/O] functions and the memory for the embedded processor core that controls the overall implementation). Recall that a Xilinx BRAM can be configured as two 9-kbit RAM blocks. This means that there are effectively 700 RAM blocks available that can hold a total of 140 tiles. Because each tile requires five 9-kbit RAM blocks and can search for at most 20 peptides, the maximum number of peptides that can be searched using this device is 700 × 20/5 = 2,800.

For the case study, reading frame data was derived using software on a standard workstation by concatenating all the chromosomes in the human genome (separated by sequences of 100 "N" characters). Next, the amino acid sequences were derived from each of the six resulting reading frames and were concatenated together. This resulted in 6,160,844,220 bytes of text to be searched for a given set of peptides. In-silico trypsin digestion was used to construct 400 different sets of peptides from chromosome 1 of the human genome for this experiment. In particular, 100 sets, each containing 2,800 randomly selected peptides ranging in lengths from 5 to 30 amino acids were constructed. Similarly, 100 sets of 2,800 peptides with lengths ranging from 10 to 30, 15 to 30, and 20 to 30 amino acids were also created. Results of the bit-split implementations for each of the 400 sets of peptides are summarized below.

12.4.1 Storage Utilization

Table 12.3 summarizes the results from generating the Aho–Corasick implementation for the various peptide sets. Most of the peptide sets where the minimum peptide size is 5 and having an average length of just more than 11 amino acids were accommodated using 140 tiles. Two of these peptide sets require an additional tile because for each of these sets one of the tiles can only accommodate 19 peptides within the 256 state limits. The average storage utilization in the tiles is approximately 52.7% because many of the bit-split FSMs require significantly fewer than the available 256 states.

The number of tiles required for the peptide sets with the minimum peptide length of 10 amino acids (average length of 15.40) varies between 141 and 142 with an average of 19.8 peptides detected per tile. The average storage utilization is a much higher 81.12%. The average number of tiles required for the peptide sets with the minimum peptide length of 15 amino acids

TABLE 12.3
Tile Packing Efficiency

Peptide Length			Average Number of Tiles Required	Average Peptides Per Tile	Average Storage Efficiency (%)
Min	Max	Average			
5	30	11.28	140.02	19.99	52.70
10	30	15.40	141.41	19.80	81.12
15	30	19.79	178.40	15.70	81.53
20	30	23.70	277.23	12.32	72.96

(average length of 19.79) is 178.40 with an average of 15.7 peptides detected per tile. The average storage utilization is relatively high at 81.53%. This efficiency is comparable to the efficiency of the peptide sets with average size of 15.40. However, while in the case of the shorter peptides, underutilization of row storage is the main cause of inefficiency, for longer peptides, underutilization of the peptide match vector storage has a larger contribution to the overall inefficiency.

The average number of tiles required for the peptide sets with the minimum peptide length of 20 amino acids (average length of 23.70) is 277.23 with an average of 12.32 peptides detected per tile. The number of tiles required is significantly larger than in the previous cases because the bit-split FSMs have more states. The storage efficiency is also reduced to 72.96% because of significant underutilization of match vector storage.

12.4.2 Implementation Performance

Maximizing the clock frequency is an important goal of many digital designs because higher operating frequencies result in faster execution. In general, the implementations of large designs in FPGAs typically have lower operating frequencies as compared with smaller designs because the signals must traverse greater distances on the chip. Therefore, to study the practical limits of implementations with large numbers of tiles, Xilinx's FPGA application development tool, XST 10.1.03, was used to implement a number of designs with varying number of Aho–Corasick tiles on a Virtex-4 FX-140 speed grade-11 device. Table 12.4 lists the performance statistics of the designs reported by XST. The smallest design composed of 40 tiles, requiring 100 BRAM blocks with a capacity of 800 peptides, operates at a frequency of 198.649 MHz. The design composed of 200 tiles, requiring 500 BRAM blocks with a capacity of 4,000 peptides, operates at a frequency of 134.971 MHz.

The runtime performance of the FPGA-based bit-split Aho–Corasick implementation was also compared with the performance of the Aho–Corasick implementation on a standard workstation. The bit-split Aho–Corasick design with supporting elements such as an embedded PowerPC processor,

TABLE 12.4

Operating Frequencies of Aho–Corasick Designs with a Variety of Tiles

Peptides	Tiles	BRAMs	Frequency (MHz)
			Virtex-4 FX140-11
800	40	100	198.649
1,600	80	200	169.866
2,400	120	300	169.635
3,200	160	400	150.648
4,000	200	500	134.971

an ATA hard disk controller, an RS232 link, system busses, and memory are synthesized to run at a clock frequency of 100 MHz on a board with a Virtex-4 FX-100 device. In this design, an ATA disk controller module implemented on the FPGA is used for reading data at a peak rate of 100 MB/s (i.e., one character from the reading frame is streamed to the Aho–Corasick tiles every clock cycle). The PowerPC is responsible for initializing the disk drive and initiating the read operations. The PowerPC also monitors the peptide match indications from the tiles and reports match data (e.g., peptide and location) to the host workstation over the RS232 link.

Previous results show that the Aho–Corasick tiles can operate at frequencies more than 150 MHz, resulting in input rates exceeding 150 Mbps. Although the Aho–Corasick tiles can operate at faster frequencies, in this series of experiments the clock frequency was restricted to execute at 100 MHz system clock to eliminate the complexity that arises with designs containing multiple clock domains. Essentially, the Aho–Corasick tiles operate at 100 MHz to match the ATA controller's peak data delivery rate of 100 MB/s and the system bus that is restricted to run at 100 MHz. Note that the tiles do not introduce any processing delays (i.e., the disk drive is the primary performance bottleneck in this implementation and increasing the implementation's clock frequency to 150 MHz produces no tangible improvements in processing time). Furthermore, to minimize processing and concomitant delays associated with a file system, the reading frame data is stored on consecutive sectors on a raw disk (i.e., the disk is not formatted using a well-known, operating system-supplied file system). Essentially, the 6,160,844,220 bytes of text derived from the human genome is written to 12,032,899 consecutive sectors on an IDE disk drive at a known starting location. The disk drive is connected to the FPGA board and a flash RAM module containing the Aho–Corasick implementation FPGA configuration file and RAM block content implementing the DFA tables is also connected to the board. On bootup, the FPGA board reads the configuration information from the flash RAM and begins executing the ACA.

For these experiments, a set of 2,800 peptides that fit in exactly 140 tiles (i.e., a set with minimum and average peptide lengths of 5 and 11.5257 amino acids, respectively) was selected. Note that storage efficiency of the selected peptide set has no bearing on the runtime performance of the bit-split Aho–Corasick implementation. This is because the Aho–Corasick tiles can each search for a subset of 20 peptides in parallel. The performance of the FPGA-based implementation was compared with the performance of a software implementation employing a traditional table-driven Aho–Corasick organization in which a single large table represents a single FSM with all the states for all 2,800 peptides. The software implementation was executed on a Windows XP workstation having a 2.67 GHz Intel Core2 Duo processor, 2 GB RAM, and a pair of Serial ATA disks configured as a RAID 0 disk drive (i.e., striped data for fast disk I/O), formatted as a new technology file system (NTFS) volume.

Five runs each of the FPGA and workstation implementations were performed. The FPGA implementation takes, on average, 94.17 seconds to process the entire 6 gigabytes of reading frame data. The workstation implementation takes an average of 1870.18 seconds to complete the search. This means that the FPGA-based implementation is nearly 20 times faster than the workstation implementation.

12.5 Conclusions

This chapter describes a technique for accelerating string set matching implementation using FPGAs for use in computational biology applications. FPGAs provide a large number of embedded memory blocks that enable more efficient implementation of DFAs than possible using the FPGA logic fabric. Furthermore, the synthesized tile-based design can be reused for different peptide sets by simply initializing the RAM block content with appropriate bit-split DFA state data. This is much faster than having to rerun the significantly time-consuming "placing and routing" synthesis stages required for logic-based implementations in FPGA fabric for each new peptide set.

Empirical results presented here show that the FPGA-based implementation of Aho–Corasik outperforms the workstation implementation by a factor of 20. This demonstrates that using specialized hardware to solve the string set matching problem can make a significant impact on the runtime of a number of computational biology processes where exact string matching is commonly required. The throughput of FPGA implementation described here is essentially limited by the data transfer speed of the ATA disk drives. Higher frequency implementations utilizing Serial ATA (SATA) disk drives, parallel disk arrays, and gigabit Ethernet interfaces are under investigation as part of ongoing implementation efforts and future research.

This chapter also demonstrates that the significantly smaller string alphabets found in computational biology enable more space-efficient designs for string matching as compared to previously published implementations focused on network intrusion detection. Although the case study focused on exact string matching, the ACA can also accommodate regular expressions. The implementation described can easily be adapted for other types of search using, for example, spaced seeds.

A number of techniques can be used to increase the number of peptides that can be searched. A simplest approach is to utilize several devices in parallel (note that this also requires the input corpus to be replicated for each FPGA). The cost of replicating the corpus for a large number of FPGAs can be eliminated by implementing a data streaming interface between the separate FPGA boards. Tools to facilitate building such interfaces are typically provided by the FPGA vendors [31, 32]. Using such an interface, only one board needs to be connected to a single disk drive containing the corpus while the other boards are connected to each other in a chain using multi-gigabit-per-second serial links available on many modern FPGA devices. This way, the reading frame data can be streamed from the disk drive to each board (i.e., as soon as an FPGA board receives a byte of data, it forwards the data to the next board in the chain). The runtime complexity of this latter implementation is essentially $O(m + k + \lambda)$, where λ represents the cumulative latency of transmitting a single character over the entire chain.

12.6 References

1. Jaffe, J.D., H.C. Berg, and G.M. Church, Proteogenomic mapping as a complementary method to perform genome annotation. *Proteomics*, 2004. 4(1): p. 59–77.
2. Jaffe, J.D., et al., The complete genome and proteome of *Mycoplasma mobile. Genome Research*, 2004. 14(8): p. 1447–1461.
3. Kalume, D.E., et al., Genome annotation of *Anopheles gambiae* using mass spectrometry-derived data. *BMC Genomics*, 2005. 6: p. 128.
4. Kuster, B., et al., Mass spectrometry allows direct identification of proteins in large genomes. *Proteomics*, 2001. 1(5): p. 641–650.
5. McCarthy, F.M., et al., Modeling a whole organ using proteomics: The avian bursa of Fabricius. *Proteomics*, 2006. 6(9): p. 2759–2771.
6. Boyer, R.S. and J.S. Moore, A fast string searching algorithm. *Communications of the ACM*, 1977. 20: p. 762–772.
7. Knuth, D.E., J.H. Morris, and V.B. Pratt, Fast pattern matching in strings. *SIAM Journal of Computing*, 1977. 6: p. 323–350.
8. Aho, A. and M. Corasick, Efficient string matching: An aid to bibliographic search. *Communications of the ACM*, 1975. 18: p. 333–340.
9. Tan, L. and T. Sherwood, A high throughput string matching architecture for intrusion detection and prevention, in *32nd Annual International Symposium on Computer Architecture*. 2005: Madison, Wisconsin US. p. 112–122.

10. Dandass, Y.S., et al., Accelerating string set matching in FPGA hardware for bioinformatics research. *BMC Bioinformatics*, 2008. 9(197).
11. Brudno, M. and B. Morgenstern, Fast and sensitive alignment of large genomic sequences. *Proceedings/IEEE Computer Society Bioinformatics Conference*. 2002. 1: p. 138–147.
12. Brudno, M., R. Steinkamp, and B. Morgenstern, The CHAOS/DIALIGN WWW server for multiple alignment of genomic sequences. *Nucleic Acids Research*, 2004. 32(Web Server issue): p. W41–44.
13. Castelo, A.T., W. Martins, and G.R. Gao, TROLL—Tandem repeat occurrence locator. *Bioinformatics*, 2002. 18(4): p. 634–636.
14. Farre, D., et al., Prediction of transcription factor binding sites with PROMO v. 3: Improving the specificity of weight matrices and the searching process, in *5th Annual Spanish Bioinformatics Conference*. 2004: Barcelona, Spain.
15. Hyyro, H., M. Juhola, and M. Vihinen, On exact string matching of unique oligonucleotides. *Computers in Biology and Medicine*, 2005. 35(2): p. 173–181.
16. Michael, M., C. Dieterich, and M. Vingron, SITEBLAST—Rapid and sensitive local alignment of genomic sequences employing motif anchors. *Bioinformatics*, 2005. 21(9): p. 2093–2094.
17. Buhler, J., U. Keich, and Y. Sun, Designing seeds for similarity search in genomic DNA. *Journal of Computer and System Sciences*, 2005. 70(3): p. 342–363.
18. Boeva, V., et al., Exact *p*-value calculation for heterotypic clusters of regulatory motifs and its application in computational annotation of *cis*-regulatory modules. *Algorithms Molecular Biology*, 2007. 2(1): p. 13.
19. Li, I.T., W. Shum, and K. Truong, 160-fold acceleration of the Smith–Waterman algorithm using a field programmable gate array (FPGA). *BMC Bioinformatics*, 2007. 8: p. 185.
20. Mak, T.S.T. and K.P. Lam, Embedded computation of maximum-likelihood phylogeny inference using platform FPGA, in *2004 IEEE Computational Systems Bioinformatics Conference (CSB'04)*, K.P. Lam, Editor. 2004. p. 512–514.
21. Lokhov, P.G., et al., Database search post-processing by neural network: Advanced facilities for identification of components in protein mixtures using mass spectrometric peptide mapping. *Proteomics*, 2004. 4(3): p. 633–642.
22. Oliver, T., et al., Using reconfigurable hardware to accelerate multiple sequence alignment with ClustalW. *Bioinformatics*, 2005. 21(16): p. 3431–3432.
23. Alex, A.T., et al., Hardware-accelerated protein identification for mass spectrometry. *Rapid Commun Mass Spectrom*, 2005. 19(6): p. 833–837.
24. Bogdan, I., et al., Hardware acceleration of processing of mass spectrometric data for proteomics. *Bioinformatics*, 2007. 23(6): p. 724–731.
25. Jung, H.-J., Z.K. Baker, and V.K. Prasanna, Performance of FPGA implementation of bit-split architecture for intrusion detection systems, in *Reconfigurable Architectures Workshop at IPDPS (RAW '06)*. 2006.
26. Sidhu, R. and V.K. Prasanna, Fast regular expression matching using FPGAs, in *The 9th Annual IEEE Symposium on Field-Programmable Custom Computing Machines*, V.K. Prasanna, Editor. 2001. p. 227–238.
27. Lin, C., et al. Optimization of regular expression pattern matching circuits on FPGA, in *Conference on Design, Automation and Test in Europe: Designers' Forum*, 2006. Munich, Germany.

28. Fide, S. and S. Jenks, A survey of string matching approaches in hardware. Technical Report SPDS 06-01, 2006, University of California, Irvine.
29. Xilinx, I. Virtex-4 Family Overview, 2007.
30. Altera Stratix III Device Handbook, Volume 1, 2007.
31. Altera SerialLite II Protocol Reference Manual, 2005.
32. Xilinx, DS128: Aurora v2.8. 2007.

13

Reconfigurable Neural System and Its Application to Dimeric Protein Binding Site Identification

Feng Lin and Maria Stepanova

13.1 Introduction

Computational approaches to the genome-wide identification of transcription factor binding sites (TFBSs) have attracted bioinformatics researchers [1–3]. Since the UIPAC matching algorithm was first reported in 1991 [4], the computational methods have evolved to complex multiple-feature prediction frameworks that include phylogenetic footprintings [5], gene expression data analysis [6], Bayesian trees and graphs, and so on [7]. Nevertheless, although the previous computational approaches to binding site identification have achieved high sensitivity, they have not demonstrated the desired high specificity comparable to that of experimental methods [8].

Artificial neural networks (ANNs) have been used as an effective tool in pattern recognition [9]. Different ANN architectures have been developed to date, each with its own domain of applicability and requirements [10]. A special ANN architecture, recurrent neural network, is an intelligent computational method of classification especially suitable in the case of partially overlapping classes. The applicability of the recurrent neural networks for biological sequence analysis has recently been reviewed by Hawkins and Boden [11] with the examples of motif recognition and prediction of subcellular localization of peptides. Empirical results obtained by the authors indicate that though the network architecture reflects the presence of a certain bias because of recurrence, properly designed recurrent neural networks indeed provide access to the patterns of biological significance. Hopfield neural network (HNN) is an excellent example of the recurrent neural networks. However, as the complexity of the network grows, computational time for network training and operation becomes prohibitively long, which is often the reason for inaccuracy of such systems. The bottleneck usually lies in certain stages of the neural adaption. Reconfigurable computing, with this respect, allows a hybrid architecture with an application-specific hardware unit, which outperforms general-purpose processors [12].

In a HNN, the states of the recurrent neural models depend on previously processed data; hence, massive data parallelism can hardly be implemented. Some parallelism may be achieved when neurons, or groups of them, are mapped to different processors for independent computation, but such an implementation will incur extremely high communication overheads because of frequent function calls for transmitting neuron outputs between the processors. The problem of overcommunication and bus contentions can be avoided by using the field-programmable gate array technology (FPGA), which, in addition to the on-chip parallelism, provides application-specific logic interconnections for data flow [13]. In contrast to that of distributed computing technologies, the communication cost within an FPGA chip is low. A recent review on the state-of-the-art in development of on-chip neural networks using FPGA was done by Zhu and Sutton [14].

In this chapter, we describe a two-phase neural system for recognition of dimeric DNA motifs, and we demonstrate its power by applying the hybrid system into genome-wide identification of hormone response elements (HREs) in DNA. The first phase is used for sequence-based selection of putative motifs by the feed-forward neural network (FFNN), and the second phase is for structure-based prediction of functional dimers by HNN. We have also invented a dynamic adaptation procedure for HNN for robust prediction of the dimeric structure, yielding highly sensitive and reliable motif recognition.

13.2 Design of the Neural System

For design of multiple-feature frameworks and highly specific ensemble models [15], two or more pattern recognition methods can be designed for different properties of the underlying object, and one may outperform each single method operating solely by aggregating their predictions. In the proposed two-phase neural system for recognition of dimeric DNA motifs, a trained FFNN recognizes putative binding sites that are similar to the preselected set of experimentally confirmed functional motifs. This first phase could be considered as a sequence-based selection. Second, a recurrent neural network assigns each putative binding site predicted at the first step with one of possible dimeric structures (either functional or not functional). The rationale behind this structure-based phase is that a sequence is unlikely to be involved in dimeric DNA-protein binding if it cannot be assigned with a functional dimeric structure [16]. In the following section, the data flow for modeling dimeric DNA motifs is described.

13.2.1 Numerical Representation of DNA Sequences

The set of putative binding sites to be processed by the two-phase neural model consists of a number of DNA sequences in the four-letter alphabet, $\Omega = \{A, C, G, T\}$. Let $\vec{\theta} = [x_1, \ldots, x_L], x_j \in \Omega \; \forall j = 1, \ldots, L$ be a DNA sequence of length L, and $\Theta = \{\vec{\theta}_1, \ldots, \vec{\theta}_n\}$ be a set of such sequences.

Although we are dealing with the DNA alphabet, the neural models require numerical representation of the input data. The space of real numbers is one dimensional, and four nucleotide bases cannot be equidistantly mapped onto it without introducing artifacts to the model. Thus, for numerical representation of the DNA alphabet, we use the *one-hot* encoding scheme that operates as follows. The encoding module Σ is a function on the space of DNA sequences $\Theta - \Sigma : \Theta \rightarrow \Xi_\Theta$, where Ξ_Θ is the space of vectors $\vec{\xi}$ of length $4L$. The elements of this vector are defined as follows:

$$\xi_{4(i-1)+k} = \begin{cases} 1 & if \quad x_i = \omega_k \\ -1 & otherwise \end{cases} \tag{13.1}$$

where the DNA sequence under transformation is $\vec{\theta} = [x_1, \ldots, x_L], x_j \in \Omega$ $\forall j = 1, \ldots, L$, and the DNA alphabet elements are $\omega_1 = 'A'$, $\omega_2 = 'C'$, $\omega_3 = 'G'$, $\omega_4 = 'T'$, $\Omega = \{\omega_1, \omega_2, \omega_3, \omega_4\}$. For example, if $x_u = G = \omega_3$, then the uth element of the sequence is transformed into a four-vector $(-1,-1,1,-1)^T$. Thus, the entire DNA sequence $\vec{\theta}$ undergoes a transformation according to that rule: $\vec{\xi} = \Sigma(\vec{\theta})$. Thus, the two-phase neural system is forestalled by the encoding module $\Sigma : \Theta \rightarrow \Xi_\Theta$. Reverse encoding procedure of the $4L$-vectors $\hat{\Sigma} : \Xi_\Theta \rightarrow \Theta$ is performed using Equation 13.1 as well.

A dimeric motif can be represented in the form of two half-sites with an optional spacer in the middle. In the case of partially symmetric motif composition, the two half-sites are similar to each other and are usually believed to have arisen due to DNA duplication [17]. Moreover, in addition to the dimeric nature of the protein–DNA interaction, the dimerization may have a different structure, namely, head-to-head, head-to-tail, or tail-to-tail composition [18], thus providing us a reason to consider different orientations of the half-sites for modeling TFBS dimers.

Consider the consensus TFBS dimer with the half-site $\vec{y}$ of b base pairs of length $\vec{y} = [y_1, \ldots, y_b]$, $y_j \in \Omega$, $j = 1, \ldots, b$, and the spacer δ_c of c nucleotides. The possible dimeric structures can be represented as follows:

$$\begin{aligned} s_{DR} &= \vec{y} \cup \delta_c \cup \vec{y} \\ s_{IR} &= \vec{y} \cup \delta_c \cup \overleftarrow{y} \\ s_{PR} &= \overleftarrow{y} \cup \delta_c \cup \vec{y} \\ s_{ER} &= \overleftarrow{y} \cup \delta_c \cup \overleftarrow{y} \end{aligned} \tag{13.2}$$

where $\overleftarrow{y} = [y_b, \ldots, y_1]$ represents the reverse of $\vec{y}$. Also, s_{DR} is called *a direct repeat*, s_{IR} is an inverted repeat, s_{PR} is a palyndromic repeat, and s_{ER} is an everted repeat. The subspace of dimeric structures is therefore defined as $\Psi = \{s_{DR}, s_{IR}, s_{PR}, s_{ER}\}$, and the HNN $\aleph$ that predicts TFBS dimeric structures actually performs the transformation from the space of DNA sequences Θ into the space of structures Ψ: $\aleph : \Theta \rightarrow \Psi$.

The two-phase neural system works as follows:

$$\aleph : \Theta \xrightarrow{\Sigma} \Xi_\Theta \xrightarrow{DP-HNN} \Xi_\Psi \xrightarrow{\hat{\Sigma}} \Psi \tag{13.3}$$

where DP-HNN stands for dynamically programmed HNN, described later.

13.2.2 The FFNN

For prior sequence-based selection of DNA motifs, we do not consider their dimeric structure. Instead, we train an FFNN in a supervised manner to distinguish putative binding sites from surely neural sequences. The output of the FFNN phase is a binary signal that indicates whether the input sequence is a motif similar to those from the training set or not, although the definition of required similarity is subject to certain adjustment by a threshold.

For training the FFNN model, back-propagation learning algorithm is used [10]. The learning procedure is stopped if either the 99.9% accuracy level is exceeded or the maximum number of back-propagation cycles (which is at most 5,000) or the error plateau is reached, meaning that no improvement of the training error has been detected by more than 100 of consecutive operations.

13.2.3 The HNN

Hopfield network is a neural network where the connections between units form a directed cycle, and the output of the network becomes its input in an iterative manner [10]. Hopfield networks were initially developed for sequence recognition tasks in the domain of natural language processing [19]. They were also shown to serve effectively as associative memories [20] and pattern classifiers [21–23].

HNNs, like other recurrent neural networks, must be treated differently from FFNNs both when analyzing their behavior and when training them. The goal of HNN training is to design a discrete system that possesses a specific set of equilibrium points such that when an initial condition is provided the network eventually converges at one of its equilibrium points. The network is recursive in that the output is fed back as the input once the network is in operation. Hopefully, the network output eventually settles in one of the original design points. Usually, the theory of dynamical systems is used to model and analyze HNNs [20].

Let the space of functional dimeric TFBSs be a subspace of all DNA sequences. This subspace should be distinguishable from other sequences, at least because the respective transcription factors are capable of distinguishing their binding sites from neutral DNA. Thus, we find a set of basis vectors (which should serve as major indicators of a functional dimeric TFBS) in this subspace, and then use a HNN with the same basis vectors as stable states, for dimeric structure prediction. Convergence of the network from its initial condition to a stable state corresponding to a functional dimeric structure serves as an indicator whether the input vector belongs to the subspace of functional dimers or not. A functional dimeric TFBS that is different from the consensus dimer can be considered as a disturbance in the space of TFBSs (whose diversity is usually explained by evolution and neutral mutations), and thus should be smoothly converted into one of the basis vectors by the trained HNN. Neutral DNA sequences are considered as the elements of the complimentary subspace.

The behavior of recurrent neural networks is more like that of an iterative process or a dynamical system rather than a neural network in its conventional feed-forward form. The equations that describe the HNN operations are defined as follows:

$$\vec{a}(0) = \vec{\xi} \tag{13.4}$$

$$\vec{a}(t+1) = F(W \cdot \vec{a}(t) + \vec{b}) \tag{13.5}$$

where $\vec{\xi}$ is the input vector to the HNN, F is the activation function, $\vec{a}(t)$ is the output vector of the network after t^{th} cycle, W is the matrix of neuron weights for the recurrent layer, and $\vec{b}$ is the vector of biases of the network. Additional requirement that weights are symmetric is set to guarantee that the error function decreases monotonically while following the activation rules, and the whole system reaches a stable state after a finite number of learning cycles [24]:

$$\frac{d\vec{a}(t)}{dt} = 0 \tag{13.6}$$

For the problem of dimeric TFBS recognition, the equilibrium points for the HNN design are the dimeric structures represented by the Equation 13.2 and nonfunctional structures taken from literature, and the HNN is expected to reach a stable state, which corresponds to one of those structures with any input DNA sequence used as an initial condition. If the network reaches a stable state other than that from the set of equilibrium points, or causes infinite oscillations, then the input DNA sequence is marked as misclassified. Otherwise, a certain dimeric structure is attributed to the input DNA sequence.

When the equilibrium points are orthogonal, the problem of describing the corresponding HNN can be easily solved by Hebb rule [10]. Otherwise, a more complex procedure of HNN synthesis must be used. Accurate numerical approximation of an iterative discrete machine by a dynamic system was proposed by Li et al. [25]. We adapt the proposed approach to learning of the HNN that is designed for recognition of dimeric DNA motifs.

Suppose we are given m vectors that represent the asymptotically stable equilibrium points for an N-dimensional dynamic system: $\vec{\xi}_1,\ldots,\vec{\xi}_m \in \Re^N$. We proceed to the HNN learning as follows:

1. Compute $N \times (m-1)$ matrix Z.

$$Z = \left[\vec{\xi}_1 - \vec{\xi}_m, \ldots, \vec{\xi}_{m-1} - \vec{\xi}_m\right] \tag{13.7}$$

2. Perform a singular value decomposition of Z and obtain the matrices U, V, and S such that $Z = USV^T$, where U and V are unitary matrices, and S is a diagonal matrix.

$$U = [\vec{u}_1, \ldots, \vec{u}_L] \tag{13.8}$$

   Let k be the rank of Z (and U correspondingly).

3. Compute the auxiliary matrices.

$$T^+ = \sum_{i=1}^{k} u_i u_i^T \tag{13.9}$$

$$T^- = \sum_{i=k+1}^{L} u_i u_i^T \tag{13.10}$$

$$T_\tau = T^+ - \tau \cdot T^- \tag{13.11}$$

$$E_\tau = \vec{\xi}_m - T_\tau \cdot \vec{\xi}_m \tag{13.12}$$

where τ is a learning parameter. According to the proposed approximation theory, the larger the value of τ, the less spurious stable states the reconstructed HNN has. For synthesis of HNNs with more than three stable states, a value $\tau \geq 10$ is usually used.

4. Compute parameters of the approximating dynamic system $\vec{a}((t+1)h) = F(W \cdot \vec{a}(t \cdot h) + \vec{b})$, namely, W and $\vec{b}$, as follows:

$$W = U \begin{bmatrix} e^h I_k & 0 \\ 0 & e^{-\tau h} I_{L-k} \end{bmatrix} U^T \tag{13.13}$$

$$\vec{b} = U \begin{bmatrix} (e^h - 1) I_k & 0 \\ 0 & \frac{1}{\tau}(e^{-\tau h} - 1) I_{N-k} \end{bmatrix} U^T E_\tau \tag{13.14}$$

where I_k is the $k \times k$ identity matrix. The parameter $0 < h < 1$ reflects the asymptotic nature of the procedure, as discrete states of the neural network are replaced by continuous values of the dynamic system.

13.2.4 Adaptation of the HNN

In a series of preliminary tests performed using dimeric HREs, we found that the HNN model allows eliminating significant amount of false-positive predictions [26]. However, the need for exact match in this approach implies a rigid correspondence between the input and output vector elements, which adds serious artifacts into the modeling process. That is, in the exact-match HNN (EM-HNN) we look for the recurrent convergence between the putative motif and its dimeric structure considering the input and the output vectors explicitly matched against each other, similarly to the exact-match sequence comparison. However, real biological sequences are exposed to short mutations including frame-shifting insertions and deletions (indels) that are not always critical for further interaction with proteins [27]. For higher flexibility of modeling, we incorporate short indels into the dimeric structure prediction.

The incitement for the model enhancement comes from the following observation: while single nucleotide mismatches are carefully handled by the HNN recurrence process, short indels can bring significant distortion into the model. In particular, as a result of an insertion of a random nucleotide into the HRE spacer (with no other changes in the rest of the HRE sequence), the predicted dimeric structure varies for 14% of functional response elements. In addition, when a single nucleotide is deleted from the spacer (also with no other changes), the predicted dimeric structure varies for 9% of the training data. Finally, the dimeric structure predicted by the trained HNN is affected by a single nucleotide indel within the HRE spacer for 18% of functional HREs (119 out of 661 response elements used for tests).

These observations motivate us to consider the reliability of the recurrent system for sequence analysis more carefully, because from a biological point of view such oscillating predictions are meaningless. In particular, the lattice of dimeric protein–DNA interaction is expected not to be influenced by 1-bp difference in the spacer length, even though the binding affinity may indeed change [28]. Thus, the necessity of developing a more flexible model for dimeric structure prediction is confirmed by the observed lack of robustness when modeling the complex pattern of dimeric motifs. On the other hand, single nucleotide substitutions within the half-sites or spacers (1,500 random nucleotide substitutions in 661 HRE sequences) do not cause predicted dimeric structures to vary for 99% of functional HREs.

For sequence comparison that has a similar limitation, numerous sequence alignment methods are proposed. Dynamic programming provides a global optimum for the sequence alignment problem. Although this technique is computationally expensive, it guarantees finding an optimal solution for any given scoring scheme. For the case of long sequences, certain heuristic optimizations are used to decrease space and time complexity. However, when dealing with relatively short sequences like protein-binding motifs in DNA, we usually can afford exponential space and computational time. That is, this approach is particularly useful when a few short sequences need to be aligned accurately.

During the HNN operation, single nucleotide substitutions within motif sequences are successfully absorbed by the convergence process. Per contra, every insertion or deletion inside the sequence causes a frame shift of a half of the sequence on average; thus, convergence for the rest of the sequence becomes misleading. We overcome this exact-match limitation by inventing the method of recurrent dynamic programming similar to that used for sequence alignment.

For the problem of dimeric structure prediction by the HNN, the exact matching between input and output is per se implied by the weight matrix by the input vector multiplication procedure (refer to the Equation 13.5).

Extending this multiplication, we have the following identity:

$$W \cdot \vec{a}(t) = \left\| \begin{matrix} \cdots \\ \sum_{j=1}^{4L} w_{i,j} a_j(t) \\ \cdots \end{matrix} \right\|_{i=1\ldots N} \tag{13.15}$$

where the assumption of jth element of the input vector $\vec{a}(t)$ being converted into j^{th} element of either S_{DR} or S_{IR}, or S_{PR} or S_{ER} dimeric structure is implied by the multiplication $w_{i,j}a_j(t)$ of the jth weight by exactly the jth element of the input vector for each ith neuron of the HNN. Thus, every nucleotide insertion or deletion can be treated as a submission of the jth element of the input vector to the $(j \pm shift)^{\text{th}}$ neuron and, therefore, can be represented in the form of index shifts for weight vectors for the procedure of multiplication. The entire procedure of establishing the best matching between the current input sequence and its dimeric prototype is referred to as the *recurrent alignment*.

For the problem of recurrent alignment, the input vector $\vec{a} = \vec{a}_1 \cup \omega \cup \vec{a}_2$ is composed of the two half-sites $\vec{a}_1$ and $\vec{a}_2$ of length $4(b + 1)$, and a single nucleotide ω located in the center. The elements of the half-site vectors are numbered starting from the internal nucleotides and, unlike the previously described representation of dimeric motifs, we merge two nucleotides of the spacer each to the nearest half-site.

The alignments with the two possible orientations of the consensus half-site $\vec{y}$ and $\bar{y}$ are performed using predefined gap penalties, and the best alignment is then selected as a prototype for the input half-site convergence. For accomplishing the alignment procedure, for each $\vec{a}_2$ and $\bar{a}_1$, two scoring matrices are generated

$$M_{\vec{y}}(k_1,k_2) = \begin{cases} M_{\vec{y}}(k_1-1,k_2-1)+s(a_{k_1},y_{k_2}) \\ M_{\vec{y}}(k_1-1,k_2)+g_a \\ M_{\vec{y}}(k_1,k_2-1)+g_h \end{cases} \tag{13.16}$$

(and similar $M_{\bar{y}}$ for the reverse consensus half-site $\bar{y}$), where

$$s(a_{k_1},y_{k_2}) = \begin{cases} s>0, & \arg\max_{i=1..4} \xi^{\vec{a}}_{4(k_1-1)+i} = \arg\max_{i=1..4} \xi^{\vec{y}}_{4(k_2-1)+i} \\ 0, & otherwise \end{cases} \tag{13.17}$$

and $g_a < 0$, $g_h < 0$ are the gap penalties for deletion and insertion in the input vector, respectively.

Vector $\vec{\xi}^{\bar{y}}$ is calculated as follows:

$$\vec{\xi}^{\bar{y}} = \Sigma(\vec{y}) \tag{13.18}$$

where Σ is the DNA encoding module (refer to Section 13.2.1). In Equation 13.17, equivalence of $\arg\max_i$ functions denotes that the two 4-vectors for the considered nucleotide positions have maximums at the same internal positions. If the two $\arg\max_i$ functions are equal, the nucleotide element of the input sequence is considered as a match with the considered nucleotide of the dimeric structure.

In the case study of steroid HREs, the initial conditions are selected so that the spacer length of three base pairs is preferred to any other length, as it has been shown in numerous experiments for steroid hormone receptors [29]. Thus,

$$\begin{aligned} M_{\bar{y}}(0,0) &= 0 \\ M_{\bar{y}}(k,0) &= sp + g_s \cdot (k-1) \\ M_{\bar{y}}(0,k) &= g_s \cdot k \end{aligned} \tag{13.19}$$

(and the same for the matrix $M_{\bar{y}}$) where $sp > 0$ represents a positive score for one nucleotide insertion at the beginning of the alignment for each half-site. This insertion, when joined with the central nucleotide ω and a similar insertion from the other half-site alignment, results in an exactly 3-bp long spacer. In addition, $g_s < 0$ is the gap penalty for a longer spacer.

The resulting alignments for the input half-sites are obtained by tracking back the scoring matrices (one of two calculated for each half-site) that have the largest values of their last columns and last rows. The resulting alignment of the input vector constructed from the alignments of its half-sites is then denoted as $\vec{al}(t)$. Its length N_{al} holds the inequality $L_{al} \leq 2L$ where L is the length of the dimeric motif.

That is, instead of the exact correspondence "j^{th} element of the input $\leftrightarrow$ j^{th} element of the output," we obtain a dynamically adaptable correspondence between input sequence and its target output structure. The current output of the HNN is thus calculated as follows:

$$W \otimes \vec{a}(t) = \left\| \begin{matrix} \cdots \\ \sum_{j=1}^{2b+1} w_{i,j+4deletion(j)} a_{j+4insertion(j)}(t) \\ \cdots \end{matrix} \right\|_{i=1\,\ldots\,N} \tag{13.20}$$

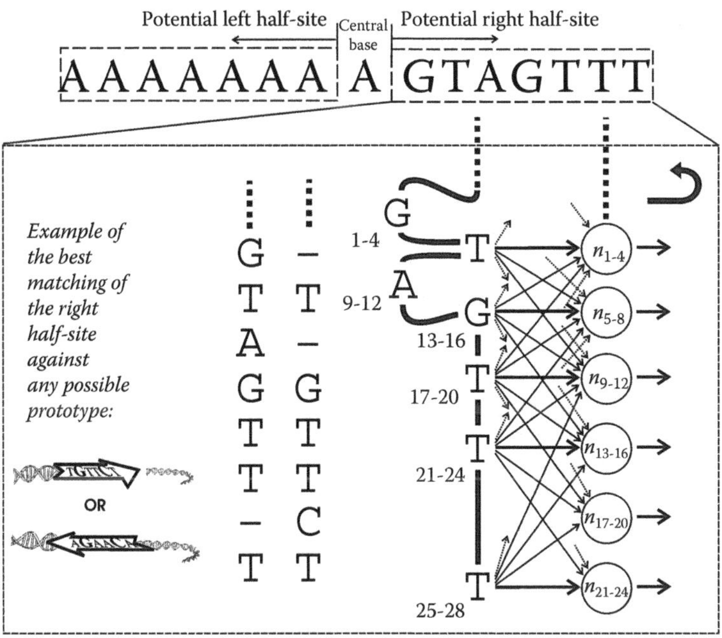

FIGURE 13.1
Recurrent alignment for HRE input sequence.

where, the operation ⊗ denotes the "aligned" multiplication; that is, multiplication involving corresponding weight shifts.

Shift indices *deletion*(j) and *insertion*(j) for the neuron weights are calculated from the alignment $\vec{al}(t)$ of the input vector $\vec{a}(t)$ using the following procedure:

```
deletion(0) = 0;
insertion(0) = 0;
for (i = 1, i < = L, i++) {
        deletion(i) = deletion(i-1);
        insertion(i) = insertion(i-1);
        if al(i + deletion(i-1)) == 'deletion'
                 deletion(i)++;
        else if al(i + deletion(i-1)) == 'insertion'
                 insertion(i)++;}
```

In Equation 13.20, we multiply each index shift by 4, as in the one-hot notation, each nucleotide corresponds to exactly four elements of the network input vector.

Figure 13.1 illustrates the procedure of recurrent alignment by the example of the right half-site of the input sequence "AAAAAAAAGTAGTTT" being aligned with the consensus HRE half-site "TGTTCT." In this example, the insertion and deletion penalties are $g_a = g_h = -2$, the long spacer penalty is $g_s = -1$, and the consensus spacer and the matching score are $s = sp = 2$. The alignment with one deletion and one insertion, in addition to one insertion in the expected spacer, is shown. For clarity, the nucleotide bases instead of the four vectors are shown. With weight shifts, after the first insertion, the second element of the aligned right half-site, namely, T (the fifth to eighth elements of the transformed vector), is an input to the first quartet of neurons for the given half-site (neurons 1–4). After the second insertion, the fourth element of the aligned right half-site, namely, T (13th–16th elements), is an input to the second quartet of neurons, and so on.

Thus, each HNN operation cycle is accompanied with the recurrent alignment procedure of the current input sequence and its targeted structure. This architecture provides us with a flexible dimeric structure modeling scheme and is expected to result in robust and hence reliable predictions.

Overall, the procedure of recognizing dimeric motifs in DNA is defined as follows:

1. Design of the two-phase neural system
   (a) The FFNN with one hidden layer of 40 neurons and the two-neuron output layer is trained using the set of experimentally validated dimeric motifs and the ten-fold set of neutral DNA sequences. The output of this network is a two-vector with probabilities of being a functional and a nonfunctional dimer for each input. The balance of these two probabilities for the following decision-making system is subject to a threshold.
   (b) Given the set of dimeric structures $\Psi = \{s_{DR}, s_{IR}, s_{PR}, s_{ER}\}$ transformed by the encoding module $\Xi_\Psi = \Sigma(\Psi)$, we construct the DP-HNN system for dimeric structure prediction. Only positive outputs of the previous sequence-based motif selection scheme returned by the FFNN are processed by the HNN (unless stated otherwise, for example, for speed testing purposes). The output of the DP-HNN system for each putative TFBS input is either one of four possible dimeric structures, or the "non-dimer" output.
2. Recognition of motifs

   To recognize dimeric motifs in the DNA sequence, we use a sliding window of a fixed length to obtain the stream of DNA subsequences that are then processed by the two-phase system.

   The recognition process operates as shown in Figure 13.2. First, a transformed DNA subsequence $\vec{\xi} = \Sigma(\vec{\theta})$ is submitted to the FFNN module, which returns the probability of this subsequence to be a motif of interest to the decision-making module. Second, the list of putative motifs returned as a result of operation of the FFNN is processed by the HNN, and for each sequence, its dimeric or nondimeric structure is predicted. The output of the system is either a binary answer for each input sequence or a list of predicted dimeric motifs with their annotations if the screening of a long DNA region is performed.

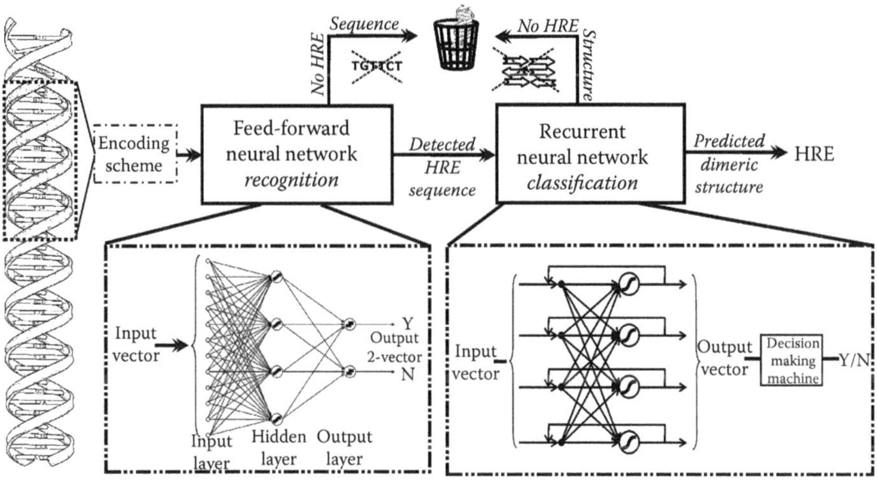

FIGURE 13.2
Two-phase neural model for recognition of dimeric motifs.

13.3 Reconfigurable DP-HNN

The bottleneck of our two-phase neural system is identified in the operation of the DP-HNN. Although the FFNN is trained once and forever, and then its operation is a straightforward pass through the sequence of its layers, the recurrent model requires hundreds of iterations for each input vector. Software for both EM-HNN and DP-HNN was developed for the second phase and run on a multicore IBM server (4 × 3.16 GHz CPUs, 3.25 GB RAM, 667 MHz system bus). It took approximately 40 minutes to screen only 1 Mb of DNA with four parallel threads by the HNN that consisted of 60 neurons without involvement of the recurrent alignment, and nearly 50 minutes when recurrent alignment procedure was involved. It becomes evident that the software implementation of HNNs for genome-wide motif recognition will need prohibitive long running time. The dynamic adaptation in the neural operations further suggests hardware acceleration for the applicable solutions.

We mapped the DP-HNN for HRE structure prediction on an FPGA chip, and then used that chip as a coprocessing unit for the two-phase neural system. In our implementation, the FPGA coprocessor communicates with the host PC via the local bus. The Alpha Data Virtex-4LX160 PCI chip with 135 168 logic elements (8 M gates) and 288 × 18 kbit RAM blocks was used. Alpha Data SDK 4.6.0 API was used for programming the communication layer between the front-end application and the on-chip HNN unit. DNA sequences that contains either putative motifs predicted by the FFNN module or functional motifs from the training set were fed into the DP-HNN, through the DNA encoding module Σ; the resulting vector of bit values, representing the one-hot encoded DNA sequence, were sent to the configured FPGA board. It also obtained the output from the board, and passed these data to the decision-making module.

In the following subsections, we describe the technical aspects of our HNN circuit design addressing the methods used for efficient data transfer and computations.

13.3.1 Representation of Numerical Values and Operations on FPGA

For implementation of numerical values on FPGA we adapt a fixed-point representation in two's complement notation. The values of neuron weights and outputs are the 32-bit numbers with one sign bit, two integer bits, and 29 fractional bits. This representation is sufficient to describe operation of the HNN with acceptable precision.

To not exhaust the limited number of logic gates, we use the 32 bit × 32 bit multiplications for generation of the synaptic inputs to the neurons, which operate in the form of finite state machines. Each of these machines regulates the operation of two dedicated hardware 18 bit × 18 bit multiplier units. Henceforth, the 32-bit arithmetic operations will be referred to as an adder and multiplier, respectively.

13.3.2 Control and Matching Units

The control unit is used for serial processing of the input vector by the HNN unit. For this purpose, the control unit dispatches control signals to all other units. In particular, the control unit is responsible for sequential processing of data flow. At its initial state, the network input is set up. Then, the recurrent alignment procedure is performed as described in Section 13.2.4. Resulting alignments of the two half-sites are then processed in parallel by the two groups of the neuron units. Each group contains four physical neuron units (a neuron quartet) and performs computations for four neurons at a time. Each neuron quartet corresponds to a single nucleotide encoded using one-hot notation for the DNA alphabet, so the depth of the cycle is equal to the length of the half-site of interest.

The verification unit contains an array of 20 × 60 32-bit registers, an oscillation detection module, and a counter of iterations. Current DP-HNN output, which is an array of sixty 32-bit fixed-point numbers, is placed into one column of 20 registers in a cyclic order, so that the set of registers always stores the network outputs from the 20 most recent iterations. The oscillation calculation module thus computes the relative oscillation of the HNN output during the 20 consecutive iterations. The HNN iterations are directed to stop by that module when the total absolute deviation is less than $1/2^{10} = 1/1{,}024 \sim 0.1\%$ of the output value for all neurons, so we conclude that a stable state is reached. The counter of iterations signals a stop when the maximum number of iterations is

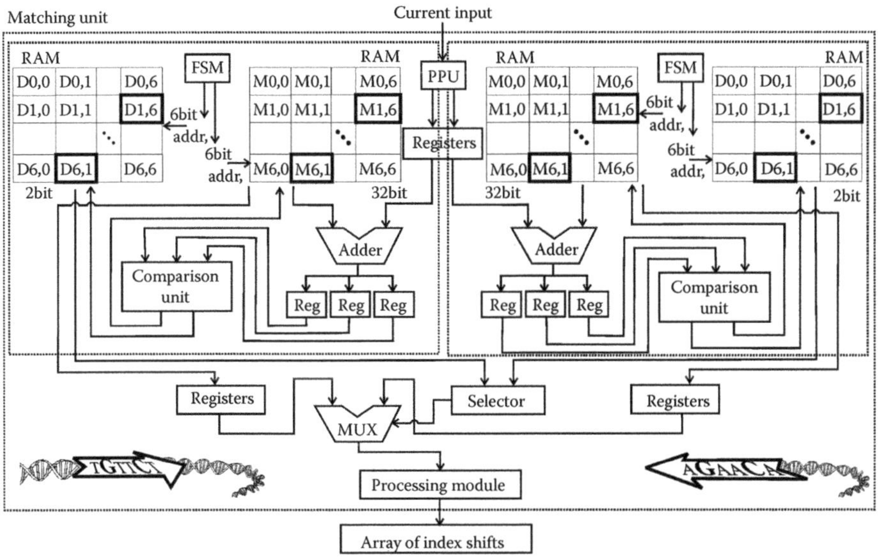

FIGURE 13.3
Matching unit for the recurrent alignment procedure.

exceeded. After testing the network operation, we set the maximum number of iterations to 10,000, which is a very conservative estimation. Even without dynamic programming, which provides a better match and therefore faster convergence, we found it to be enough for the HNN to converge with majority of the inputs unless a specific pattern caused substantial oscillations in the network. The stop signal produced by the verification module causes the whole DP-HNN unit to finish the processing of the current input. Current output is then sent to the PCI bus to the awaiting front-end application, which conveys to it the user interface or to the decision-making scheme.

The recurrent alignment procedure is performed by the matching unit shown in Figure 13.3. It obtains the input vector, performs the procedure of alignment for its half-sites, and returns an array of weight index shifts that are used for further processing of the vector by proper HNN neurons. An input vector is preprocessed by the preprocessing unit (PPU on the figure), which defines the indices of maximum elements for each consecutive four elements (thus defining a particular nucleotide base encoded by them), and only then these indices are used.

Inside the matching unit, the two identical half-site processing modules perform the recurrent alignment procedure for the two consensus HRE half-sites, each for one of its orientation, namely, the direct (left part of the figure) or the inverted (right part of the figure). To decrease the number of registers for sequence representation, consensus vectors are stored in the chip memory and never changed during the network operation. Each half-site processing module uses two 6-bit addressed RAM sections. One RAM section stores the alignment scoring matrix M and the other one stores the trace-back array D for reconstructing the resulting optimal alignment.

A finite state machine regulates the succession of scoring matrix calculations. Although it is possible to implement alignment score matrix in linear space [32], we are dealing with the iterative process so it is more important to minimize the time latency. That is, our implementation of the recurrent alignment procedure consumes space quadratic to the length of the half-site, as it is affordable for the case of relatively short HRE motifs. However, such an implementation permits simultaneous calculation of two matrix elements at a time: the cell (i,j) and its symmetric cell (j,i), which are contoured with bold in Figure 13.3. In addition, the initial matrix values, which are stored in its first row and first column, can be filled simultaneously if we use the gap penalties of degrees of two, thus avoiding costly multiplications.

The total latency of the half-site alignment procedure for an input of length n is therefore $2+\frac{n(n+1)}{2}+2n-1$. Specifically, two sequential operations are required for the input preprocessing, and they are parallelized with one operation of initialization of the scoring matrices; $n(n+1)/2$ operations are required to compute both alignment scoring matrices of size $n \times n$, and $2n-2$ operations are needed to select the maximum values from their last columns and rows. Finally, at most $2n$ operations are required to reconstruct the resulting alignment using the trace-back matrix D. For the HRE half-site of length 6, the latency of the alignment module is at most 52 calculation cycles.

Outputs of each half-site processing module are the best alignments of the half-sites and their scores. Then, a multiplexer controlled by a selector picks out the alignment with the highest score, and the downstream processing module returns the best alignment in the form of index shifts for neuron weights.

13.3.3 Neuron and Memory Units

The main part of the on-chip HNN consists of the neuron units that are connected to the memory unit. The memory unit stores the neuron weights, as well as current and initial input and output vectors.

The neuron unit is implemented using several types of calculations, namely, the adder, the multiplier, and the register. Inside the neuron unit, two consecutive elements of the input vector and the appropriate weight values are multiplied at a time using two 32 bit × 32-bit multiplier modules operating in parallel. Thus, the latency of the neuron unit operation is half of the input vector length. The unit's addressing scheme required for communication with the memory unit uses the index shifts for neuron weights returned by the matching unit. The computed weighted inputs for the neurons are stored in the memory registers, and then summed up into the synaptic input of the neuron. The output of the neuron is computed from its synaptic input using its activation function. In our implementation, we use a linear approximation of the sigmoid function $F_0(x)=\frac{1-e^{-x}}{1+e^{-x}}$. The approximation is used to avoid calculations of actual sigmoid, which requires series of multiplications and at least one division. Instead, we use a combination of linear curves with slopes of 1/2, 1/4, and 1/8. Thus, the resource-consuming multiplication and division operations are replaced by "cheap" register shift operations.

Four instances of the neuron unit are used in parallel in each of the two parts of the on-chip HNN, so the total amount of neuron units involved is eight. Each part of HNN processes the input half-site. Inside each part, after a neuron quartet is calculated (the end of calculation is reported to the control unit by a handshake signal), it is replaced by the next four of a total 24 for each of the HRE half-sites.

The memory unit stores the weight values for the neurons and the outputs of the neurons. It is also used for storing the 20 recent HNN outputs used for oscillation measurements, the initial input, weight index shifts, and the iteration counts. These data are read or written according to the commands of the control unit. As we have 48 neurons in the HNN, there are 48 × 48 32-bit weight values. Therefore, we need the 12-bit addressing scheme. Verification unit stores 20 × 60 32-bit values and requires 11-bit addressing scheme.

13.3.4 Operation of DP-HNN

The operation of the on-chip DP-HNN system is mainly divided into three functional units: the matching unit, the actual HNN unit, and the verification unit. Figure 13.4 shows the configuration of the DP-HNN from the point of view of digital data processing.

The top-level control is performed by the counter of iterations. The top-level control is performed by the counter of iterations, which determines the following:

- When the counter of neurons should be reset to zero,
- Whether a vector has to be put into the system as its initial input,
- Which register of the verification unit contains the oldest output, and
- Whether the maximum number of iterations is exceeded and the iteration is terminated.

The counter of neurons is reset before the HNN iteration starts, and then it regulates the succession of neuron quartets processing, as well as decides which values of weights must be selected from memory using the corresponding weight index shifts.

The operation of the HNN is implemented by eight identical physical neuron units grouped into two sections, and the memory. The neuron unit performs sum-of-products operations for calculation of neuron synaptic inputs at its run mode. Two groups of four input elements are processed in series. That is, eight physical neuron units emulate 48 neurons as required by our

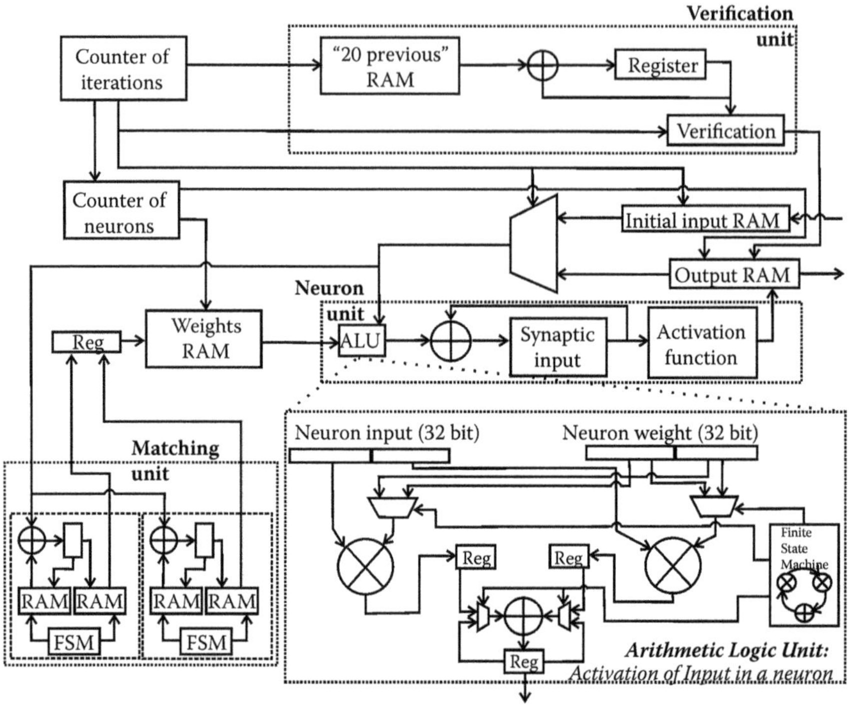

FIGURE 13.4
Configuration and control of the DP-HNN on FPGA.

HNN-based HRE recognition model. The weight values, inputs, and outputs are read from or written to the distributed memory.

Summary of the ensued implementation for the ADM Virtex-4 chip is as follows:

Logic elements:	101,696 of 135,168 (75%)
RAM:	960 Kbit of 5,184 Kbit (19%)
I/O pins:	101 of 960 (11%)
DSP slices:	48 of 96 (50%)

13.4 Application to Dimeric Protein Binding Site Identification

13.4.1 The Biological Problem

Steroid hormone receptors are transcription factors that exist in the cytoplasm or nucleus [33]. Connection of a hormone molecule results in an allosteric change of conformation of the receptor (the process known as "an activation of a receptor") that raises affinity of the receptor's DNA-binding domain to DNA, thus allowing the receptor to bind to specific parts of DNA molecule inside a nucleus and to adjust transcription of *cis*-linked genes. Cellular mechanisms are described in detail for only a modest number of known target genes of steroid hormone receptors [34,35]. However, steroid hormones are clearly involved in the expression regulation of a considerable number of genes about which not enough is known [36].

With a few exceptions [33], DNA-binding domain of a steroid hormone receptor molecule interacts with an HRE that is composed of two half-sites separated by a short spacer. Response element's half-sites can occur in different orientations while interacting with zinc-fingers of a hormone receptor's DNA-binding domain [37–39]. Consensus DNA sequence of the half-site is known to be $\vec{h}_{HRE}$ = TGTTCT. With reference to notation from Section 13.2, the four possible structures of HRE are

$$\begin{aligned} HRE_{DR} &= [\vec{h}_{HRE} \cup \omega_{sp1}\omega_{sp2}\omega_{sp3} \cup \vec{h}_{HRE} \\ HRE_{IR} &= [\vec{h}_{HRE} \cup \omega_{sp1}\omega_{sp2}\omega_{sp3} \cup \overleftarrow{h}_{HRE}. \\ HRE_{PR} &= [\overleftarrow{h}_{HRE} \cup \omega_{sp1}\omega_{sp2}\omega_{sp3} \cup \vec{h}_{HRE} \\ HRE_{ER} &= [\overleftarrow{h}_{HRE} \cup \omega_{sp1}\omega_{sp2}\omega_{sp3} \cup \overleftarrow{h}_{HRE} \end{aligned} \tag{13.21}$$

where $[\omega_{sp1}\omega_{sp2}\omega_{sp3}]$ stands for the 3-bp-long spacer ($\omega_{sp,i} \in \Omega$, $\forall i = 1,2,3$). These consensus HRE structures and six non-HRE sequences (taken from the experimental papers by Thackray et al. [40] and by Lieberman et al. [27]) were used as ten equilibrium points for the HNN design.

We use the on-chip implementation of DP-HNN for two different purposes. First, it is used to classify functional HREs from the collected dataset to find any interesting trends and to test the applicability of the approach for the general problem of modeling symmetrically structured weak TFBS signal. Second, the on-chip DP-HNN is used as a part of a two-phase neural system for TFBS recognition with the aim to estimate its ability to eliminate false-positive predictions.

The collection of progesterone, glucocorticoid, and androgen response elements used for the current project has been described earlier [29]. In short, it contains seven hundred experimentally verified binding sites for androgen, glucocorticoid, and progesterone nuclear receptors collected from biomedical literature. For an HRE to be accepted into the collection, a convincing experimental evidence was required, namely, validated binding in vitro or confirmed mediation of gene expression by transfection assay.

13.4.2 Dimeric Structure of HREs

When estimating predictive capabilities of the developed neural system, we tested both the dynamically adaptable and the exact-match versions of the

HNN. In addition, as the set of possible dimeric and nondimeric structures is predefined, the HNN model can be considered as a sequence classifier; thus, we performed *k*-means classification of the same set of HREs for comparison. Unlike the HNN classifier with fixed stable states, the *k*-means procedure iteratively adjusts the set of its centroids, so that the cumulative variance for these centroids and the points in the dataset is eventually minimized. That is, for the procedure of *k*-means classification, instead of fixing the class centroids we set the possible dimeric structures as starting points for centroid adjustment.

The results of the three classification procedures for the set of functional HRE sequences are shown in Table 13.1. In this table, the first column is the total number of HREs for a given steroid hormone receptor (namely, progesterone, glucocorticoid, or androgen receptor). In the second column, we list the possible HRE dimeric structures for each of the groups. The next columns show the results of dimeric structure prediction by a given classifier for the HRE group.

As shown in Table 13.1, the three hormone receptors demonstrate different preferences toward the dimeric structure of their response elements on DNA. The difference between the distributions of predicted HRE structures for ARE, PRE, and GRE is statistically significant (p value $< 10^{-4}$) for *k*-means clustering, and for Hopfield-based classification as well (p value = .007). However, this finding is not unexpected, as a similar observation that AREs have stronger preferences toward the IR structure has already been reported by Reid et al. [41] and Claessens et al. [42].

To estimate the robustness of the dynamically adaptable neural model, we performed a series of tests where single nucleotides were inserted to or deleted from the spacers and half-sites of the HRE sequences. We observed that one indel within the HRE spacer caused variation of structure prediction returned by the EM-HNN for 18% of HREs, while for the DP-HNN this variation was 3%. In addition, one indel within half-sites was critical for 7% of predictions made by the EM-HNN, and for 1% of predictions made by the DP-HNN.

However, in spite of these findings, there still exists a small group of HREs that are highly different from other known HREs and cannot be robustly assigned with the conserved configuration, that is, the six nucleotide repeats with a three nucleotides spacer, although all of them have been convincingly proved to be functional (reviewed in [43]). If those outlying HREs are not false positives, then the nature of their interaction with the hormone receptor's DNA-binding domains should be considered more carefully.

In addition, for both EM-HNN and DP-HNN, we have estimated the median numbers of iterations required for reaching a stable state (medians were used instead of means because absence of oscillations during the convergence process could not be guaranteed while the mean value would be heavily affected by a single instability). For the EM-HNN, the average median number of iterations is 480, and for the DP-HNN it is 400 for all experimentally validated HREs.

TABLE 13.1

Predicted Structures for Androgen (ARE), Glucocorticoid (GRE), and Progesterone (PRE) Response Elements with Exact-Match Hopfield Neural Network (EM-HNN), Dynamically Adaptable Hopfield Network (DP-HNN), and *k*-Means Classifier

		EM-HNN		DP-HNN		*k*-Means	
HRE	**Structure**	*N*	%	*N*	%	*N*	%
	DR	35	53.0	32	48.5	26	39.4
PRE	IR	1	1.5	2	3.0	5	7.6
total:	PR	24	36.4	27	40.9	20	30.3
66	ER	3	4.5	4	6.1	12	18.2
	non-HRE	3	4.5	1	1.5	3	4.5
	DR	225	59.7	210	55.7	134	35.5
GRE	IR	3	0.8	7	1.9	26	6.9
total:	PR	90	23.9	124	32.9	127	33.7
377	ER	28	7.4	22	5.8	62	16.4
	non-HRE	31	8.2	14	3.7	28	7.4
	DR	94	43.1	93	42.7	33	15.1
ARE	IR	1	0.5	5	2.3	55	25.2
total:	PR	65	29.8	67	30.7	69	31.7
218	ER	52	23.9	51	23.4	40	18.3
	non-HRE	6	2.8	2	0.9	21	9.6
	DR	354	53.6	335	50.7	193	29.2
$\cap$	IR	5	0.8	14	2.1	86	13.0
total:	PR	179	27.1	218	33.0	216	32.7
661	ER	83	12.6	77	11.6	114	17.2
	non-HRE	40	6.1	17	2.6	52	7.9

Note: N denotes number, and % is percentage of total.

13.4.3 Two-Phase Neural System for HRE Prediction

The receiver operating characteristic (ROC) curves were tracked on each step of HRE prediction; ten-fold cross-validation was used for estimating the prediction accuracies and their variances. For the two-phase prediction and for the FFNN itself, the point of the ROC curve with the smallest Euclidean distance from the 100% accuracy point was selected as a cutoff where the sensitivity and the specificity values were collected. However, when the FFNN was used as a first phase of prediction followed by the HNN, lower threshold values were used to enlarge the set of putative HREs for further validation. The results of HRE prediction tests are summarized in Table 13.2.

For HRE prediction, the first phase of machine learning is awakening the trained FFNN that selects HRE-like sequence patterns. For the FFNN step, the prediction sensitivity value was found to be as high as 98% (i.e., 15 among

TABLE 13.2

Accuracy of Two-Phase Hormone Response Element Prediction Tool

Neural Network	Misclassified HREs	Sensitivity, %	Specificity, kbp^{-1}	Prediction Rate on Human Genome (NCBI Genbank #36.2), $kbps^{-1}$	AUC
FFNN	15	98 ± 4.4	5.84 ± 0.78	7.28	0.98 ± 0.04
FFNN™EM-HNN	52	92 ± 2.3	7.29 ± 1.13	8.15	0.92 ± 0.03
FFNN™DP-HNN	25	96 ± 2.6	7.08 ± 1.21	8.14	0.96 ± 0.03

661 HREs were always misclassified), combined with the specificity of 1:6 Kb. The DP-HNN allowed increasing the specificity level to 1:7.1 Kb, while the sensitivity was kept at the reliably high level of 96% (6% or 9% of PREs, 37% or 10% of GREs, and 9% or 4% AREs, or total 52 HREs, were misclassified).

Note the two interesting observations in Table 13.2. First, the area under curve (AUC) for HRE prediction by solely FFNN is better than that for the two two-phase systems. Although the AUC metric treats type-I and type-II errors equally, if this is the case for a particular task, the user certainly would prefer the approach with better AUC. However, for the TFBS prediction problem, a high false-positive rate is always a big challenge, so we prefer to consider the approach with possibly lower AUC, as it provides lower false-positive rate, rather than a generally superior one.

Second, for HRE prediction by the neural networks, we have the AUC values that are nearly equal to the products of corresponding sensitivity and specificity. That is, the ROC curves have rectangular shapes, and therefore, there is no actual trade-off between the type-II and type-I errors. Indeed, neural networks usually converge to particular answers within machine precision for most inputs, so there is little chance for any threshold-mediated balance. If the selection of one particular error type is of greater importance, then it may be reasonable to use another pattern recognition method that allows for more user-defined accuracy trade-off, such as those exploiting the statistic models [29].

13.4.4 Performance of the Hardware-Accelerated System

To evaluate the speedup gained because of using the FPGA-based hardware acceleration, we developed a software version of the same HNNs (for both its exact-match and dynamically adaptable versions). C applications were tested using a four-core IBM server. We also implemented the software versions of the HNNs with both one thread and four threads being processed in parallel by four central processing units (CPUs). The FPGA on-chip clock frequency was set to 100 MHz.

TABLE 13.3

Performance of HNNs Implemented with Use of Virtex-4 FPGA and 4-Way IBM PC

Hopfield Neural network	Implementation		Processing Time (sec) of Training Set of 661 HREs	Processing Time (sec) of 1Mb of DNA
	V-4 FPGA		0.46	485
EM-HNN	C++	4 threads	2.47	3,000
	application	1 thread	10.15	10,500
	V-4 FPGA		0.49	540
DP-HNN	C++	4 threads	2.81	3,400
	application	1 thread	11.36	12,500

Table 13.3 confirms a nearly 50× speedup of the hardware-accelerated version versus the single-threaded software implementation, and a 10× speedup versus the high-performance software implementation. It is interesting to note that software implementation of an HNN may require less operation cycles to converge even though the models implemented in hardware and software are essentially the same. An explanation is that software implementations written in high-level programming languages exploit the 64-bit floating point numbers, although the FPGA solutions use fixed-point number representation with 32 bits and thus introduce additional imprecision into the calculations.

13.5 Discussions

Transcription factor binding site recognition, although is conceptually a well-understood task, has some very challenging constraints. The typically short length of TFBSs poses a big problem as it significantly increases their chances to occur randomly. Hence, high false-positive rate has always been a limiting factor for precise TFBS recognition. The possible solution for eliminating the excessive false-positive predictions is to design multiple-feature recognition schemes reflecting the specific characteristics of a particular binding site or a family of those.

Special cases of structured motifs, the inverted and direct repeats, could be recognized by both prokaryotic [44] and eukaryotic [16] transcription factors. We designed and evaluated a novel computational method for prediction of dimeric DNA motifs and developed original hardware-accelerated implementation of the proposed two-phase neural system. Using the case study of steroid HREs, it has been found that random expectation of motif prediction by our two-phase system is at least 7.1 kbp^{-1} combined with 96% sensitivity.

For real genomic sequences the prediction rate is even better, such as 8.1 kbp^{-1} for the human genome [43].

Considering the extensive dataset of HRE sequences that currently has no analogs, our findings are indeed promising. For comparison, the results of the TRANSFAC-based TFBS prediction experiments provided by Rahmann et al. [45] can be used. In that paper, the authors show that specificity level higher than 99% can be achieved for only 43 TFBS profiles (i.e., 7%) among 623 used for testing. Some profiles of high interest in practice like nuclear-receptor binding sites are not included into the high-quality group. All other profiles reside below the specificity level of 0.99, which is fairly trivial because the corresponding prediction rate is as low as 1 hit per 0.1 kb.

The most reliable algorithm for prediction of dimeric binding sites reported to date for the superfamily of vertebrate nuclear receptors is the nuclear hormone receptor binding site prediction (NHR)-scan. The algorithm exploits hidden-Markov model specifically adapted for recognizing the dimeric structure of its input motifs [46]. The authors report a specificity of one match per 5 kb versus a sensitivity of approximately 50%, and one match per 1 kb accompanied with a sensitivity of 85%; the former values are more indicative because for the problem of prediction lower false-positive output is usually of higher priority.

For the structure prediction by the recurrent neural network enhanced with dynamic adaptation, we found that it works well for most of the functional HREs, and provides robust results in case of short frame-shifting mutations. However, there exists an unsolved issue that the system may fall into oscillations if a chimerical motif is encountered. For such a motif, one-half of the sequence half-site comes from one orientation of the consensus half-site, and the other half from its reverse form. We met at least two experimentally validated examples of such chimerical HREs identified in the promoter regions of vertebrate genes: a progesterone response element with the right half-site AGTACT (compare with the HRE consensus TGTTCT and its reverse form AGAACA) is known to be involved in the regulation of rabbit *uteroferrin* gene [47], and the same androgen receptor–responsive DNA sequence acts in the promoter area of rat *probasin* gene [48]. Such cases are marked by our two-phase system as false negatives and will require more careful investigation in future.

Software implementation of the HNN enhanced with dynamic programming, when applied for genome-wide analysis, could cause prohibitively long execution time as can be seen from the performance results summarized in Section 13.4. In particular, processing of 1 Mb of DNA takes hours with one computational thread, and nearly 1 hour with four parallel threads when tested on a very powerful PC. Nowadays, having gigabytes of annotated DNA, running HNNs in software becomes impossible. Hardware-accelerated implementation of the most computationally expensive phases of motif prediction process should help to benefit by the trade-off between speed and accuracy. In particular, a well-designed parallel FPGA architecture provides access to the best possible precision and reveals the potential of the recurrent neural solutions in full.

The main challenge of our FPGA design is finding the balance between bit and node parallelism to reach the best overall performance and keep the applicability to the chosen domain given the implementation constrains. In the current implementation, we process the input vector using eight physical neuron units, and each unit involves four embedded multipliers. However, we may trade the time efficiency for more complex activation rules to improve the numerical precision. In particular, we use 29 fractional bits for number representation and thus obtain the imprecision of up to -10^{-8} per HNN iteration. For the case of steroid HREs, that error is not critical because the number of iterations does not exceed the level of 500 for most inputs. However, if we now consider more complex motif patterns, it may result in significantly worse HNN convergence. In turn, longer convergence leads to resulting error that cannot be negligible any more.

Involving dynamic adaptation of the HNN recurrence we resolve a very challenging issue of motif prediction, namely, incorporation of short indels within the half-sites and especially within the spacer of dimeric motifs. Properly designed DP-HNN carefully carries short mutations including both indels and substitutions, thus making structure prediction more sensible.

In conclusion, we have developed a novel dynamically adaptable neural architecture for recognition of dimeric DNA motifs, and demonstrated its performance using the case study of steroid HREs. Our two-phase prediction framework provides access to robust and biologically meaningful predictions, while the invented FPGA architecture guarantees the applicability of the proposed approach to genomic scale.

13.6 References

1. Tompa M, Li N, Tompa M, Li N, Bailey TL, Church GM, De Moor B, Eskin E et al. (2005) Assessing computational tools for the discovery of transcription factor binding sites. *Nat Biotechnol,* 23(1): 137–144.
2. Wasserman WW and Sandelin A (2004) Applied bioinformatics for the identification of regulatory elements. *Nat Rev Genet,* 5(4): 276–287.
3. Vavouri T and Elgar G (2005) Prediction of *cis*-regulatory elements using binding site matrices—The successes, the failures and the reasons for both. *Curr Opin Genet Dev,* 15(4): 395–402.
4. Prestridge DS (1991) SIGNAL SCAN: A computer program that scans DNA sequences for eukaryotic transcriptional elements. *Comput Appl Biosci,* 7(2): 203–206.
5. Li X, Zhong S, and Wong WH (2005) Reliable prediction of transcription factor binding sites by phylogenetic verification. *Proc Natl Acad Sci USA,* 102(47): 16945–16950.

6. Kim SY and Kim Y (2006) Genome-wide prediction of transcriptional regulatory elements of human promoters using gene expression and promoter analysis data. *BMC Bioinformatics*, 7: 330.
7. Grau J, Ben-Gal I, Posch S, and Grosse I (2006) VOMBAT: Prediction of transcription factor binding sites using variable order Bayesian trees. *Nucleic Acids Res*, 34(W): 529–533.
8. Hu J, Li B, and Kihara D (2005) Limitations and potentials of current motif discovery algorithms. *Nucleic Acids Res*, 33(15): 4899–4913.
9. Jain LC (2000) Recent advances in artificial neural networks: Design and applications. CRC Press, Boca Raton.
10. Hagan M, Demuth H, and Beale M (1996) *Neural Network Design*. PWS Pub, Boston.
11. Hawkins J and Boden M (2005) The applicability of recurrent neural networks for biological sequence analysis. *IEEE/ACM Trans Comput Biol Bioinform*, 2(3): 243–253.
12. Ormondi A and Rajapakse J (2006) *FPGA Implementations of Neural Networks*. Springer, Netherlands.
13. Deschamps JP, Bioul G, and Sutter GD (2006) *Synthesis of Arithmetic Circuits: FPGA, ASIC, and Embedded Systems*, Wiley-Interscience.
14. Zhu J and Sutton P (2003) FPGA implementation of neural networks: A survey of a decade of progress. *13th International Conference on Field-Programmable Logic and Applications (FPL 2003)*, 1062–1066.
15. Kuncheva LI and Whitaker CJ (2003) Measures of diversity in classifier ensembles and their relationship with the ensemble accuracy. *Machine Learning*, 51(2): 181–207.
16. Khorasanizadeh S and Rastinejad F (2001) Nuclear-receptor interactions on DNA-response elements. *Trends Biochem Sci*, 26(6): 384–390.
17. Freedman LP and Luisi BF (1993) On the mechanism of DNA binding by nuclear hormone receptors: A structural and functional perspective. *J Cell Biochem*, 51(2): 140–150.
18. Aranda A and Pascual A (2001) Nuclear hormone receptors and gene expression. *Physiol Rev*, 81(3): 1269–1304.
19. Pollack JB (1991) The introduction of dynamical recognizers. *Machine Learning*, 7: 227–252.
20. Haykin S (1999) *Neural Networks: a Comprehensive Foundation*. Prentice Hall, New Jersey.
21. Simpson JJ and McIntire TJ (2001) A recurrent neural network classifier for improved retrievals of areal extent of snow cover. *IEEE Trans Geosci Remote Sens*, 39(10): 2135–2147.
22. Guler I and Ibeyli ED (2006) A recurrent neural network classifier for Doppler ultrasound blood flow signals. *Pattern Recognition Letters*, 27(13): 1560–1571.
23. Hammer B (2000) *Learning with Recurrent Neural Networks*. Springer Lecture Notes in Control and Information Sciences 254, Springer-Verlag, London.
24. Hagan M, Demuth H, and Beale M (1996) *Neural Network Design*. PSW Publishing Company, Boston, MA.
25. Li JH, Michel A, and Porod W (1989) Analysis and synthesis of a class of neural networks: linear systems operating on a closed hypercube. *IEEE Trans Circuits Syst*, 36(11): 1405–1422.

26. Stepanova M, Lin F, and Lin V (2007) A Hopfield neural classifier and its FPGA implementation for identification of symmetrically structured DNA motifs. *J VLSI Sig Process S*, 48(3): 239–254.
27. Lieberman BA, Bona BJ, Edwards DP, and Nordeen SK (1993) The constitution of a progesterone response element. *Mol Endocrinol*, 7(4): 515–527.
28. Dahlman-Wright K, Siltala-Roos H, Carlstedt-Duke J, and Gustafsson JA (1990) Protein-protein interactions facilitate DNA binding by the glucocorticoid receptor DNA-binding domain. *J Biol Chem*, 265(23): 14030–14035.
29. Stepanova M, Lin F, and Lin V (2006) Establishing a statistic model for recognition of steroid hormone response elements. *Comput Biol Chem*, 30(5): 339–347.
30. International Human Genome Sequencing Consortium (2001) Initial sequencing and analysis of the human genome. *Nature*, 409(860): 921.
31. Venter JC, Adams MD, Myers EW, Li PW, Mural RJ, Sutton GG, et al. (2001) The sequence of the human genome. *Science*, 291(5507): 1304–1351.
32. Gusfield D (1997) Algorithms on strings, trees, and sequences. Cambridge University Press.
33. Alberts B, Bray D, Lewis J, Raff M, Roberts K, and Watson J (1994) Intercellular signalling. In *Molecular Biology of the Cell*. Garland Publishing, New York.
34. Richer JK, Jacobsen BM, Manning NG, Abel MG, Wolf DM, and Horwitz KB (2002) Differential gene regulation by the two progesterone receptor isoforms in human breast cancer cells. *J Biol Chem*, 277(7): 5209–5218.
35. Leo JC, Wang SM, Guo CH, Aw SE, Zhao Y, Li JM, Hui KM, and Lin VC (2005) Gene regulation profile reveals consistent anticancer properties of progesterone in hormone-independent breast cancer cells transfected with progesterone receptor. *Int J Cancer*, 117(4): 561–568.
36. Xu W (2005) Nuclear receptor coactivators: The key to unlock chromatin. *Biochem Cell Biol*, 83(4): 418–428.
37. Evans RM (1988) The steroid and thyroid hormone receptor superfamily. *Science*, 240(4854): 889–895.
38. Nelson CC, Hendy SC, Shukin RJ, Cheng H, Bruchovsky N, Koop BF, and Rennie PS (1999) Determinants of DNA sequence specificity of the androgen, progesterone, and glucocorticoid receptors: Evidence for differential steroid receptor response elements. *Mol Endocrinol*, 13(12): 2090–2107.
39. Gronemeyer H (1992) Control of transcription activation by steroid hormone receptors. *FASEB J*, 6(8): 2524–2529.
40. Thackray VG, Lieberman BA, and Nordeen SK (1998) Differential gene induction by glucocorticoid and progesterone receptors. *J Steroid Biochem Mol Biol*, 66(4): 171–178.
41. Reid KJ, Hendy SC, Saito J, Sorensen P, and Nelson CC (2001) Two classes of androgen receptor elements mediate cooperativity through allosteric interactions. *J Biol Chem*, 276(4): 2943–2952.
42. Claessens F, Verrijdt G, Schoenmakers E, Haelens A, Peeters B, Verhoeven G, and Rombauts W (2001) Selective DNA binding by the androgen receptor as a mechanism for hormone-specific gene regulation. *J Steroid Biochem Mol Biol*, 76(1–5): 23–30.
43. Stepanova M, Lin F, and Lin V (2007) A two-phase ANN method for genome-wide detection of hormone response elements. *Lecture Notes in Bioinformatics*, 4774: 19–29.

44. Favorov AV, Gelfand MS, Gerasimova AV, Ravcheev DA, Mironov AA, and Makeev VJ (2005) A Gibbs sampler for identification of symmetrically structured, spaced DNA motifs with improved estimation of the signal length. *Bioinformatics,* 21(10): 2240–2245.
45. Rahmann S, Muller T, and Vingron M (2003) On the power of profiles for transcription factor binding site detection. *Stat Appl Genet Mol Biol,* 2(1): Article 7.
46. Sandelin A and Wasserman WW (2005) Prediction of nuclear hormone receptor response elements. *Mol Endocrinol,* 19(3): 595–606.
47. Jantzen K, Fritton HP, Igo-Kemenes T, Espel E, Janich S, Cato AC, Mugele K, and Beato M (1987) Partial overlapping of binding sequences for steroid hormone receptors and DNaseI hypersensitive sites in the rabbit uteroglobin gene region. *Nucleic Acids Res,* 15(11): 4535–4552.
48. Claessens F, Alen P, Devos A, Peeters B, Verhoeven G, and Rombauts W (1996) The androgen-specific probasin response element 2 interacts differentially with androgen and glucocorticoid receptors. *J Biol Chem,* 271(32): 19013–19016.

14

Parallel FPGA Search Engine for Protein Identification

Daniel Coca, Istvan Bogdan, and Robert J. Beynon

14.1 Introduction

In the aftermath of the successful completion of the sequencing of the human genome, which highlighted the surprising fact that humans have only approximately 20,000–25,000 protein-coding genes [1], hardly enough to explain the complexity gap between humans and the lowly round worm whose genome boasts approximately 19,000 protein-coding genes, the focus of biological research has rapidly shifted to the study of the encoded proteins. Proteins, which are the main workhorses in our cells, acting as molecular motors, catalysts, structural elements, signaling messengers, and defensive agents, are the key to understanding fundamental biological mechanisms in cells and tissues, from development to aging and disease. In the postgenomic phase of molecular biology, novel high-throughput proteomic tools and technologies have been developed to study proteins expressed in tissues, cells, and organelles, leading to an explosive growth in the volume of proteomics data.

The vast diversity of proteins, dynamic range of expression, and interaction make the identification of the entire proteome a computational challenge that is more complex by several orders of magnitude than the sequencing of the genome [2]. The advances in mass spectrometry (MS) technology have played a key role in the extraordinary growth of the proteomics research field and have revolutionized protein analysis [3]. MS, which started being used for biological applications in 1950s [4], is today a fundamental tool for protein identification [3]. The state-of-the-art MS instruments available today, which have acquisition rates of up to 200 spectra per second, enable scientists to carry out large-scale proteomic studies that routinely generate tens or even hundreds of gigabytes of complex, multidimensional datasets, and allow the simultaneous identification of hundreds or even thousands of proteins present in complex protein mixtures.

Postinstrument data processing is already a major bottleneck in proteomics workflow and is expected to get worse given the importance of MS-based proteomics in modern biology. An unprecedented growth of proteomics data over the next decade is forecasted to escalate further the demand for computing power, outstripping the expansion in supply predicted by the celebrated Moore's law. Conventional workstations based on low number of multicore processors are unlikely to deliver the speed that will be required to analyze such large volume of data, so high-performance computing (HPC) resources are essential to address the analysis bottlenecks. Grid computing technology could in principle help meet this challenge [5]. However, for the grid implementation of a proteome sequence similarity search algorithm using the Ensembl database of protein sequences, the performance gains reported—60 fold speed increase using 600 CPUs [6]—are far from spectacular relative to the significant power consumption, maintenance costs, and floor space used. For comparison, the implementation of Allegro, a computer program for multipoint linkage analysis based on hidden Markov models on the same grid achieved a significantly better speedup: 455-fold for 600 grid nodes [7]. This algorithm, however, does not involve any database searching. A preliminary grid-based implementation of protein identification algorithms based on tandem MS data is presented in [8] but the study provides no information on the performance gains achieved. It is worth highlighting the generic limitations of computational grids that may have prevented, so far, the large-scale adoption of grid solutions for protein identification involving database searching: (1) the job submission process on the platform is usually complex and difficult to automate; (2) high parallelization is hampered by resource brokering and job-scheduling time; (3) to maximize performance, the database has to be split and distributed across the nodes of the cluster, which introduces a significant computational overhead; (4) there are large variations in processing times caused by variations in load and network latency. In addition, processing of MS data is best performed "near-instrument," where the end user has the option of adjusting the search strategy according to results obtained in real time. In this context, relying on grid computing or dedicated computer clusters may not be the best solution for protein identification. Equally dedicated high-performance computer clusters require a significant amount of infrastructure to deal with interconnectivity and power dissipation. It has been argued [9] that more efficient HPC solutions are necessary to mitigate the costs of housing and powering the next-generation petascale and larger high-performance computer systems, which are expected to be prohibitive for many institutions and programs.

This work advocates the use of reconfigurable computing, as an alternative approach to conventional HPC, for a specific bioinformatics problem in proteomics, namely, protein identification based on MS, using database searching.

14.2 The Reconfigurable Computing Paradigm

Reconfigurable computers consist of a standard microprocessor system coupled with hardware processors whose circuitry can be programmed (and reprogrammed) according to the algorithm that is being run. The idea of reconfigurable computing originated in the 1960s [10]. In a seminal paper [11] Estrin proposed the concept of a computer made of a standard processor and an array of "reconfigurable" hardware. The introduction of high-density field-programmable gate arrays (FPGAs) in the 1990s made reconfigurable computing possible. FPGAs are de facto the reconfigurable processors in almost all current reconfigurable computing platforms. Modern FPGAs can be programmed to run a custom digital hardware design providing the flexibility afforded by a conventional computer program. To fully understand the significant advances made in this area the reader is referred to the excellent books [12, 13] that are amongst the first comprehensive surveys and tutorials in the field of FPGA-based reconfigurable computing.

Since their introduction in 1985, FPGAs have continuously expanded their use from being the ultimate prototyping platform and providing basic "glue logic" functionality to being at the heart of complex digital systems in a wide range of application areas ranging from telecommunication, automotive, aerospace, and defense to biomedical and HPC. The remarkable success of these devices is attributed to the inherent advantages offered by the parallel programmable architecture, which allows designers to exploit algorithm and instruction-level parallelism to accelerate computations and to add or modify features and functionality provided by an existing FPGA-based system by reconfiguring the device. The tremendous increase in gate densities and lowering of unit costs and the development of more sophisticated and user-friendly design tools have also been determining factors to skyrocketing demand for FPGAs in recent years.

Because of the widening spectrum of applications for FPGA devices, modern FPGAs have evolved to include specialized programmable blocks such as embedded RAM, dedicated DSP structures, embedded microprocessors, system monitoring functions, digital clock managers, and fast serial transceivers. Moreover, to satisfy diverging user demands device manufacturers such as Xilinx have followed a new strategic route of creating a family of FPGA platforms [14] that have been optimized for particular application domains.

As the static random access memory (SRAM)-based FPGA manufacturers are amongst the earlier adopters of new digital-CMOS manufacturing processes, in recent years FPGAs have advanced at a faster pace than microprocessors, the latest devices offering unprecedented performance and density gains with speeds on average 30% faster and a logic capacity 65% greater than previous generations. The latest devices have as many as 1.2 billion transistors and allow the implementation of a few thousand conventional microcontrollers on a single FPGA chip.

Despite the availability of high-level design software, the difficulties of mapping an algorithm in hardware are still considerable, as FPGA development tools were designed for electronics hardware engineers and require in-depth knowledge of hardware design languages (VHDL, Verilog) and digital electronics. The emerging high-level design tools, whilst offering a great level of abstraction, still require a fair amount of manual optimization; hence low-level design knowledge is still essential. Moreover, because there is no standard RC architecture, most common design tools do not target specific FPGA boards and as a result designs have to be mapped manually on the chosen RC platform. This cannot be achieved without a detailed understanding of the architecture of the hardware system.

In biocomputation, early applications of FPGA devices addressed the gene sequence analysis problem [15] and have been successfully employed to speedup DNA sequencing algorithms [16–21]. FPGAs were also used in the attempt to accelerate search of substrings similar to a template in a proteome

[22]. A multiple sequence alignment solution has been implemented in FPGA hardware by [23] and FPGAs have been used to accelerate sequence database searches with MS/MS-derived query peptides [24]. FPGA-accelerated basic local alignment search tool (BLAST) search algorithms have been developed and used, for example, to perform expressed sequence tag (EST) sequencing [25]. More recently, the Aho-Chorasick string set matching algorithm was implemented in FPGA hardware and used for matching peptide sequences against a genome translated in six reading frames [26].

14.3 Protein Identification by Sequence Database Searching Using Mass Spectral Fingerprints

14.3.1 Overview of the Approach

MS is a powerful technique for chemical characterization, which has become the central tool of proteomics [3]. In protein identification, mass spectrometers do not deal with intact proteins but with their constituent peptides generated by proteolytic digestion. The mass spectrometer ionizes the peptides in the experimental sample producing charged ions that are directed to the mass analyzer where they are separated according to their mass-to-charge ratio (m/z) and ultimately detected.

Proteolytic enzymes with site-specific cleavage properties are used to produce a subset of predictable peptide fragments. The resulting peptide mixture is typically analyzed by matrix-assisted laser desorption/ionization time-of-flight (MALDI-TOF) mass spectrometer. The resulting spectral map is used to perform protein identification.

Peptide mass fingerprinting (PMF) is an established protein identification technique that is predicated on the assumption that the detected pattern of proteolytic peptide masses provides a quasi-unique signature for every protein in the database. The computations associated with the PMF approach can be divided into two distinct stages:

(a) Process the raw MALDI-TOF MS data to extract a spectral peptide fingerprint, which is a subset of the experimentally generated "peak list."

(b) Find the best matching protein by correlating the experimental mass fingerprint with theoretical peptide maps generated by *in silico* digestion of protein sequences from a database of known proteins.

Both stages of computation have been implemented as dedicated hardware processors [27–30].

14.3.2 Abstract Computational Model

Proteins can be modeled mathematically using the concept of weighted strings [31–33].

Let Σ = {A,C,D,E,F,G,H,I,K,L,M,N,P,Q,R,S,T,V,W,Y} be the 20-letter alphabet corresponding to the naturally occurring amino acids. Let $\mu:\Sigma\rightarrow IR$ be a function assigning to each character $s\in\Sigma$ its mass or weight $\mu(s)$. The pair (Σ,μ) is called a *weighted alphabet* with character mass function μ. The letters of the alphabet Σ "spell out" peptide or protein "words." A peptide or protein can be represented as length-n weighted strings $S_\mu = (\overline{s}, \overline{\mu}_s)$ over (Σ,μ) where $\overline{s} = (s_i)$ $\overline{\mu}_s = \mu(s_i)$, $i = 1,\ldots,n$. Polypeptides of specific sequence for which $n >$ 50 amino acid residues are usually known as proteins.

The mass of a peptide or protein of length n is given by

$$m(S_\mu) = \sum_{i=1} \mu(s_i) \tag{14.1}$$

After a protein has been synthesized by the cell, it often suffers further "natural" chemical (so-called *posttranslational*) modifications, such as phosphorylation and glycosylation, that often have functional roles. There are also "accidental" modifications, such as oxidation or modifications carried out deliberately during experiment phase. These modifications typically result in changes of the amino acid masses.

A modification that is applied universally, to every instance of the specified amino acid for example, is known as a *fixed modification*. Fixed modification can be described in terms of modified weighted alphabets (Σ, μ^K), where μ^K is the modified mass function corresponding to the K-type modification.

In reality, however, it is possible that not every instance of an amino acid has suffered a modification. This can be described by considering a modified weighted alphabet (Σ, μ^K) where $\mu^K:\Sigma \times \Gamma\rightarrow IR$ is a probabilistic function that assigns to every character s in Σ a random variable $\mu^K(s,\gamma)$, where $\gamma\in\Gamma$ is a possible outcome and $P(\gamma)$ denotes the probability associated with the elementary events. For example, if we consider a single modification, $\gamma = 1$ if the amino acid residue is "modified" or $\gamma = 0$ if "unmodified."

14.3.3 Cleavage Rules

Proteolytic enzymes or proteases will break down a protein by cleaving the peptide at specific points, for example, after the occurrence of a specific amino acid residue. Each proteolytic enzyme can be associated with a set of cleavage rules. Table 14.1 shows examples of cleavage site rules for some of the most commonly used proteases. In this table, the amino acid residues of the N-terminal side of the scissile bond are noted as P_1, P_2 and the residues of the C-terminal side are noted as P_1', P_2' (Figure 14.1) in line with the Schechter and Berger nomenclature for the description of the protease subsites [34].

TABLE 14.1

Examples of Cleavage Site Rules

Enzyme	P_1	P_1'	P_2'
Arg-C	R	X	X
Asp-N	X	D	X
Chymotrypsin 1	F,Y,W	X	X
Chymotrypsin 2	A,L,M,F,Y,W	X	X
Glu-C 1	E	X	X
Glu-C 2	D,E	X	X
Lys-C	K	X	X
Trypsin	R, K	not P	X
Thermolysin	X	A,I,L,M,F,V	not P
CNBr	M	X	X
Formic acid	D	P	X

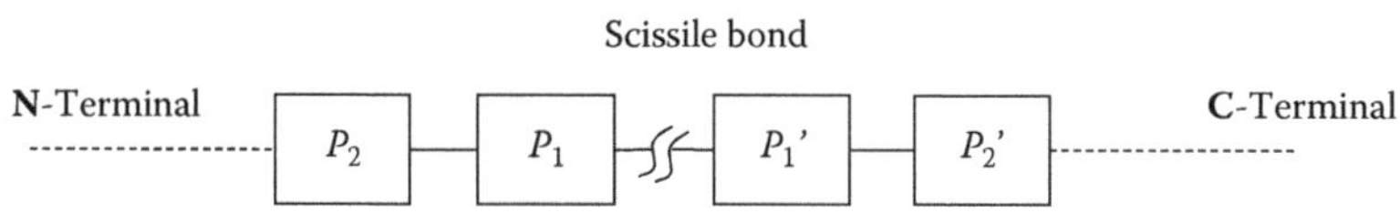

FIGURE 14.1
Illustration of the notation used to for cleavage sites.

Trypsin is by far the most commonly used enzyme in MS studies because each proteolytic fragment contains a basic arginine (R) or lysine (K) amino acid residue that is abundant and well distributed, yielding peptides of molecular weights that can be analyzed by MS. A model for the cleavage behavior of a proteolytic enzyme E has been proposed in [33]. As an alternative model, more suitable in the present context, we introduce a cleavage function defined on the space of three-letter substrings $\overline{s}_{i,i+2} = s_i s_{i+1} s_{i+2}$ of $\overline{s}$, over the alphabet Σ, such that

$$c_E(\overline{s}_{i,i+2}) = \begin{cases} 0 & \text{if } \overline{s}_{i,i+2} \in J^3 \setminus R_E \\ i & \text{if } \overline{s}_{i,i+2} \in R_E \end{cases} \tag{14.2}$$

where zero means that no cleavage occurs, i indicates that the cleavage occurs after character s_i, J^3 is the set of all possible three-character strings over the alphabet Σ, and $R_E \subset J^3$ is the subset of cleavage patterns associated with enzyme E.

The application of the cleavage function (2) to a string $S = s_{1 \ldots} s_n$ of length n, produces an ordered set of all possible cleavage points

$$C_E(S_\mu) = \left\{ l(k) \,\middle|\, l(k) \in \bigcup_{i=1}^{n-2} \{c_E(\overline{s}_{i,i+2})\}, l(k) < l(k+1), l(k) \neq 0, k = 1, \ldots, \left| \bigcup_{i=1}^{n-2} \{c_E(\overline{s}_{i,i+2})\} \right| - 1 \right\} \tag{14.3}$$

The "digestion" of the string S generates the following fragmentation set (peptide map):

$$\mathsf{F}_E(S) = \{F_\mu^k\} = \{S_\mu^{l(k),l(k+1)} = (\overline{s}_{l(k)+1,l(k+1)}, \overline{\mu}_{l(k)+1,l(k+1)}) \,|\, l(k), l(k+1) \in C_E(S), k = 1,\ldots,|C_E(S)| - 1\} \tag{14.4}$$

where $\overline{s} = s_{l(k)} \ldots s_{l(k+1)}$ denotes the k-th fragment of $\overline{s}$ (i.e., a substring $\overline{s}$ of between two successive cleavage points).

In practice, if enzymatic cleavage is not complete, not all adjacent peptides may separate. Assuming that one missed cleavage site has occurred, the corresponding fragmentation set is given by,

$$\begin{aligned} \mathsf{F}_{E,1}(S_\mu) = &\{S_\mu^{l(k),l(k+1)} \,|\, l(k), l(k+1) \in C_E(S), k = 1,\ldots,|C_E(S_\mu)| \text{-} 1\} \cup \\ &\{S_\mu^{l(k),l(k+2)} \,|\, l(k), l(k+2) \in C_E(S_\mu), k = 1,\ldots,|C_E(S_\mu)| \text{-} 2\} \end{aligned} \tag{14.5}$$

Typically, when searching the database, up to two missed cleavage sites are allowed but, for best results, one missed cleavage site seems to be the optimal choice. Each additional missed cleavage in the search increases the search space and reduces the significance level of the matches found as the probability of getting random matches increases. Using more than the two cleavage sites implies that the quality of the enzymatic digestion was really poor and really begs the question of whether the experiment should be repeated.

Assuming no PTMs- and p-missed cleavage sites to the fragmentation set of a protein string S_m digested with an enzyme E we can associate the following set of corresponding fragment masses (the peptide mass map):

$$M_{E,p}(S_\mu) = \{m(F_\mu^{k,j}) \,|\, F_\mu^{k,j} \in \mathsf{F}_{E,p}(S_\mu), j = 0,\ldots,p\} \tag{14.6}$$

If fixed modifications are specified, the peptide map is computed using

$$m(S_\mu) = \sum_{i=1}^{n} \mu^K(s_i, \gamma),\ \gamma = 1 \tag{14.7}$$

In this case, we still have $|\mathsf{F}_{E,p}(S_\mu)| = |M_{E,p}(S_{\mu^K})|$; that is, the size of mass set equals the number of protein fragments.

However, if a modification is assumed to be variable (may or may not be present at a particular location), the mass of a peptide is a random variable, so that all possible arrangements of that modification have to be considered, during searching, for every peptide fragment. For example, if an amino acid $s \in \Sigma$ is assumed to undergo a variable modification, for a peptide $m(F_\mu^k)$

of length l containing three s residues, the following masses have to be computed:

$$m(F_{\mu}^{k},0)=\sum_{i=1}^{l}\mu(s_i)$$

$$m(F_{\mu}^{k},1)=m(F_{\mu}^{k},0)-(\mu(s)-\mu^{K}(s,1))=m(F_{\mu}^{k},0)-\Delta\mu(s)$$

$$m(F_{\mu}^{k},2)=m(F_{\mu}^{k},0)-2(\mu(s)-\mu^{K}(s,1))=m(F_{\mu}^{k},0)-2\Delta\mu(s)$$

$$m(F_{\mu}^{k},3)=m(F_{\mu}^{k},0)-3(\mu(s)-\mu^{K}(s,1))=m(F_{\mu}^{k},0)-3\Delta\mu(s) \tag{14.8}$$

These correspond to the cases in which we have 0, 1, 2, or 3 modified residues. As a consequence, the size of the theoretical peptide masses corresponding to a single protein can increase dramatically, especially if more than one variable modification is specified.

14.3.4 Protein Identification by Spectral Matching

Protein mass fingerprinting involves comparing an experimental peptide mass fingerprint; that is, a list of r mass queries $X = \{x_1,\ldots,x_r\}$, with the theoretical peptide mass map $M_E(S_\mu)$ computed *in silico* for every protein $S_{\mu,j}$ string in the database.

All algorithms used to perform the protein identification task implement in essence a Bayes statistical decision rule with the goal to minimize/maximize the cost/benefit of incorrect/correct classification. A Bayes classifier may assign an unknown peptide mass fingerprint pattern X to a protein if

$$\sum_{i=1}^{N}\epsilon_{i,j}P(S_{\mu,i}\mid X)<\sum_{i=1}^{N}\epsilon_{i,k}P(S_{\mu,k}\mid X), \qquad \forall k\neq j \tag{14.9}$$

where N is the total number of proteins in the database and $\xi_{j,k}$ is the cost of wrongly assigning a mass fingerprint of a protein S_j to a protein S_k. For simplicity $\xi_{i,k}$ can be assumed to be 1 for all $k \neq i$ and 0 when $k = i$. A practical score or discriminator function can be defined by approximating the probability functions $P(S_{\mu,i}\mid X)$ for all proteins in the database.

A rough approximation is given by the number of theoretical peptide masses in a protein that match the experimental mass fingerprint. An exact match between x_i is found $m(F_{\mu}^{k})$ if

$$x_i = m(F_{\mu}^{k}) \tag{14.10}$$

If we take into account mass accuracy R (expressed in ppm) of the instrument, an approximate match is found if

$$x_i^L \le m(F_\mu^k) \le x_i^H \tag{14.11}$$

where x_i^L and x_i^H are positive lower and upper tolerances

$$x_i^L = x_i \frac{(1-2R/10^6)}{(1-R/10^6)}; x_i^H = x_i \frac{1}{(1-R/10^6)} \tag{14.12}$$

The basic score provides a distorted approximation of $P(S_{\mu,i} \mid X)$. Specifically, it does not take into account the frequency of occurrence of each peptide in the database; the measurement accuracy; the individual properties of each protein searched S_i, such as theoretical number of peptides generated by fragmentation and the total mass of the protein; or the fact that the sample that is analyzed may be a mixture of two or more proteins. Whilst the MOWSE [35] or MASCOT [36] scoring schemes, which incorporate information about the frequency of peptides from all proteins in the database within a molecular weight range, are superior to the naïve scoring scheme, it still does not account for the individual properties of the proteins analyzed. A comprehensive Bayesian scoring approach, which accounts for all factors listed earlier, can deal with protein mixtures and can incorporate additional a priori information about the experimentally observed peptides, is implemented by ProFound (http://prowl.rockefeller.edu/prowl-cgi/profound.exe). The algorithm used by ProFound is described in detail in the paper by Zhang and Chait [37].

Whilst the basic score alone is not ideal, it has to be computed to implement more sensitive scoring functions. To decide the most likely protein match, the score has to be computed for all proteins in the sequence database. To have an idea of the computational challenge, it is worth mentioning that UniProt, the largest protein database in the world, currently holds a total of almost 9 million sequence entries covering 204,373 species, with the largest known protein consisting of 34,350 amino acids.

14.4 Reconfigurable Computing Platform

The hardware processors described in this paper were implemented on a commercial off-the-shelf (COTS) multi-FPGA reconfigurable hardware platform, consisting of a BenNuey motherboard and a BenDATA DIME-II module from Nallatech Ltd. (www.nallatech.com).

The BenNUEY board is a full-length PCI DIME-II motherboard that houses a Xilinx Virtex-II XC2V8000 FPGA and 4 Mbytes on-board RAM, providing substantial on-board FPGA resource for processing and system management. In our application, this FPGA has been used to implement the mass spectra processor that is described elsewhere [28].

The communication between the PC server and the FPGA system, via a standard PCI interface (32 bits, 33 MHz), is handled by a second, smaller FPGA (Xilinx Spartan-II) on the motherboard. The motherboard has three DIME-II expansion slots, which allow users to configure additional system resources to meet processing, memory, and input/output (I/O) requirements. The BenDATA DIME-II module has one user FPGA device (Virtex-II XC2V8000) and 1 GB of DDR SDRAM memory organized in four banks with a 64-bit wide data bus each. The total data bus width is 256 bits. Each module is connected with the motherboard FPGA and with the other two modules via a 64-bit, 40-MHz local bus. This architecture enables the implementation of parallel searches at FPGA level as well as across modules. The block diagram of the FPGA system is shown in Figure 14.2.

An important factor that has to be considered when choosing an FPGA platform is the communication overhead associated with data transfer between PC and device, which should represent only a fraction of the actual execution time. For a known reconfigurable computing platform it is possible to evaluate at this stage the actual communication costs incurred by transferring data between hardware and software. This aspect is a major decision factor in the selection of the FPGA system best suited for an application.

The reconfigurable computing platform adopted in this work is well suited for this particular bioinformatics application. More specifically

(a) The FPGA module used to implement the database search engine (BenData DD) provided sufficient on-board memory to hold the entire database, allowing the entire database that is to be searched to be stored in local memory, resulting in very low communication overhead.
(b) The architecture of the module allowed data transfers between memory and FPGA on a 256-bit-wide channel at 100 MHz so that multiple proteins could be streamed out from the memory and processed in parallel by individual search processors programmed on the FPGA fabric.
(c) The hardware system offers the flexibility to implement distributed search strategies by adding additional modules.

The code selected to run in hardware should ideally have low data dependency, to facilitate parallel implementation. Fingerprint matching, for example, can be performed in parallel on different database partitions. Because the

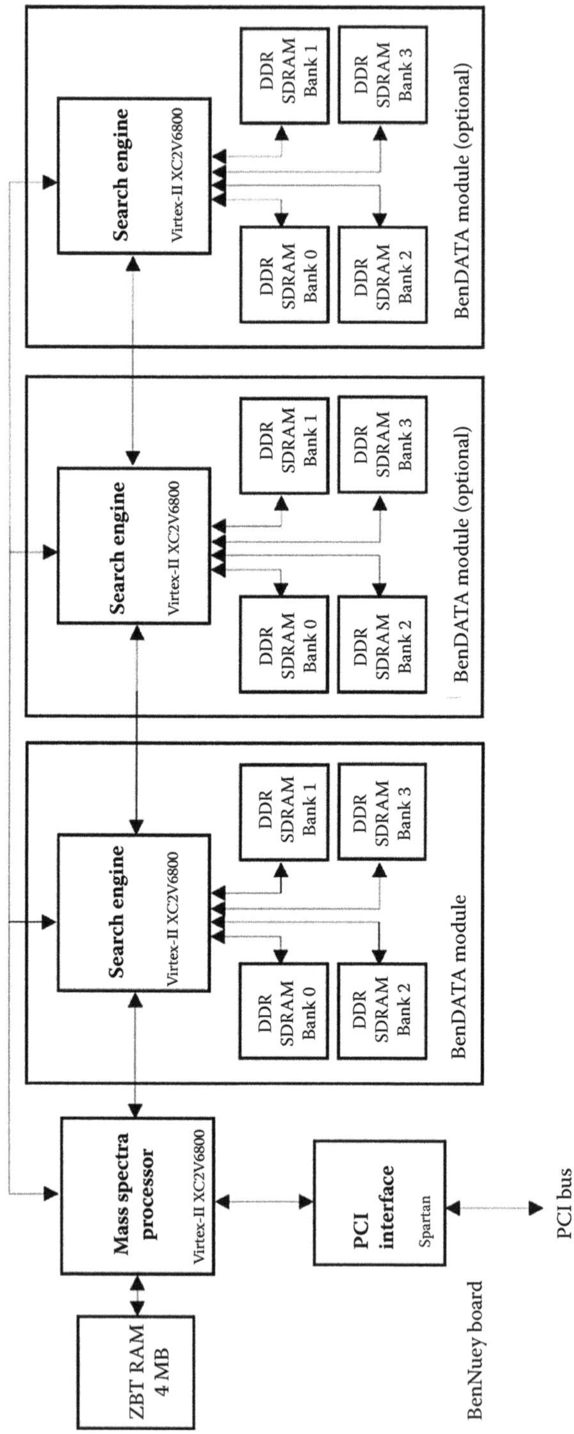

FIGURE 14.2
Block diagram of the FPGA system.

number of proteins in the database is very large, the potential for speedup is huge, given the right FPGA platform.

The design of the present implementation of the database search engine has been optimized to exploit this particular RC hardware architecture. In particular, the implementation is scalable so that the system performance can be easily increased by incorporating additional search modules and distributing the database across different modules.

Basically, a motherboard equipped with three search modules is able to deliver a match three times faster than a single module motherboard. Each system (box) can be scaled up easier by adding up additional FPGA motherboards (two to four motherboards per "box" for example). Furthermore, for the ultimate performance, a FPGA "cluster" could be set up by interconnecting two or more FPGA computing boxes. Considering the fact that single database search module can deliver the performance of hundreds or even thousand of conventional single-core microprocessors, a "cluster" FPGA protein identification system could easily deliver more computational power than even some of the largest HPC in operation today.

The complete FPGA-hardware solution for PMF, which incorporates a raw mass spectra processor and a parallel search engine, is presented in the following sections.

14.5 Protein Sequence Database FPGA Search Engine

The database search engine [28] traces the peptide fingerprint back to the originating peptide by matching it against the expected (theoretical) peptide masses obtained by digesting *in silico*—on the fly—all protein sequences in the database.

To maximize database search speed, the initial search engine has been configured as a set of 48 identical search processors that can process database records (encoded protein streams) in parallel. The search processors operate at a clock rate of 100 MHz, which is dictated by clock frequency of the 1-GB on-board DDR SDRAM.

14.5.1 Database Encoding

The protein sequence databases such as MSDB, the database used in this study, are in fact flat text files. To fully exploit the benefits of FPGA acceleration, the entire MSDB database was encoded using 28 symbols coded on five-bit words. Of the 28 symbols, 20 symbols were required to encode the constituent amino acids, 6 were additional standard symbols adopted in the FASTA format, and 2 symbols were used to mark the end of a protein sequence and the end of the database. By encoding the database using only five-bit "characters," the database size was reduced by approximately 40%. The encoded database occupies approximately 680 MB of the total 1-GB DDR SDRAM memory installed on the module.

The encoded database was loaded on the local on-board memory of the BenDATA module in a format that facilitates fast parallel searches. The database was divided into 4 × 12 = 48 data streams of consecutive records for parallel processing. Each data stream contains a variable number of complete protein sequences and the unused memory locations, which could not hold an entire protein sequence, were padded with zeroes (Figure 14.3). In this format, each memory module can supply synchronously 12 × 5-bit wide data streams to 12 search processors that connect to the output of that random access memory (RAM) bank.

Storing the database in the local module memory eliminates a significant memory access bottleneck that would be caused, if the protein database were stored in the computer memory, by the relatively slow PCI interface.

However, the most significant reason for encoding and storing the protein database in the local memory is that it enables parallel processing of protein sequences. In the current implementation there are 48 protein sequences that are streamed out, in parallel, from the memory as shown in Figure 14.4. Each protein sequence is processed sequentially by a search processor implemented in the module's FPGA.

14.5.2 Database Search Processor

Each search processor has two major functional blocks, an *in silico* protein digestion unit and a scoring module. These blocks perform the following basic operations:

(a) The digestion unit computes the theoretical peptide masses [5] for every protein in the database by *in silico* digestion [2].
(b) The scoring unit compares the user supplied experimental masses with the theoretical peptide masses [10, 11] generated by the digestion unit, computing for each protein the matching score; that is, the number of matched peptide masses.

14.5.2.1 Digestion Unit

Each search processor reads a five-bit code every clock cycle from the corresponding memory column and passes it to the digestion unit. The digestion unit is responsible for calculating the peptide masses according to the user-defined digestion rule/parameter (Figure 14.5).

For every clock cycle, the digestion unit calculates the cumulative mass of the amino acids received from the database until it encounters a cleavage pattern [2], a protein sequence delimiter or the end-of-database marker. The theoretical masses of individual amino acids, used to compute the peptide

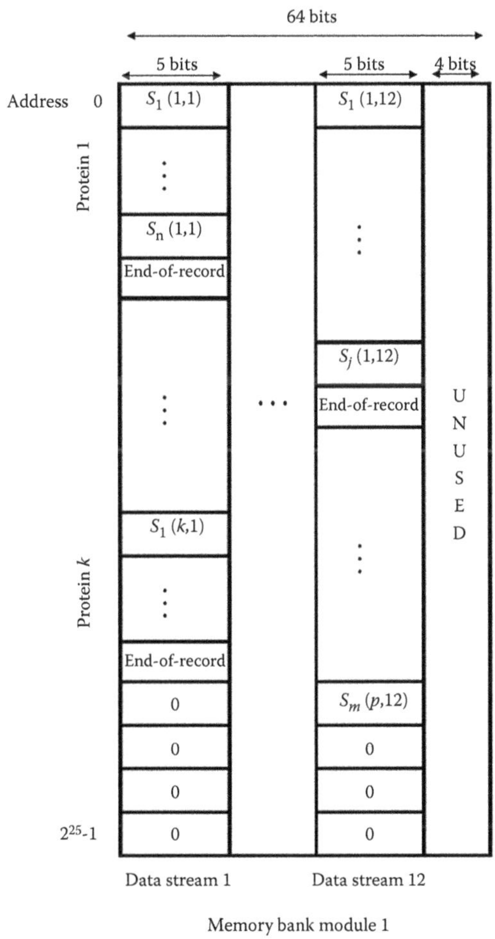

FIGURE 14.3
Structure of the encoded database stored in one DDR SDRAM memory bank.

masses, are stored in a lookup table as 32-bit (12 bits after the radix point) fixed-point numbers. In this implementation, fixed PTM rules are handled implicitly. The lookup table is loaded with the modified amino acid residue masses, according to the PTMs specified by the user.

14.5.2.2 Variable Modifications

To deal with variable modifications, the current design has additional registers and control logic compared with the original design [28].

One additional register stores the codes of the amino acids that are modified. Another register is used to store the corresponding $\Delta\mu(s)$ values [7]. If a variable modification is specified, when a processor encounters a cleavage point,

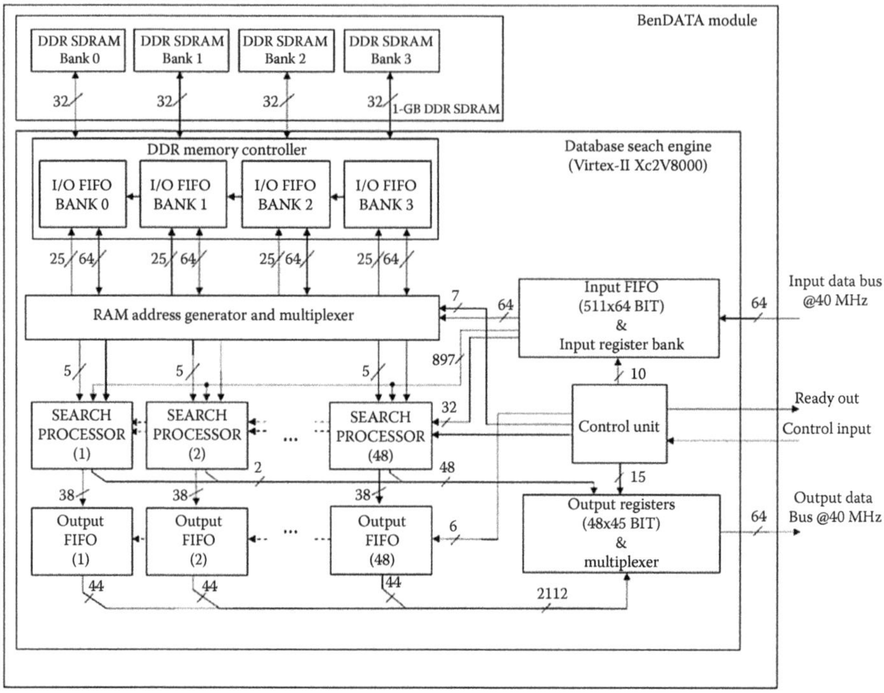

FIGURE 14.4
Block diagram of the database search engine.

the input FIFO is disabled and no more data is streamed out from it, until the digestion unit that encountered the cleavage point computes the additional masses required [7]. The number of additional clock cycles inserted equals the number of possible distinct modifications that have to be evaluated; that is, the number of affected amino acid residues in that peptide.

Because of the increased complexity of the design, only 36 processors that implement variable modifications could be fitted on the FPGA. From a computational point of view, variable modifications are costly, as the parallel computation flow is broken every time a cleavage site is encountered. This problem can be significantly alleviated by distributing the database across different search modules.

14.5.2.3 Scoring Unit

The scoring unit calculates the number of peptide masses in the peptide mass fingerprint that are matched for every digested protein in the database. The user can specify the matching error tolerance *R* (ppm) in [11], which reflects the accuracy of the mass spectrometer and other known sources of errors.

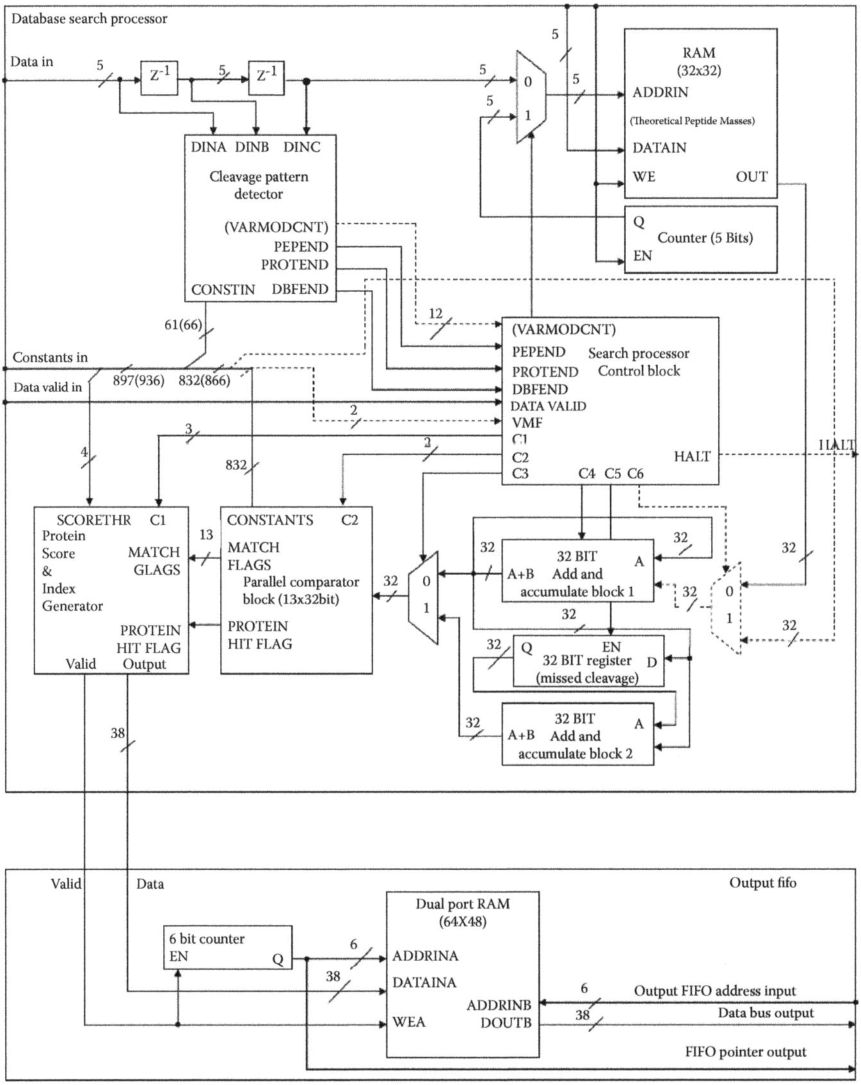

FIGURE 14.5
Block diagram of a search processor. The dashed-line is used to indicate the additional blocks and signals that are used to implement variable modification functionality.

When the digestion unit detects a cleavage point, the computed theoretical mass $m(F_{\mu}^{k})$ of a peptide fragment is transferred to a bank of comparators.

The theoretical peptide map is compared in parallel with the experimental peptide mass fingerprint $X = \{x_1, x_2, \ldots, x_r\}$, generated by the mass spectra processor.

For the implementation evaluated here r=13. However, the design could be easily modified to increase the number of m/z query values, at the expense of increasing the complexity of the design of individual search processors that will consume more FPGA resources. Effectively, the number of processors that can be allocated on current FPGA device (Xilinx XC2V8000) will be smaller. However, as the logic capacity of the latest FPGA devices (e.g., Xilinx Virtex 5 family) has increased dramatically compared with the device used in this implementation. In this situation, the complexity of the design can be increased without compromising on performance. Moreover, because of the parallel nature of the computations, the entire database can be divided into distinct subsets and loaded on separate BenDATA modules.

If a match is found, the score counter is incremented by one. The position of a match is also recorded in an n-bit match index word. When the end of a record is found, the record index counter, the score counter, and the match index register outputs are stored in intermediate registers.

Each search processor has four outputs: (1) the index of the matched protein, (2) the corresponding protein score, (3) a flag indicating that the index and score output can be written in the output FIFO, and (4) a flag that indicates that the processor has reached the end of the corresponding database segment. The user can specify a score threshold τ_s so that the record and match indexes are stored in the output FIFO only if the score is higher than the specified threshold τ_s.

Results of the 48 search processor are collected in dual port RAM devices organized as FIFO structures of 64 words of 38 bits each. When all search processors reached the end of the database segment, a global flag indicating that the search engine has completed processing is set and the results are transferred on the PC server side.

The basic score is normally used to implement more sensitive scoring schemes that account for peptide frequency distributions such as MOWSE [35], PIUMS [38], or more comprehensive Bayesian scoring approaches that also account for the individual properties of the proteins analyzed such as ProFound [37]. Because of the low speedup gain expected from a hardware implementation, these scoring methods have been implemented in software and run on the PC server post-FPGA processing. The externalization of the scoring statistics means that the output of the search can be rapidly evaluated using different scores and even developed into a consensus score validation scheme.

The design includes all necessary control and FIFO structures that implements a 64-bit wide data transfer between the FPGA devices at a rate of 320 MB/s. In general, the development of a complete reconfigurable computing solution involves significant low-level programming for designing control and synchronization modules to manage data transfers between the hardware processors running on different FPGAs, between the FPGA system and the host PC, and between FPGAs and the on-board memory modules.

The 48-processor search engine occupies 99% of the FPGA's logic resources, 99% of the FPGA's internal RAM resources, and 53% of the FPGA's I/O resources.

14.6 Performance Evaluation

Performance evaluations of the FPGA implementation were carried out using both single and dual (single core) 3.06 GHz Xeon processor servers with 4-GB RAM under Windows XP Professional [28]. In this study, the performance of the 48-processor database search engine is compared with that of a dual quad 3.16 GHz Intel Xeon server with 4 GB of RAM running Windows 2003 Server operating system.

The MSDB database was encoded and loaded in the local 1-GB DDR SDRAM module memory. The database contains 3,239,079 records with 1,079,594,700 effective code letters. If the additional separator codes are included the encoded database requires 1,082,833 779 symbols. It is important to emphasize that the MSDB database used here is no longer actively maintained, being last updated on August 8, 2006. It now contains only a fraction of the currently known proteins.

Performance evaluation was carried out using a reference C program that models the exact computational flow implemented by the hardware design. In tests, the output results of both the software and FPGA implementation of the database search engine were identical. For the current comparative study, the C program was run on a Dual Quad Core 3.16 GHz Intel Xeon PC server as well as on a single (single core) 3.06 GHz Xeon server.

The performance of both software and hardware (FPGA) designs were assessed using randomly selected database records that were digested *in silico* using trypsin digestion rules. In each case, the search was carried out using 13 query peptides (m/z values) selected randomly from the theoretical protein digests. The processing time for the software implementation accounts only for the main computational loop after all variables have been initialized.

The FPGA system performs a complete database search in 240(±0.02) ms while the completed average processing time for the C implementation is approximately 3.92 minutes.

As seen in Figure 14.6, the speed gain of the FPGA over the C software implementation running on a dual quad processor server is still significant. Figure 14.7 shows comparatively the speed gains relative to single and dual quad processor systems. The average speed gain on the dual quad processor system is 982.77 (standard deviation = 14.9). This compares with an average of 1,805 (standard deviation = 50.66) when the software was run on the single-core processor system.

The results show that the latest multiprocessor systems have narrowed the performance gap but, although the number of cores has increased from one to eight, the dual system manages to be only twice as fast as the single-core machine. Moreover, the comparison is somehow "unfair" as the FPGA platform used in this study is based on very "old" 2004 Virtex-II technology.

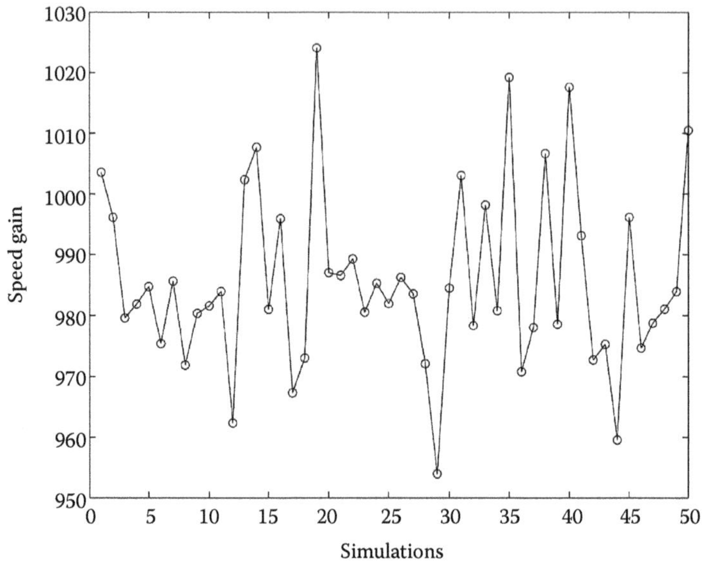

FIGURE 14.6
Speed gains of FPGA versus C implementation.

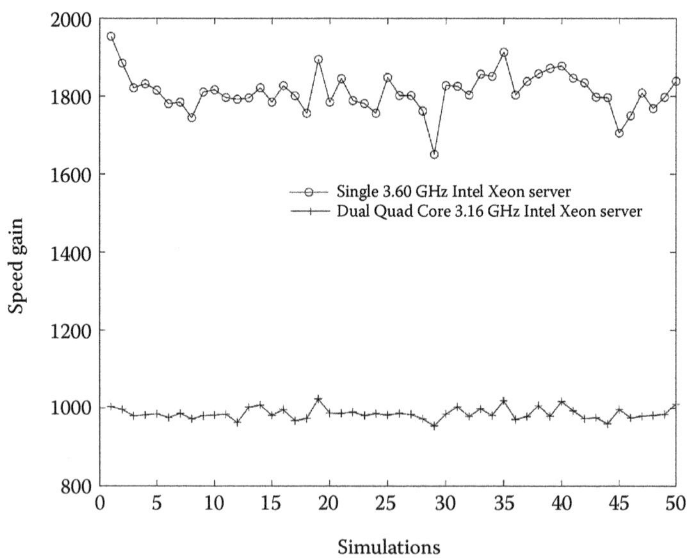

FIGURE 14.7
Speed gains of FPGA versus C implementation run on single/dual quad processor servers.

Acknowledgments

The authors gratefully acknowledge that this work was funded by BBSRC. The authors are also grateful for the support received from Xilinx Inc. who donated the devices and design tools used in this study.

14.7 References

1. Stein L. D. (2004), Human genome: End of the beginning, *Nature* 431, 915–916.
2. Naaby-Hansen S., Waterfield M.D., Cramer R. (2001), Proteomics—Post-genomic cartography to understand gene function, *Trends in Pharmacological Sciences*, 22, 376–384.
3. Hugh L., Arthur J.W. (2008), Computational methods for protein identification from mass spectrometry data, *PLoS Computational Biology*, 4, 1553–7358.
4. Beynon J.H. (1956), The use of the mass spectrometer for the identification of organic compounds, *Microchimica Acta*, 44, 437–453.
5. Krishnan A. (2004), A survey of life sciences applications on the grid, *New Generation Computing*, 22, 111–126.
6. Andrade T., Berglund L., Uhlén M., Odeberg J. (2006), Using Grid technology for computationally intensive applied bioinformatics analyses, *In Silico Biology*, 6, 495–504.
7. Andrade J., Andersen M., Sillen A., Graff C., Odeberg J. (2007), The use of grid computing to drive data-intensive genetic research, *European Journal of Human Genetics*, 15, 694–702.
8. Quandt A., Hernandez P., Kunzst P., Pautasso C., Tuloup M., Hernandez C., Appel R.D. (2007), Grid-based analysis of tandem mass spectrometry data in clinical proteomics, *Studies in Health Technology and Informatics*, 126, 13–22.
9. El-Ghazawi T., Bennett D., Poznanovic D., Cantle A., Underwood K., Pennington R., Buell D., George A., Kindratenko V. (2006), Is high-performance reconfigurable computing the next supercomputing paradigm? *Proceedings of the 2006 ACM/IEEE Conference on Supercomputing*, Tampa, Florida. doi:10.1109/SC.2006.38.
10. Estrin G. (1960), Organization of computer systems—The fixed plus variable structure computer, *Proc. Western Joint Computer Conf.*, New York, pp. 33–40.
11. Estrin G. (2002), Reconfigurable computer origins: The UCLA fixed-plus-variable (F+V) structure computer, *IEEE Ann. Hist. Comput*, 24, 3–9.
12. Gokhale M.B., Graham P.S. (2005), *Reconfigurable computing: Accelerating computation with Field-Programmable Gate Arrays*, Springer.
13. Hauck S., Dehon A. (2008), *Reconfigurable computing: The theory and practice of FPGA-Based Computation*, Elsevier.
14. Xilinx. (2009), Virtex 5 family overview. DS100, Xilinx Inc.
15. Fagin B., Watt J.G., Gross R. (1993), A special-purpose processor for gene sequence analysis, *Computer Applications in the Biosciences*, 9, 221–226.

16. Hughey R. (1996), Parallel hardware for sequence comparison and alignment, *Computer Applications in the Biosciences*, 12, 473–479.
17. Wozniak A. (1997), Using video-oriented instructions to speed up sequence comparison, *Comput Applied Bioinformatics*, 13, 145–150.
18. Guerdoux-Jamet P., Lavenier D. (1997), SAMBA: hardware accelerator for biological sequence comparison, *Computer Applications in the Biosciences*, 13, 609–615.
19. Lavenier D. (1998), Speeding up genome computations with systolic accelerator, *SIAM News* 31, 1–8.
20. Guccione A.S., Keller E. (2002), Gene matching using Jbits, *Proceedings of the Reconfigurable Computing Is Going Mainstream, 12th International Conference on Field-Programmable Logic and Applications*, 1168–1171.
21. Simmler H., Singpiel H., Männer R. (2004), Real-time primer design for DNA chips, *Interscience Concurr. Comput.: Pract. Exper.* 16, 855–872.
22. Marongiu A., Palazzari P., Rosato V. (2003), Designing hardware for protein sequence analysis, *Bioinformatics*, 19, 1739–1740.
23. Oliver T., Schmidt B., Nathan D., Clemens R., Maskell D. (2005), Using reconfigurable hardware to accelerate multiple sequence alignment with ClustalW, *Bioinformatics*, 21, 3431–3432.
24. Anish T.A., Dumontier M., Rose J.S., Hogue C.W.V. (2005), Hardware-accelerated protein identification for mass spectrometry, *Rapid Communications in Mass Spectrometry*, 19, 833–837.
25. Panitz F. et al. (2007), SNP mining porcine ESTs with MAVIANT, a novel tool for SNP evaluation and annotation, *Bioinformatics*, 23, 387–391.
26. Dandass Y.S., Burgess S.C., Lawrence M., Bridges S.M. (2008), Accelerating string set matching in FPGA hardware for bioinformatics research, *BMC Bioinformatics*, 9, doi: 10.1186/1471-2105-9-197.
27. Bogdan I.A., Coca D., Rivers J., Beynon J.R. (2007), Hardware acceleration of processing of mass spectrometric data for proteomics, *Bioinformatics*, 23, 724–731.
28. Bogdan I.A., Rivers J., Beynon J.R., Coca D. (2008), High-performance hardware implementation of a parallel database search engine for real-time peptide mass fingerprinting, *Bioinformatics*, 24, 1498–1502.
29. Bogdan I.A., Coca D., Beynon J.R. (2009), Peptide mass fingerprinting using field-programmable gate arrays, *IEEE Transactions on Biomedical Circuits and Systems*, 3, 142–149.
30. Coca D., Bogdan I.A., Beynon R.J. (2009), A high-performance reconfigurable computing solution for peptide mass fingerprinting, In *Proteome Bioinformatics*, Series: Methods in Molecular Biology, 604, Hubbard, Simon J.; Jones, Andrew R. (Eds.), Humana Press.
31. Edwards N., Lippert R. (2002), Generating peptide candidates from amino-acid sequence databases for protein identication via mass spectrometry. In *Proc. of the 2nd International Workshop on Algorithms in Bioinformatics (WABI)*, 68–81.
32. Cieliebak M., Erlebach T., Liptak Z., Stoye J., Welzl E. (2004), Algorithmic complexity of protein identification: combinatorics of weighted strings, *Discrete Applied Mathematics*, 137 (1), 27–46.
33. Kaltenbach H.M., Sudek H., Böcker S., Rahmann S. (2005), Statistics of cleavage fragments in random weighted strings. Tech. Rep. TR-2005-06, Technische Fakultät der Universität Bielefeld, Abteilung Informationstechnik, http://bieson.ub.unibielefeld.de/volltexte/2006/900/.

34. Schechter I., Berger A. (1967), On the size of the active site in proteases. *Biochemical and Biophysical Research Communications*, 27, 157–162.
35. Pappin D.J., Hojrup P., Bleasby A.J. (1993), Rapid identification of proteins by peptide-mass fingerprinting, *Current Biology*, 3, 327–332.
36. Perkins D.N., Pappin D.J.C., Creasy D.M., Cottrell J.S. (1999), Probability-based protein identification by searching sequence databases using mass spectrometry data, *Electrophoresis*, 20, 3551–3567.
37. Zhang W., Chait B.T. (2000), ProFound: An expert system for protein identification using mass spectrometric peptide mapping information, *Analytical Chemistry*, 72, 2482–2489.
38. Samuelsson J., Dalevi D., Levander F., Rögnvaldsson T. (2004), Modular, scriptable and automated analysis tools for high-throughput peptide mass fingerprinting, *Bioinformatics*, 20, 3628–3635.

Index

Note: *Italicized* page numbers refer to figures and tables.

G

P

Q

R

For Product Safety Concerns and Information please contact our EU
representative GPSR@taylorandfrancis.com
Taylor & Francis Verlag GmbH, Kaufingerstraße 24, 80331 München, Germany

www.ingramcontent.com/pod-product-compliance
Ingram Content Group UK Ltd.
Pitfield, Milton Keynes, MK11 3LW, UK
UKHW021504240926
475702UK00035B/355
* 9 7 8 1 4 3 9 8 1 4 8 8 8 *